LIQUEFACTION AROUND MARINE STRUCTURES

ADVANCED SERIES ON OCEAN ENGINEERING

Series Editor-in-Chief
Philip L-F Liu (*Cornell University*)

---

*For the complete list of titles in this series, please write to the Publisher.

Advanced Series on Ocean Engineering — Volume 39

LIQUEFACTION AROUND MARINE STRUCTURES

B. Mutlu Sumer

Technical University of Denmark, Denmark

World Scientific

NEW JERSEY · LONDON · SINGAPORE · BEIJING · SHANGHAI · HONG KONG · TAIPEI · CHENNAI

Published by

World Scientific Publishing Co. Pte. Ltd.

5 Toh Tuck Link, Singapore 596224

USA office: 27 Warren Street, Suite 401-402, Hackensack, NJ 07601

UK office: 57 Shelton Street, Covent Garden, London WC2H 9HE

British Library Cataloguing-in-Publication Data
A catalogue record for this book is available from the British Library.

Advanced Series on Ocean Engineering — Vol. 39
LIQUEFACTION AROUND MARINE STRUCTURES
(With CD-ROM)

ISBN 978-981-4329-31-6

Typeset by Stallion Press
Email: enquiries@stallionpress.com

Printed in Singapore.

Contents

Preface

In geotechnical-engineering terminology, liquefaction refers to the state of the soil in which the effective stresses between individual soil grains vanish and the water–sediment mixture as a whole, therefore, acts like a fluid. Under this condition, the soil fails, therefore precipitating failure of the supported structure such as pipelines, sea outfalls, breakwaters, seawalls, pile structures, gravity structures, rock berms, etc.

Although a substantial amount of knowledge had accumulated on flow and scour processes around marine structures, comparatively little was known about the impact of wave-induced liquefaction on these structures until recently. Indeed, the topic had received little coverage in research, which had substantially advanced the design of coastal structures but not the design of their foundations with regard to soil liquefaction.

The European Union (EU) supported a three-year (2001–2004) research program called LIquefaction around MArine Structures (LIMAS), which was preceded by another EU research program (1997–2000) called SCour ARound COastal STructures (SCARCOST) in which liquefaction around coastal structures was one of the two focus areas. The main results of LIMAS were published in two special issue volumes in the *Journal of Waterway, Port, Coastal and Ocean Engineering* by the American Society of Civil Engineers (ASCE) (see the editorials by Sumer, 2006 and 2007). Those from SCAR-COST were summarized in a paper in *Coastal Engineering* by Sumer *et al.* (2001).

The topic has continued to receive much attention, which has led to a substantial number of recent publications in journals and conference proceedings.

The present book intends (1) to collect state-of-the-art knowledge based on the above work, (2) to build content, and (3) to create interest in this currently popular area.

The primary aim of the book is to describe wave-induced liquefaction processes and their implications for marine structures. A clear hydrodynamic/geodynamic understanding makes it relatively easier for a consulting engineer to assess liquefaction potential, to make engineering predictions, and to make recommendations as to how to avoid potential risks. This book is essentially intended for professionals and researchers in the area of Coastal, Ocean and Marine Civil Engineering, and as a textbook for graduate/postgraduate students.

Soil liquefaction caused by earthquakes has been studied quite extensively in the past 30 years or so. This has culminated into a substantial body of literature, including books by Seed and Idriss (1982), Kramer (1996, Chapter 9), and most recently, Jefferies and Been (2006) and Idriss and Boulanger (2008). The application to port structures of soil liquefaction caused by earthquakes is covered in a book titled "Seismic Design Guidelines for Port Structures", published by PIANC (The World Association for Waterborne Transport Infrastructure) in 2001.

The present book differs from the aforementioned literature in the sense that it focuses on wave-induced liquefaction, although one chapter (Chapter 10) is devoted solely to the topic of earthquake-induced liquefaction and its implications for marine structures.

The author believes that the present book and the existing body of literature on seismic-induced liquefaction with special reference to marine structures form a complementary source of information on liquefaction around marine structures.

The author has interacted in his research on liquefaction with:

- Many colleagues: Professor Jørgen Fredsøe (Technical University of Denmark (DTU)), Professor Dong Jeng (University of Dundee, UK), Professor Liang Cheng (University of Western Australia), and colleagues from the LIMAS project (in alphabetical order): Mr. J. S. Damgaard (COWI, UAE), Mr. M. B. de Groot (GeoDelft, the Netherlands), Dr. S. Dunn (HR Wallingford, UK), Professor P. Foray (Domaine Universitaire, Grenoble, France), Dr. N.-E. O. Hansen (LICengineering, Denmark), Mr. M. Kudella (Technische Universitat Braunschweig, Germany), Professor M. Mory (University of Pau, France), Professor H. Oumeraci (Technische Universitat Braunschweig,

Germany), Professor A. C. Palmer (University of Cambridge, UK, currently at National University of Singapore), Professor R. Sandven (NTNU, Norway), and Professor A. Sawicki (Inst. of Hydroengineering, Poland);

- Former research associates: Professor Nian-Sheng Cheng, Dr. Waldemar Magda, Dr. Figen Hatipoglu Dixen, Mr. S. Kaan Sumer, Dr. Abidin Kaya, and Dr. Ozgur Kirca; and

- Former students: Mr. Morten T. Lind, Mr. Steffen Christensen, Dr. Christoffer Truelsen, Mr. Søren Juhl Andersen and Mr. Jørgen Bang Jensen.

A draft of the manuscript was sent to Dr. Eng. Shinji Sassa (Head, Soil Dynamics Group, Geotechnical Engineering Field, Port and Airport Research Institute (PARI), Yokosuka, Japan), and Dr. Ozgur Kirca (Istanbul Technical University, Department of Civil Engineering) for their review and comments.

Dr. Sassa went through the entire manuscript, and provided the author with invaluable comments, which have been incorporated into the final version of the manuscript. The author is truly grateful to Dr. Sassa for this.

Dr. Kirca (who was a former postdoctoral research fellow with the author at DTU from 2009–2011) also went through the manuscript, and provided the author with equally invaluable comments. Dr. Kirca also produced excellent videos (shot during his stay at DTU) which are collected on the CD-ROM accompanying this book. The author is also truly grateful to Dr. Kirca for these.

The author would like to express his appreciation for the support of Professor Henrik Carlsen (Head, Department of Mechanical Engineering, DTU) and for the excellent working environment at DTU Mekanik. The author's thanks go also to Professor Jørgen Juncher Jensen (Head, Section for Fluid Mechanics, Coastal and Maritime Engineering, Department of Mechanical Engineering, DTU), who gave support to the author; to Mr. Henning Jespersen and Mr. Jan Larsson who skillfully provided technical assistance to the author and to his research associates and his students for the "endless" wave-induced liquefaction tests in the laboratory; and to Mr. Hans Jørn Poulsen, who prepared the figures for the book with patience and skill.

A significant part of the author's research on wave-induced liquefaction during the last 20 years has been supported by various research programs, notably:

- "Scour Around Coastal Structures (SCARCOST)", 1997–2000, Contract No. MAS3-CT97-0097 of the Commission of the European Communities, Directorate-General XII for Science, Research and Development (Program Marine Science and Technology, MAST III);

- "Liquefaction Around Marine Structures (LIMAS)", 2001–2004, Contract No. EVK3-CT-2000-00038, of the same commission (FP5 specific program "Energy, Environment and Sustainable Development");

- "Exploitation and Protection of Coastal Zones (EPCOAST)", 2005–2007, Sagsnr. 26-00-014 of the Danish Research Agency, Danish Councils for Independent Research, The Danish Research Council for Technology and Production Sciences (FTP);

- "Seabed Wind Farm Interaction", 2008–2012, Sagsnr. 2104-07-0010 of the Danish Agency for Science Technology and Innovation, Danish Council for Strategic Research (DSF), Programme Commission on Energy and Environment; and

- "Innovative Multi-purpose Offshore Platforms: Planning, Design and Operation (MERMAID)", 2012–2016, Grant Agreement No. 288710 of the European Commission, 7th Framework Programme for Research.

The author has been the program leader of SCARCOST, LIMAS, EPCOAST, and Seabed Wind Farm Interaction.

References

1. Idriss, I.M. and Boulanger, R.W. (2008): Soil Liquefaction During Earthquakes. Earthquake Engineering Research Institute, Oakland, CA, 244 p.

2. Jefferies, M. and Been, K. (2006): Soil Liquefaction. A Critical State Approach. Taylor & Francis, Abingdon, 479 p.

3. Kramer, S.L. (1996): Geotechnical Earthquake Engineering. Prentice Hall, Upper Saddle River, NJ, 653 p.

4. PIANC (2001): Seismic Design Guidelines for Port Structures. Working Group No. 34 of the Maritime Navigation Commission, International Navigation Association (PIANC). A.A. Balkema Publishers, Lisse/Abingdon/Exton (PA)/Tokyo.

5. Seed, H.B. and Idriss, I.M. (1982): Ground Motions and Soil Liquefaction During Earthquakes. Earthquake Engineering Research Institute, Oakland, CA, 134 p.

6. Sumer, B.M. (2006): Special issue on liquefaction around marine structures: Processes and benchmark cases. Journal of Waterway, Port, Coastal and Ocean Engineering, ASCE, vol. 132, No. 4, 225–226.

7. Sumer, B.M. (2007): Special issue on liquefaction around marine structures: Miscellaneous. Journal of Waterway, Port, Coastal and Ocean Engineering, ASCE, vol. 133, No. 1, 1–2.

8. Sumer, B.M., Whitehouse, R.J.S. and Tørum, A. (2001): Scour around coastal structures: A summary of recent research. Coastal Engineering, Vol. 44, No. 2, 153–190.

Chapter 1

Introduction and Physics of Liquefaction

1.1   Introduction

In geotechnical-engineering terminology, liquefaction refers to the state of the soil where the effective stresses between the individual grains vanish, and the water–sediment mixture as a whole, therefore, acts like a fluid. Under this condition, the soil fails, thus precipitating failure of the supported structure. Some such failures have been catastrophic. With the soil liquefied, buried pipelines may float to the surface of the seabed, pipelines laid on the seabed may sink (penetrating to greater depths), large individual blocks (like those used for scour protection) may penetrate into the seabed, sea mines may enter into the seabed and eventually disappear, and gravity structures (such as caisson breakwaters, offshore platforms, box caissons, etc.) may fail.

Marine soils that can be liquefied under wave action are basically limited to fine soils such as silt or fine sand, or, in composite soils, to silty sand, or clayey sand, as will be discussed in greater details later in the book. From Table B.1 in Appendix B, these kinds of soil components correspond to grain sizes $d = 0.074 - 0.4\,\mathrm{mm}$ (for fine sand), and $d < 0.074\,\mathrm{mm}$ (for silt).

Figs. 1.1 and 1.2 display distributions of sediments in the North Atlantic Ocean and in the North Pacific Ocean, respectively (Keller, 1967). As seen from these figures, these types of sediments, sand-silt and silt-clay, are almost invariably present in the areas where most marine and offshore construction takes place.

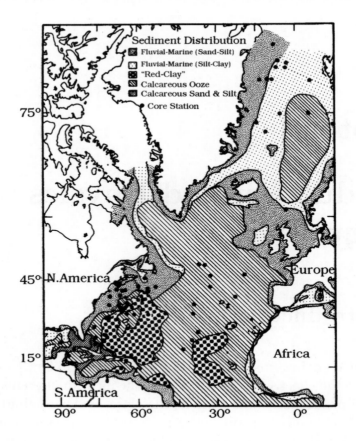

Figure 1.1:   Distribution of sediments in the North Atlantic Ocean. Adapted from Keller (1967).

Table 1.1 gives examples of soil types (with corresponding grain sizes) in projects where the author has been involved as a consultant (a partial list). In these projects, either wave-induced liquefaction has occurred, or there has been large potential.

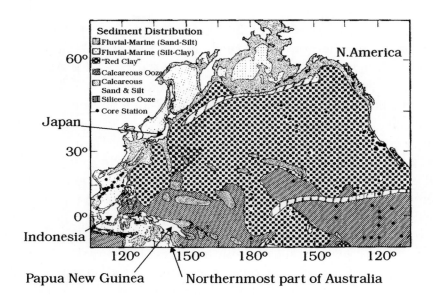

Figure 1.2:   Distribution of sediments in the North Pacific Ocean. Adapted from Keller (1967).

Table 1.1. Examples of soil types taken from projects where either liquefaction has occurred, or where there has been large potential.

Location	Soil type	Grain size (mm)
Australia	Fine sand	0.135
Bangladesh	Fine sand	0.080
	Silt	0.003–0.070
Canada	Silty sand	0.050–0.150
Denmark	Fine sand	0.150
UK	Silty sand	0.080–0.120
	Silt	0.060–0.070

Several examples of liquefaction have been reported in the literature in conjunction with marine engineering.

Christian, Taylor, Yen and Erali (1974) and Herbich, Schiller, Dunlap and Watanabe (1984) report incidents where sections of pipelines floated to the surface of the soil during storms.

   Floatation is just one of possible other consequences of the seabed liquefaction as far as pipelines are concerned. Fig. 1.3 illustrates various scenarios
(Damgaard, Sumer, Teh, Palmer, Foray and Osario, 2006). Case (a) shows
a pipeline that is lighter than the surrounding liquefied soil. In this case the
pipeline will float to the surface of the seabed and become exposed to the
wave and current forces, as well as to impacts from dropped objects, trawl
collision, etc. Case (b) where the pipeline is heavier than the surrounding
liquefied soil, is probably not very usual; normally the pipe weight will be
reduced as much as possible to save on costs of construction and installation,
but there are situations where other factors dictate a heavy pipeline. In this
case the seabed liquefaction will cause the pipe to sink. Pipelines may, for
various reasons, be protected by a rock cover (Sumer and Fredsøe, 2002). In
that case seabed liquefaction can result in sinking of the rock cover (Case (c)).
The fourth case (Case (d)) involves a bottom-seated pipeline designed to
remain stable on the surface of the bed, assuming that the seabed itself will
remain stable. If the seabed is liquefied before the pipeline stability is lost,
then the pipeline will remain probably partially buried in the seabed, but
exposed to the action of current and waves. This will cause large horizontal
and vertical displacements of the pipeline.

   Dunlap, Bryant, Williams and Suheyda (1979) describe storm-induced
pore pressures in soft, clayey sediments in the Mississippi Delta where sinking
of several of the measuring instruments up to 6–14 ft was noted.

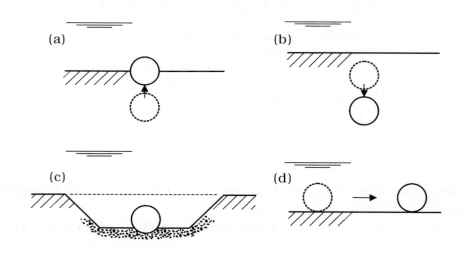

Figure 1.3:   Possible consequences of liquefaction. Damgaard *et al.* (2006).

Figure 1.4: Pipeline floatation accident. Taken from Damgaard *et al.* (2006). Courtesy of Mr. D. Osorio.

Miyamoto, Yoshinaga, Soga, Shimizu, Kawamata, and Sato (1989) report the subsidence of offshore breakwaters composed of concrete blocks at Niagata Coast, Japan (see also Goda, 1994).

Fig. 1.4 illustrates a pipeline floatation accident near Fao Island in Shatt Al Arab Delta in 1960 (Damgaard *et al.*, 2006). The pipeline was installed in a predredged trench by a bottom pull operation. The seabed material was described as "very soft sandy silt" with $d_{30} = 0.02\,\text{mm}$. Due to waves and tidal motion the seabed soil around the trench turned into a dense liquid that flooded the trench and, as a result, the pipe floated.

Liquefaction occurs due to cyclic effects, such as waves and earthquakes, although other effects (shocks, blasts, etc.) may also cause the liquefaction phenomenon. The remainder of this chapter will briefly review the physics of liquefaction (the mechanisms) caused by these effects.

1.2   Wave-Induced Liquefaction

Marine soils are constantly exposed to wave action. The liquefaction phenomenon is essentially associated with large waves. These waves may have a period of $O(5-15\,\mathrm{s})$, with 50- or 100-year-return-period wave heights as much as $O(1-2\,\mathrm{m})$ or larger in coastal areas, and $O(10-20\,\mathrm{m})$ in offshore areas. Here the symbol $O(\ )$ means order of magnitude.

Liquefaction is generated mainly by two different mechanisms in waves, namely:

1. By buildup of pore pressure (residual liquefaction); and

2. By an upward-directed vertical pressure gradient in the soil during the passage of a wave trough (momentary liquefaction).

Each mechanism is now considered individually.

1.2.1   Residual liquefaction

This mechanism can best be described by reference to a progressive wave over a horizontal seabed (Fig. 1.5).

The seabed, in the case of a progressive wave will, undergo a periodic pressure variation, as sketched in Fig. 1.5b. Owing to the increased bed pressure under the wave crest, and the opposite effect under the wave trough, the soil will be compressed under the wave crest, and expanded under the wave trough. Therefore, the water–soil interface will be (nearly) 180° out of phase with the water surface elevation Fig. 1.5c. This will result in the generation of shear stresses in the soil, as illustrated in Fig. 1.5c. These shear stresses will vary periodically in time, as the wave continues. (Normal stresses will also be generated in the soil. However, for the time being, we put aside these latter stresses for the sake of simplicity.)

Now, if the grains are initially loosely packed, the previously mentioned periodic shear stresses and their associated shear deformations in the soil will gradually rearrange the soil grains at the expense of the pore volume of the soil, as illustrated in Fig. 1.6 (Frames 1–4). The latter effect will "pressurize" the water in the pores, and presumably lead to a buildup of pore-water pressure in the case of an undrained soil (i.e., in the case of silt, for example). As the wave action continues, the pore-water pressure will continue to accumulate.

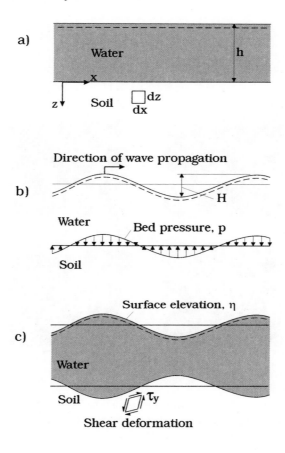

Figure 1.5: Elastic deformation of the seabed soil under a progressive wave.

During this progressive buildup, the pore-water pressure may reach such levels that it may exceed the value of the overburden pressure. In this latter situation, the soil grains will become unbound and completely free, and the soil will begin to act like a liquid; see Video 1 on the CD-ROM accompanying the present book. This process is called the *residual liquefaction*.

1.2.2 Momentary liquefaction

The second mechanism generating soil liquefaction is related to the phase-resolved component of the waves. This kind of liquefaction occurs during the passage of the wave trough. Under the wave trough, the pore pressure (in excess of the hydrostatic pressure) has a negative sign (Fig. 1.5b). Therefore

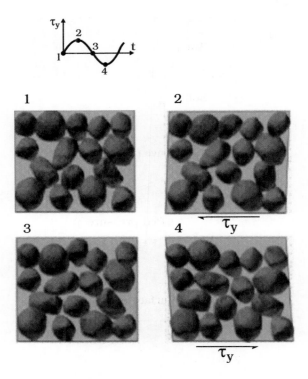

Figure 1.6:   Periodic shear stresses and their associated shear deformations in the soil will gradually rearrange the soil grains at the expense of the pore volume. Adapted from an animation posted at http://www.ce.washington.edu/~liquefaction/html/what/what1.html

the pressure distribution across the soil depth will be as sketched in Fig. 1.7a. This figure describes the pressure distribution in the case of a completely saturated soil, whereas Fig. 1.7b describes that in the case of an unsaturated soil. In the latter case, the soil contains some air/gas and therefore the pore pressure is "dissipated" at a very fast rate with the depth, as sketched in Fig. 1.7b.

Now, in the case of the completely saturated soil, the pressure gradient is not tremendously large (Fig. 1.7a). However, in the case of the unsaturated soil (Fig. 1.7b), the pressure gradient can be very large, particularly at small values of z, meaning that quite a substantial amount of lift can be generated at the top layer of the soil during the passage of the wave trough. If this lift exceeds the submerged weight of the soil, the soil will fail, and as a

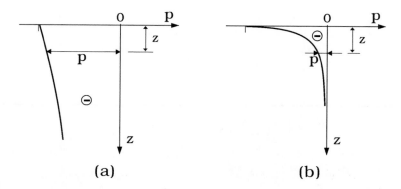

Figure 1.7: Typical distributions of pore pressure (in excess of hydrostatic pressure) during the passage of a wave trough. (a) Saturated soil. (b) Unsaturated soil.

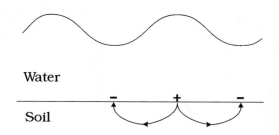

Figure 1.8: Seepage flow under a progressive wave.

result, it will be liquefied. This type of liquefaction is termed the *momentary liquefaction*. (The liquefaction here occurs over a short period of time during the passage of the wave trough; for the rest of the wave period, the soil will be in the no-liquefaction regime. We note that the term momentary liquefaction, widely used in the literature, implies that the liquefaction has only a very "brief life".) It may be noted that the seepage flow under the wave trough (Fig. 1.8) may help enhance the momentary liquefaction due to the upward drag acting on the individual grains.

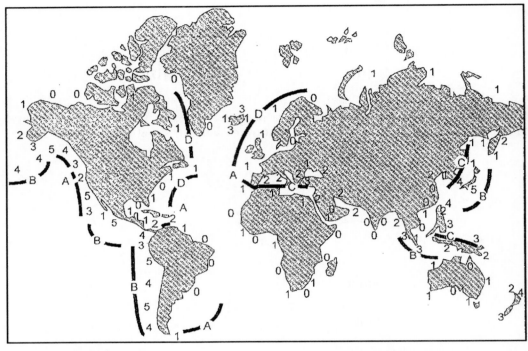

Earthquake intensity	Seismotectonic type
Zone 0 = 0.00~0.05 g Zone 1 = 0.05~0.15 g Zone 2 = 0.15~0.25 g Zone 3 = 0.25~0.35 g Zone 4 = 0.35~0.45 g Zone 5 = 0.45~0.55 g	A = Shallow crustal fault zones B = Deep subduction zones C = Mixed shallow crustal fault and deep subduction zones D = Intraplate zones

Figure 1.9:   Zone averaged earthquake intensity.   Where g is the acceleration due to gravity.   Values of acceleration correspond to a return period of 475 years.   Note that some areas of low average seismic hazard have historically experienced major destructive earthquakes. Adapted from PIANC (2001), GSHAP (1999), and Bea (1997).

1.3   Earthquake-Induced Liquefaction

Fig. 1.9 (adapted from PIANC, 2001, GSHAP, 1999, and Bea, 1997) displays zone average earthquake intensity across the world. The indicated earthquake zones lie in areas with sediment prone to liquefaction, sand-silt and silt-clay; see Figs. 1.1 and 1.2.

Fig. 1.10 (PIANC, 2001) presents four examples of acceleration response spectra in which S is the normalized acceleration response spectrum defined by $S = SA/PGA$ with SA being the acceleration response spectrum, and PGA the peak ground acceleration, which may vary from practically nil to $0.6\,g$ where g is the acceleration due to gravity. Fig. 1.10 indicates that the peak period may vary over the range $O(0.1\,\mathrm{s}) - O(1\,\mathrm{s})$, certainly an order of magnitude smaller than the range of wave period referred to in the preceding paragraphs, indicating that even coarser sediment may be subject to (residual) liquefaction. (Owing to the smaller periods, pore-water pressure accumulates faster than in the case of waves.)

The physics of earthquake-induced liquefaction can be explained as follows. During earthquakes, the ground experiences strong, cyclic accelerations, $a(t)$. The equation of motion for a soil column (Fig. 1.11) in its simplest form reads:

$$\tau(t) = \frac{1}{g}\gamma_t z\, a(t) \qquad (1.1)$$

in which γ_t is the total specific weight of the soil, z the depth of the soil column, τ the shear stress at the bottom face of the soil column, and t the time. This equation implies that the cyclic ground acceleration "translates" to a cyclic shear stress in the soil (Fig. 1.12). These cyclic shear stresses cause the soil to undergo cyclic shear deformations, in exactly the same way as in the case of waves, leading to pressure buildup, and eventually liquefaction if the soil is "undrained" (silt, fine sand, and even medium sand).

1.4   Other Mechanisms

Liquefaction in the marine (or otherwise) environment occurs, not only, by waves or earthquakes but also by other effects as well.

One such effect is shocks or blasts. These may be caused by a sudden failure of a slope, or blasting effects. Chaney and Fang (1991) give a comprehensive review of case histories experienced in coastal areas. The results

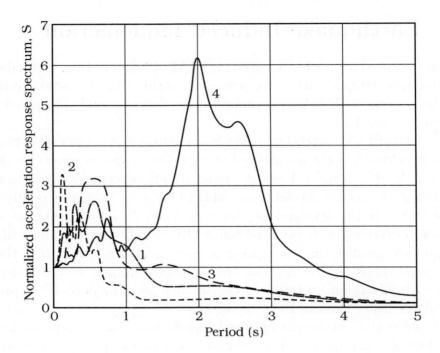

Figure 1.10:   Examples of acceleration response spectra. 1: El Centro, 1940; 2: Lima, 1966, 3: Parkfield, 1966; and 4: Mexico City, 1985. Adapted from PIANC (2001).

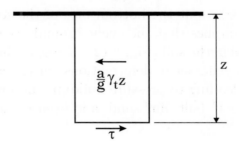

Figure 1.11:   Forces on a soil column with dimensions $1 \times 1 \times z$.

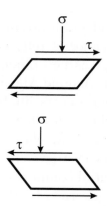

Figure 1.12:   Ground acceleration in earthquakes "translates" into cyclic shear stress.

of a recent study (Hatzor, Gvirtzman, Wainshtein and Orian, 2009) (where a controlled blast-induced liquefaction experiment at the field scale was conducted) showed that liquefaction may be induced by the blast effect in dense silty clayey sands with the relative density as high as 63–89% whereas liquefaction induced by cyclic effects such as waves and earthquakes is normally associated with loose sediment, as will be detailed later.

Another effect is solitary waves. A solitary wave can induce liquefaction on a coastal slope, as demonstrated by Young, White, Xiao and Borja (2009) and Xiao, Young, and Prevost (2010). Pore-water pressure measurements of Sumer *et al.* (2011) show that sediment at or near the surface of a sloping bed experiences upward-directed pressure gradient forces during the rundown stage of a solitary wave. This force is caused by the delay in the pore pressure in responding to the fluid loading (infiltration and exfiltration). If this upward-directed pressure gradient force is larger than the submerged weight of the soil column, the soil at these locations will be liquefied, as revealed by the Young *et al.* (2009) and Xiao *et al.* (2010) work.

A third effect is rocking motions that structures may execute under cyclic loadings (rocking motion of vertical-wall breakwaters under waves, or that of monopiles, or pipelines). In this case, soil undergoes shear strains/ deformations in much the same way as in the case of wave- and earthquake-induced liquefaction. Hence, the end result is the buildup of pore pressure, and eventually liquefaction around the structure.

For further reading on liquefaction around marine structures (furnished with several examples), the reader is referred to de Groot *et al.* (2006), a paper published in the special issue "Liquefaction Around Marine Structures", devoted to Liquefaction Around Marine Structures (LIMAS) (2001–2004), a European research program coordinated by the author. An extensive review of the subject can also be found in Sawicki and Mierczynski (2006).

Returning to the wave-induced liquefaction, as seen from the preceding discussion, the wave-induced shear stresses in the soil, the pore pressure, and the ground-water flow are essential components of the liquefaction processes. Basically, these quantities are governed by the Biot consolidation equations. The following chapter will describe these equations and their solutions.

1.5   References

1. Bea, R.G. (1997): Background for the proposed International Standards Organization reliability based seismic design guidelines for offshore platforms, Proceedings of Earthquake Criteria Workshop: Recent Developments in Seismic Hazard and Risk Assessments for Port, Harbor, and Offshore Structures, Port and Harbour Research Institute, Japan, and University of California, Berkeley, CA, USA, 40–67.

2. Chaney, R.C. and Fang, H.Y. (1991): Liquefaction in the coastal environment: An analysis of case histories. Marine Geotechnology, vol. 10, No. 3, 343–370.

3. Christian, J.T., Taylor, P.K., Yen, J.K.C. and Erali, D.R. (1974): Large diameter underwater pipeline for nuclear plant designed against soil liquefaction. Offshore Technology Conference, May 6–8, 1974, Houston, TX, OTC 2094, 597–606.

4. Damgaard, J.S., Sumer, B.M., Teh, T.C., Palmer, A.C., Foray, P. and Osorio, D. (2006): Guidelines for pipeline on-bottom stability on liquefied noncohesive seabeds. Journal of Waterway, Port, Coastal and Ocean Engineering, ASCE, vol. 132, No. 4, 300–309.

5. de Groot, M.B., Bolton, M.D., Foray, P., Meijers, P., Palmer, A.C., Sandven, R., Sawicki, A. and Teh, T.C. (2006): Physics of liquefaction phenomena around marine structures. Journal of Waterway, Port, Coastal and Ocean Engineering, ASCE, vol. 132, No. 4, 227–243.

6. Dunlap, W., Bryant, W.R., Williams, G.N. and Suheyda, J.N. (1979): Storm wave effects on deltaic sediments — Results of SEASWAB I and II. Port and Ocean Engineering Under Arctic Conditions (POAC 79), Norwegian Institute of Technology, vol. 2, 899–920.

7. Goda, Y. (1994): A plea for engineering-minded research efforts in harbor and coastal engineering. International Conference on Hydro-Technical Engineering for Port and Harbor Construction, Hydro-Port'94, October 19–21, 1994, Yokosuka, Japan, 1–21.

8. GSHAP (1999): Global Seismic Hazard Assessment Program. Available at: http://www.seismo.ethz.ch/static/GSHAP/. July 25, 2013.

9. Hatzor, Y.H., Gvirtzman, H., Wainshtein, I. and Orian, I. (2009): Induced liquefaction experiment in relatively dense, clay-rich sand deposits. Journal of Geophysical Research, vol. 114, No. B2, 1–22.

10. Herbich, J.B., Schiller, R.E., Dunlap, W.A. and Watanabe, R.K. (1984): Seafloor Scour, Design Guidelines for Ocean-Founded Structures, Marcel Dekker, Inc., New York and Basel.

11. Keller, G.H. (1967): Shear strength and other physical properties of sediments from some ocean basins. Civil Engineering in the Oceans, ASCE Conference, San Francisco, September 1967, Proceedings, 1968, 391–417.

12. Miyamoto, T., Yoshinaga, S., Soga, F., Shimizu, K., Kawamata, R. and Sato, M. (1989): Seismic prospecting method applied to the detection of offshore breakwater units settling in the seabed. Coastal Engineering in Japan, vol. 32, No. 1, 103–112.

13. PIANC (2001): Seismic Design Guidelines for Port Structures. Working Group No. 34 of the Maritime Navigation Commission, International Navigation Association (PIANC). A book published by A.A. Balkema Publishers, XV + 474 p. Lisse/Abingdon/Exton (PA)/Tokyo.

14. Sawicki, A. and Mierczynski, J. (2006): Developments in modelling liquefaction of granular soils, caused by cyclic loads. Applied Mechanics Reviews, vol. 59, March issue, 91–106.

15. Sumer, B.M. and Fredsøe, J. (2002): The Mechanics of Scour in the Marine Environment. World Scientific, Singapore, 552 p.

16. Sumer, B.M., Sen, M.B., Karagali I., Ceren, B., Fredsøe, J., Sottile, M., Zilioli, L. and Fuhrman, D.R. (2011): Flow and sediment transport induced by a plunging solitary wave. Journal of Geophysical Research, vol. 116, No. C1, 1–15.

17. Xiao, H., Young, Y.L. and Prevost, J.H. (2010): Parametric study of breaking solitary wave induced liquefaction of coastal sandy slopes. Ocean Engineering, vol. 37, No. 17, 1546–1553.

18. Young, Y.L., White, J.A., Xiao, H. and Borja, R.I. (2009): Liquefaction potential of coastal slopes induced by solitary waves. Acta Geotechnica, vol. 4, No. 1, 17–34.

Chapter 2

Biot Equations and their Solutions

As pointed out in the previous chapter, the wave-induced shear stresses in the soil, the pore pressure, and the ground-water flow are essential components of the liquefaction processes. Basically, these quantities are governed by the Biot consolidation equations. This chapter will describe these equations, and their solutions for some "benchmark" cases.

2.1 Biot Equations

It is common practice that the soil stresses induced by waves are calculated using the classical elasticity theory in which the soil is assumed to be a poro-elastic medium (e.g., Terzaghi, 1948, p.265, and Biot, 1941). The voids of the elastic skeleton are filled with water. A good example of such a model is a rubber sponge saturated with water (Biot, 1941). In the following paragraphs, the governing equations related to a poro-elastic soil will be derived.

1) Equilibrium conditions for a stress field

The equilibrium conditions for a stress field shown in Fig. 2.1 (where only the stress components in the x-direction are shown, to keep the figure relatively simple) are as follows:

$$\frac{\partial \sigma_x}{\partial x} + \frac{\partial \tau_z}{\partial y} + \frac{\partial \tau_y}{\partial z} = 0 \tag{2.1}$$

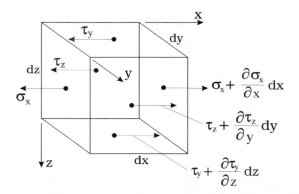

Figure 2.1:   Stresses in the x-direction on the surface of a small soil element.

$$\frac{\partial \tau_z}{\partial x} + \frac{\partial \sigma_y}{\partial y} + \frac{\partial \tau_x}{\partial z} = 0 \tag{2.2}$$

$$\frac{\partial \tau_y}{\partial x} + \frac{\partial \tau_x}{\partial y} + \frac{\partial \sigma_z}{\partial z} = 0 \tag{2.3}$$

in which the normal stresses are denoted by the symbol σ, and the shear stresses by τ. (Note that we adopt here Biot's (1941) original notation; for example, σ_x is the normal stress in the x-direction, while τ_y is the shear stress in the x-direction acting on the side perpendicular to the z-direction, or τ_z is the shear stress in the x-direction acting on the side perpendicular to the y-direction.)

2) The stress–strain relationships

The stress–strain relationships for an isotropic, elastic material are given as follows (Hooke's law):

$$e_x = \frac{\sigma_x}{E} - \frac{\nu}{E}(\sigma_y + \sigma_z) \tag{2.4}$$

$$e_y = \frac{\sigma_y}{E} - \frac{\nu}{E}(\sigma_x + \sigma_z) \tag{2.5}$$

$$e_z = \frac{\sigma_z}{E} - \frac{\nu}{E}(\sigma_x + \sigma_y) \tag{2.6}$$

$$\gamma_x = \frac{\tau_x}{G} \tag{2.7}$$

$$\gamma_y = \frac{\tau_y}{G} \tag{2.8}$$

$$\gamma_z = \frac{\tau_z}{G} \tag{2.9}$$

in which E is Young's modulus, and G the shear modulus. The quantity ν is Poisson's ratio which links Young's modulus to the shear modulus by:

$$G = \frac{E}{2(1+\nu)}. \tag{2.10}$$

The quantities e_x, e_y, and e_z represent the strains (namely, the linear deformations per unit length), while γ_x, γ_y, and γ_z represent the shear (angular) deformations in the x-, y-, and z-directions, respectively:

$$e_x = \frac{\partial u}{\partial x}, \qquad e_y = \frac{\partial v}{\partial y}, \qquad e_z = \frac{\partial w}{\partial z} \tag{2.11}$$

$$\gamma_x = \frac{\partial w}{\partial y} + \frac{\partial v}{\partial z}, \qquad \gamma_y = \frac{\partial u}{\partial z} + \frac{\partial w}{\partial x}, \qquad \gamma_z = \frac{\partial v}{\partial x} + \frac{\partial u}{\partial y} \tag{2.12}$$

in which u, v, and w are the x-, y-, and z-components of the soil displacement, respectively.

Now, for later use, solving σ_x, σ_y, and σ_z, and τ_x, τ_y, and τ_z from Eqs 2.4–2.6, we get:

$$\sigma_x = 2G\left(e_x + \frac{\nu\epsilon}{1-2\nu}\right) \tag{2.13}$$

$$\sigma_y = 2G\left(e_y + \frac{\nu\epsilon}{1-2\nu}\right) \tag{2.14}$$

$$\sigma_z = 2G\left(e_z + \frac{\nu\epsilon}{1-2\nu}\right) \tag{2.15}$$

and

$$\tau_x = G\gamma_x, \qquad \tau_y = G\gamma_y, \qquad \tau_z = G\gamma_z \tag{2.16}$$

in which ϵ is termed the volume expansion per unit volume of soil, and given by:

$$\epsilon = \frac{\partial u}{\partial x} + \frac{\partial v}{\partial y} + \frac{\partial w}{\partial z}. \tag{2.17}$$

3) The stress–strain relationships in the case of a poro-elastic soil

In the case of a poro-elastic soil where the voids are filled with water (Fig. B.1a), the normal stresses will be apportioned by the soil skeleton and the pore water. For example, σ_x, the normal stress in the x-direction, will be

σ_x = normal stress carried by soil + normal stress carried by water.
$$\text{(2.18)}$$

The first part is (from Eq. 2.13):

$$\text{normal stress carried by soil} = 2G\left(e_x + \frac{\nu\epsilon}{1 - 2\nu}\right) \qquad \text{(2.19)}$$

while the second part is:

$$\text{normal stress carried by water} = -p \qquad \text{(2.20)}$$

in which p is the pore-water pressure. (Note that tensile stresses have positive signs while pressures have negative signs for convenience).

Hence, from Eqs 2.18, 2.19, and 2.20,

$$\sigma_x = 2G\left(e_x + \frac{\nu\epsilon}{1 - 2\nu}\right) - p. \qquad \text{(2.21)}$$

This is the stress–strain relationship for the x-direction for a poro-elastic soil. Similarly, the other normal stresses (Eqs 2.14–2.16) will be:

$$\sigma_y = 2G\left(e_y + \frac{\nu\epsilon}{1 - 2\nu}\right) - p \qquad \text{(2.22)}$$

$$\sigma_z = 2G\left(e_z + \frac{\nu\epsilon}{1 - 2\nu}\right) - p \qquad \text{(2.23)}$$

in which ϵ is given by Eq. 2.17. The shear stresses, on the other hand, will remain unchanged, since they are carried only by the soil (Eq. 2.16):

$$\tau_x = G\gamma_x \qquad \text{(2.24)}$$
$$\tau_y = G\gamma_y \qquad \text{(2.25)}$$
$$\tau_z = G\gamma_z \qquad \text{(2.26)}$$

The soil part of the stress (i.e., Eq. 2.19, and similar equations for the other two directions, namely y-, and z-directions) is termed the effective stress:

$$\sigma'_x = 2G\left(e_x + \frac{\nu\epsilon}{1 - 2\nu}\right) \qquad \text{(2.27)}$$

$$\sigma_y' = 2G \left(e_y + \frac{\nu \epsilon}{1 - 2\nu} \right) \qquad (2.28)$$

$$\sigma_z' = 2G \left(e_z + \frac{\nu \epsilon}{1 - 2\nu} \right). \qquad (2.29)$$

4) The equations of equilibrium for a poro-elastic soil

Inserting the stress–strain relationships in Eqs 2.21–2.26 into Eqs 2.1–2.3, the following equations are obtained.

$$G\nabla^2 u + \frac{G}{1 - 2\nu} \frac{\partial \epsilon}{\partial x} = \frac{\partial p}{\partial x} \qquad (2.30)$$

$$G\nabla^2 v + \frac{G}{1 - 2\nu} \frac{\partial \epsilon}{\partial y} = \frac{\partial p}{\partial y} \qquad (2.31)$$

$$G\nabla^2 w + \frac{G}{1 - 2\nu} \frac{\partial \epsilon}{\partial z} = \frac{\partial p}{\partial z} \qquad (2.32)$$

in which

$$\nabla^2 = \frac{\partial^2}{\partial x^2} + \frac{\partial^2}{\partial y^2} + \frac{\partial^2}{\partial z^2}. \qquad (2.33)$$

5) Darcy's law

The variation in the pore-water pressure will drive a flow in the pores. The flow velocities are related to the pressure gradient through Darcy's law in the following way:

$$V_x = -\frac{k}{\gamma} \frac{\partial p}{\partial x}, \qquad V_y = -\frac{k}{\gamma} \frac{\partial p}{\partial y}, \qquad V_z = -\frac{k}{\gamma} \frac{\partial p}{\partial z} \qquad (2.34)$$

in which V_x, V_y, and V_z are the velocity components in the x-, y- and z-directions, respectively, k is the coefficient of permeability of the soil, and γ is the specific weight of water.

6) Continuity equation for the pore water

Finally, from the conservation of mass of pore water, the following equation is obtained:

$$\frac{\partial}{\partial t} \left(\epsilon + \frac{n}{K'} p \right) + \frac{\partial V_x}{\partial x} + \frac{\partial V_y}{\partial y} + \frac{\partial V_z}{\partial z} = 0 \qquad (2.35)$$

in which n is the porosity of the soil (see Appendix B for the relationships among various soil quantities), and K' is the *apparent* bulk modulus of elasticity of water (Biot, 1941).

Regarding the first term in the preceding equation, there are two contributions: first, $\frac{\partial}{\partial t}(\epsilon)$, which represents the increase in the volume of water due to the expansion of the soil skeleton per unit volume of soil per unit time, and second, $\frac{\partial}{\partial t}(\frac{n}{K'}p)$, which represents the increase in the volume of water due to the compressibility of water itself (including the effect of gas/air content in water).

The quantity K' is related to the *true* bulk modulus of elasticity of water, K, by (Verruijt, 1969):

$$\frac{1}{K'} = \frac{1}{K} + \frac{1 - S_r}{p_0} \tag{2.36}$$

in which S_r is the degree of saturation,

$$S_r = \frac{V_w}{V_v} \tag{2.37}$$

with V_w being the volume of water and V_v the total pore volume (Fig. B.1, Appendix B), and p_0 in Eq. 2.36 is the absolute (not excess) pore-water pressure and can be taken equal to the initial value of pressure. When the pore water is gas/air free, S_r will be unity, therefore, in this case, K' will be equal to the true bulk modulus of elasticity of water, K.

Now, inserting Eq. 2.34 in Eq. 2.35, one gets

$$\frac{k}{\gamma}\nabla^2 p = \frac{n}{K'}\frac{\partial p}{\partial t} + \frac{\partial \epsilon}{\partial t}. \tag{2.38}$$

This equation is known as the storage equation, while the entire set of equations (namely, Eqs 2.30–2.32, and 2.38) are known as the Biot consolidation equations (Biot, 1941).

The Biot consolidation equations are to be solved to get the four unknown quantities, namely u, v, w, and p. Once the solution is obtained, then the stresses in the soil can be found from Eqs 2.21–2.26.

Discussion of inertia effects

The Biot equations do not contain the inertia terms. It can be shown that the inertia effect can be ignored in most engineering problems (Cheng and

Liu, 1986). Mei and Foda (1981) can be consulted for the full version of the previous equations including the inertia effect.

Recently, Ulker, Rahman and Jeng (2009) made a systematic study of the inertia effect. They developed three formulations; Ulker *et al.* termed these formulations as fully dynamic, partly dynamic, and quasi static:

1. Fully dynamic: in this formulation, they included both the acceleration of soil skeleton and the acceleration of pore water (relative to that of soil skeleton);

2. Partly dynamic: in this formulation, they included only the acceleration of soil skeleton; and

3. Quasi static: here, both inertia terms associated with soil skeleton and pore water were neglected, resulting essentially in quasi-static coupled flow and strain formulation, the Biot equations formulation above.

Ulker *et al.* (2009) developed analytical solutions for the above three formulations. One set of results obtained by Ulker *et al.* (2009) is particularly interesting: they illustrated the effect of inertia in a real-life scenario in which the depth of seabed was taken as $d = 30$ m, the water depth $h = 25$ m, the wave height $H = 4.5$ m, the wave period $T = 10$ s, the wave length $L = 130.4$ m, Young's modulus $E = 14,000$ kPa, Poisson's ratio $\nu = 0.35$, the degree of saturation $S_r = 1$, and the porosity $n = 0.333$. In order to observe the effect of inertia, they run three kinds of numerical experiments with three different values of the coefficient of permeability, namely $k = 0.001$, 0.1, and 1 cm/s, the smallest value, $k = 0.001$ cm/s, representing silt or fine sand, and the largest value very coarse sand or gravel (see Fig. B.2 in Appendix B).

Their results (Fig. 3 in Ulker *et al.*, 2009) clearly demonstrated that:

1. The solutions for the soil mechanic quantities for all three formulations practically collapse for the case of $k = 0.001$ cm/s (no inertia effect).

2. The solutions for the fully-dynamic and partially-dynamic formulations collapse even for the case of the largest permeability, $k = 1$ cm/s; illustrating that the inertia effect associated with the acceleration of water is nil.

3. Although the solution for the fully- (or partly-) dynamic formulation deviates from that for the quasi-static formulation for the case of the largest permeability, $k = 1\,\text{cm/s}$, the difference appears to be less than 5% at most.

The above results are not unexpected as the cyclic loading involves low frequencies (the wave period being $T = O(1-10\,\text{s})$), and therefore the forces induced by the soil and water accelerations will be small.

Ulker *et al.* (2009, Figs 11–14) presented charts for what they termed domains of applicability of the previously mentioned three formulations in terms of permeability (or soil type) and the wave type (from the deep-water limit to the wave-breaking limit). Caution must be observed in using these charts, however, as the maximum discrepancy between the results of two neighbouring formulations is considered as 3% in categorizing the domains of applicability. For larger, acceptable, discrepancy values, the boundaries between the domains of applicability will move towards the smaller permeabilities (towards coarser sediment). (Ulker and Rahman, 2009, carried out a similar study with various other applications including blast-loading-induced response of a soil layer, earthquake response of soils, and blood-flow-induced response of a porous tissue.)

These results revealed those of Cheng and Liu (1986) conclusively in that the inertia effects can be ignored for most engineering problems, and particularly for the problems involving fine sediment (silt and fine sand), the sediment relevant for liquefaction processes.

2.2   Solutions to Biot Equations

2.2.1   Stresses in soil under a progressive wave

The case of infinitely large soil depth

We now consider a soil, with an infinitely large depth, subject to a progressive wave (Fig. 1.5b). (Appendix A can be consulted for a summary review of small amplitude linear progressive waves.) As discussed in the preceding chapter in conjunction with Fig. 1.5, the waves induce a pressure distribution on the bed as sketched in Fig. 1.5b, and this will, in turn, cause an elastic deformation in the soil (Fig. 1.5c), resulting in the generation of shear stress, τ_y, and pore-water pressure, p, in the soil (Fig. 1.5c). Our objective is to describe τ_y and p. As will be seen later, we need these quantities to describe the wave-induced liquefaction processes, i.e., the process of residual liquefaction and that of momentary liquefaction.

The governing equations are the Biot consolidation equations, i.e.,

1. The x- and z-components of the equations of equilibrium, Eqs 2.30 and 2.32; and

2. The storage equation, Eq. 2.38 (the 2-D case):

$$GV^2u + \frac{G}{1-2\nu}\frac{\partial\epsilon}{\partial x} = \frac{\partial p}{\partial x} \tag{2.39}$$

$$GV^2w + \frac{G}{1-2\nu}\frac{\partial\epsilon}{\partial z} = \frac{\partial p}{\partial z} \tag{2.40}$$

$$\frac{k}{\gamma}\nabla^2 p = \frac{n}{K'}\frac{\partial p}{\partial t} + \frac{\partial\epsilon}{\partial t}. \tag{2.41}$$

There are three unknowns: (1) the two components of the soil displacement u and w, and (2) the pore-water pressure p. Once the soil displacements u and w are obtained, then the stresses τ_y, σ'_x, and σ'_z in the soil can be found from Eqs 2.25, 2.27, and 2.29, respectively.

The solution to these equations is to be sought under the following three boundary conditions.

1) Pressure at the bed surface:

At the bed surface, the excess pore pressure generated by a small amplitude, linear, progressive wave (Fig. 1.5b)

$$\eta = \frac{H}{2}\exp[i(\lambda x + \omega t)] \tag{2.42}$$

is given by

$$z = 0: \qquad p = p_b \exp[i(\lambda x + \omega t)]. \tag{2.43}$$

Here, η is the surface elevation measured from the Mean Water Level (see Fig. A.1 in Appendix A for the definition sketch), λ is the wave number

$$\lambda = \frac{2\pi}{L}, \tag{2.44}$$

L is the wave length, ω is the angular frequency of the waves

$$\omega = \frac{2\pi}{T},\tag{2.45}$$

T is the wave period, i the imaginary unit, $i = \sqrt{-1}$, and p_b, the maximum value (the amplitude) of the pressure exerted on the bed by the progressive wave (Fig. 1.5b), given by

$$p_b = \gamma \frac{H}{2} \frac{1}{\cosh(\lambda h)}\tag{2.46}$$

(Eq. A.17, Appendix A. See also e.g., Dean and Dalrymple, 1984, p. 89). γ is the specific weight of water.

2) Stresses at the bed surface:

At the bed surface, the vertical effective stress (the vertical stress carried by the soil) must be zero (from Eq. 2.29):

$$z = 0: \qquad \sigma'_z = 2G \left[\frac{\partial w}{\partial z} + \frac{\nu}{1 - 2\nu} \left(\frac{\partial u}{\partial x} + \frac{\partial w}{\partial z} \right) \right] = 0\tag{2.47}$$

and the shear stress is (from Eqs 2.25 and 2.12):

$$z = 0: \qquad \tau_y = G \left(\frac{\partial u}{\partial z} + \frac{\partial w}{\partial x} \right) = \tau_0 (= \rho U_f^2)\tag{2.48}$$

in which τ_0 is the bed shear stress due to the wave boundary-layer flow, U_f is the corresponding friction velocity, and ρ is the water density (see, e.g., Fredsøe and Deigaard, 1992, for general concepts of wave boundary-layer flows). However, normally, $\tau_0 \ll \tau_y$. Hence, the right-hand side of Eq. 2.48 can be put equal to zero.

3) The boundary conditions at large depths:

At large depths, obviously no soil displacement will be experienced, and also no pore-water pressure will develop. Therefore

$$z \to \infty: \qquad u, w \quad \text{and} \quad p \to 0.\tag{2.49}$$

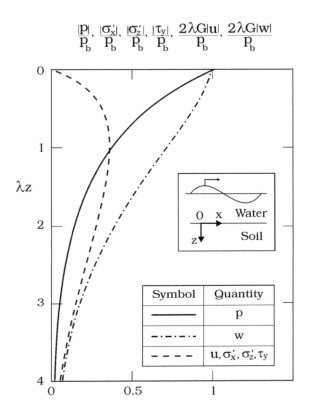

Figure 2.2:   Distribution of the amplitudes of pore pressure, effective stresses and displacements for a soil (with infinitely large depth) exposed to a progressive wave. Yamamoto *et al.* (1978).

Solution:

The solution to the present set of equations (Eqs 2.39–2.41) under the preceding boundary conditions (Eqs 2.43, 2.47, 2.48, and 2.49) was obtained by Yamamoto, Koning, Sellmeijer and van Hijum (1978) for the general case where the soil is not completely saturated, i.e., $S_r < 1$.

In the solution, the ratio G/K' emerges as a key parameter. In the case of a *completely saturated soil*, however, it can be shown that this parameter becomes practically zero for most soils except for dense sand. For completely saturated soil, $K' = K = 1.9 \times 10^6 \, \text{kN/m}^2$ and $G = 4.8 \times 10^2$ (silt and clay)-4.8×10^5 (dense sand) kN/m^2 (Yamamoto *et al.*, 1978), and therefore G/K' will be extremely small.

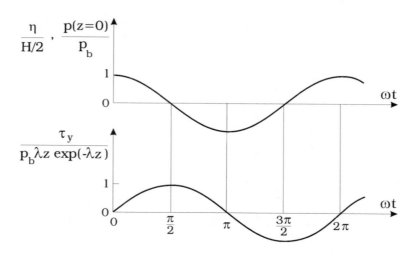

Figure 2.3:   The surface elevation, the bed pressure and the shear stress in the soil for a soil with infinitely large depth.

In the limit when $G/K' \to 0$, Yamamoto $et\ al.$'s (1978) solution reduces to

$$u = -i\lambda z \exp(-\lambda z)(p_b/2\lambda G) \exp[i(\lambda x + \omega t)] \tag{2.50}$$

$$w = [\exp(-\lambda z) + \lambda z \exp(-\lambda z)](p_b/2\lambda G) \exp[i(\lambda x + \omega t)] \tag{2.51}$$

$$p = p_b \exp(-\lambda z) \exp[i(\lambda x + \omega t)]. \tag{2.52}$$

The stresses in the soil can then be calculated by inserting these solutions into Eqs 2.27, 2.29, and 2.25. The effective stress in the x-direction will be

$$\sigma'_x = 2G\left(e_x + \frac{\nu \epsilon}{1 - 2\nu}\right) = p_b \lambda z \exp(-\lambda z) \exp[i(\lambda x + \omega t)] \tag{2.53}$$

that in the z-direction

$$\sigma'_z = 2G\left(e_z + \frac{\nu \epsilon}{1 - 2\nu}\right) = -p_b \lambda z \exp(-\lambda z) \exp[i(\lambda x + \omega t)] \tag{2.54}$$

and the shear stress τ_y,

$$\tau_y = -ip_b \lambda z \exp(-\lambda z) \exp[i(\lambda x + \omega t)]. \tag{2.55}$$

Fig. 2.2 illustrates how the maximum values of the various quantities vary with respect to the depth. As seen, the quantities asymptotically go to zero for larger depths, as dictated by Eq. 2.49. Also, while the pore

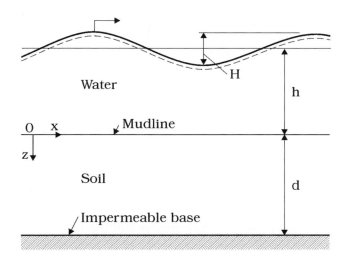

Figure 2.4:   Definition sketch. Finite soil depth.

pressure is dissipated in a monotonous manner with the depth, the soil stresses appear to first increase, attain their maximum values, and then decrease with z.

Furthermore, from Eqs 2.42, 2.43, and 2.55, it is seen that the shear stress is zero below the wave crest and trough, while it attains its maximum value below the zero-crossing points of the surface elevation (Fig. 2.3), a result anticipated from the physical arguments given in Section 1.2.

Yamamoto *et al.* (1978, Fig. 5) compared the poro-elastic theory (in terms of the pore-water pressure) with experiments from the same study, and found very good agreement. Other studies such as Tzang (1998, Fig. 4), for example, also came to the same conclusion.

The case of finite soil depth

We now consider a soil with a finite depth. The soil is subject to a progressive wave, as in the previous case.

Hsu and Jeng (1994) have developed an analytical solution to the Biot equations (Eqs 2.39–2.41) for this case. The boundary conditions (Fig. 2.4) are:

$$z = 0 : \qquad p = p_b \cos(\lambda x - \omega t) \tag{2.56}$$

$$z = 0 : \qquad \sigma'_z = 0 \tag{2.57}$$

$$z = 0 : \qquad \tau_y = 0 \tag{2.58}$$

$$z = d : \qquad u \text{ and } w = 0 \tag{2.59}$$

$$z = d : \qquad \frac{\partial p}{\partial z} = 0. \tag{2.60}$$

The last two conditions imply zero displacements of the soil and no vertical seepage flow at the impermeable base, respectively.

As discussed in the previous section, in the case of the *completely saturated soil*, the parameter G/K' is extremely small (also discussed by Hsu and Jeng, 1994, Section 5 in their paper). For this case $(G/K' \to 0)$, Hsu and Jeng's (1994, p. 793) solution gives the soil displacements as

$$u = \frac{ip_b}{2G\lambda} U(z) e^{i(\lambda x - \omega t)} \tag{2.61}$$

$$w = \frac{p_b}{2G\lambda} W(z) e^{i(\lambda x - \omega t)} \tag{2.62}$$

the pore pressure

$$p = \frac{p_b}{(1 - 2\nu)} P(z) e^{i(\lambda x - \omega t)} \tag{2.63}$$

and the effective normal and shear stresses

$$\sigma'_x = -p_b \Xi_x(z) e^{i(\lambda x - \omega t)} \tag{2.64}$$

$$\sigma'_z = p_b \Xi_z(z) e^{i(\lambda x - \omega t)} \tag{2.65}$$

$$\tau_y = ip_b \Upsilon(z) e^{i(\lambda x - \omega t)} \tag{2.66}$$

in which the functions $U(z), W(z), P(z), \Xi_x(z), \Xi_z(z)$, and $\Upsilon(z)$ are given as

$$U(z) = (C_1 - C_2\lambda z)e^{-\lambda z} + (C_3 - C_4\lambda z)e^{\lambda z} \tag{2.67}$$
$$+\lambda^2 C_5 e^{-\delta z} + \lambda^2 C_6 e^{\delta z}$$

$$W(z) = [C_1 - (1 + \lambda z)C_2]e^{-\lambda z} - [C_3 + (1 - \lambda z)C_4]e^{\lambda z} \tag{2.68}$$
$$+\lambda\delta(C_5 e^{-\delta z} - C_6 e^{\delta z})$$

$$P(z) = (1 - 2\nu)(C_2 e^{-\lambda z} - C_4 e^{\lambda z}) \tag{2.69}$$
$$+(1 - \nu)(\delta^2 - \lambda^2)(C_5 e^{-\delta z} + C_6 e^{\delta z})$$

$$\Xi_x(z) = (C_1 - C_2\lambda z)e^{-\lambda z} + (C_3 - C_4\lambda z)e^{\lambda z} \qquad (2.70)$$

$$+ \left[\lambda^2 - \frac{(\delta^2 - \lambda^2)\nu}{1 - 2\nu}\right](C_5 e^{-\delta z} + C_6 e^{\delta z})$$

$$\Xi_z(z) = (C_1 - C_2\lambda z)e^{-\lambda z} + (C_3 - C_4\lambda z)e^{\lambda z} \qquad (2.71)$$

$$+ \frac{1}{1 - 2\nu}[\delta^2(1 - \nu) - \lambda^2\nu](C_5 e^{-\delta z} + C_6 e^{\delta z})$$

$$\Upsilon(z) = (C_1 - C_2\lambda z)e^{-\lambda z} - (C_3 - C_4\lambda z)e^{\lambda z} \qquad (2.72)$$

$$+ \lambda\delta(C_5 e^{-\delta z} - C_6 e^{\delta z})$$

in which δ is given by

$$\delta^2 = \lambda^2 - \frac{i\omega\gamma(1 - 2\nu)}{2k(1 - \nu)G} \qquad (2.73)$$

and i is the imaginary unit.   The coefficients C_1 to C_6 are given in Appendix D.

It can easily be shown that the above solution converges to the solution of Yamamoto *et al.* (1978) (given in the previous section, Eqs 2.50–2.55) for large soil depths. Incidentally, Cheng, Sumer and Fredsøe (2001) numerically solved the Biot equations subject to the boundary conditions given in Eqs 2.56–2.60, and found that their numerical solution and Hsu and Jeng's analytical solution are in good agreement.

In the next chapter, we shall return to Hsu and Jeng's solution in conjunction with the shear stress in the soil given in Eq. 2.66.

The present subject, namely soil stresses under waves, has been investigated quite extensively over the years. Table 2.1 presents a partial review of these studies (the review has been adapted mainly from McDougal, Tsai, Liu and Clukey, 1989).

Table 2.1.   A partial list of the past work regarding stresses and pore pressure in soil under waves.

Author	Soil	Pore water	Note
Putnam (1949)	Rigid skeleton	Incompressible	See (1)
Reid and Kajiura (1957)	,,	,,	,,
Hunt (1959)	,,	,,	,,
Murray (1965)	,,	,,	,,
Sleath (1970)	,,	,,	,,
Moshagen and Tørum (1975)	Rigid skeleton	Compressible	See (2)
Gade (1958)	See (3)	-	-
Mallard and Dalrymple (1977)	,,	-	-
Dawson (1978)	,,	-	-
Dalrymple and Liu (1978)	,,	-	-
MacPherson (1980)	,,	-	-
Hsiao and Shemdin (1980)	,,	-	-
Dawson et al. (1981)	,,	-	-
Yamamoto (1977)	See (4)	Incompressible/ Compressible	See (5)
Madsen (1978)	,,	,,	,,
Yamamoto (1978, 1981a, 1981b)	,,	,,	,,
Yamamoto et al. (1978)	,,	,,	,,
Yamamoto and Suzuki (1980)	,,	,,	,,
Mei and Foda (1981)	,,	,,	,,
Dalrymple and Liu (1982)	,,	,,	,,
McDougal and Sollitt (1984)	,,	,,	,,
McDougal et al. (1989)	,,	,,	,,
Hsu and Jeng (1994)	,,	,,	,,
Cheng et al. (2001)	,,	,,	,,
Liu and Garcia (2007)	,,	,,	,,

(1) The governing equation is the Laplace equation satisfied by the pore pressure.

(2) An additional term (time-derivative of the pore pressure) appears in the Laplace equation satisfied by the pore pressure.

(3) Soil behaves as a viscous liquid, as an elastic soil or as a combined viscoelastic medium.

(4) Soil behaves as a linearly elastic medium.

(5) The Biot consolidation equations are the governing equations.

2.2.2 Stresses in soil under a standing wave

Consider a vertical, rigid wall (Fig. 2.5), subjected to a progressive wave (the incident wave). As the incident wave impinges on the wall, a reflected wave moves in the offshore direction. The superposition of these two waves results in a standing wave. The surface elevation of this standing wave (Fig. 2.5), measured from the mean water level, is given by

$$\eta(x,t) = \frac{(2H_i)}{2}\cos(\lambda x)\cos(\omega t) \tag{2.74}$$

$$\omega^2 = gk\tanh(kh) \tag{2.75}$$

in which ω is the angular frequency ($\omega = 2\pi/T$), λ is the wave number ($\lambda = 2\pi/L$), T is the wave period, L is the wave length, and h is the water depth. The quantity $2H_i$ is the height of the standing wave, and is twice the wave height (H_i) of each of the two progressive waves forming the standing wave (Dean and Dalrymple, 1984, Chapter 4).

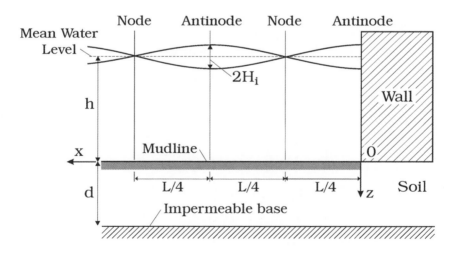

Figure 2.5: Definition sketch. Standing wave in front of a wall.

In the case of the *completely saturated soil*, the parameter G/K' is extremely small, as discussed in the preceding paragraphs. For this case ($G/K' \to 0$), Hsu and Jeng (1994, p. 794) have developed an analytical

solution to the Biot equations. Hsu and Jeng's solution for a soil with a
finite depth gives the soil displacements as

$$u = \frac{-p_b}{2G\lambda}U(z)\sin(\lambda x)e^{-i\omega t} \tag{2.76}$$

$$w = \frac{p_b}{2G\lambda}W(z)\cos(\lambda x)e^{-i\omega t} \tag{2.77}$$

the pore pressure

$$p = \frac{p_b}{(1-2\nu)}P(z)\cos(\lambda x)e^{-i\omega t} \tag{2.78}$$

and the effective normal and shear stresses

$$\sigma'_x = -p_b\Xi_x(z)\cos(\lambda x)e^{-i\omega t} \tag{2.79}$$
$$\sigma'_z = p_b\Xi_z(z)\cos(\lambda x)e^{-i\omega t} \tag{2.80}$$
$$\tau_y = -p_b\Upsilon(z)\sin(\lambda x)e^{-i\omega t} \tag{2.81}$$

in which p_b is given by Eq. 2.46 where the wave height H is to be replaced
with $2H_i$, the height of the standing wave:

$$p_b = \gamma\frac{(2H_i)}{2}\frac{1}{\cosh(\lambda h)}. \tag{2.82}$$

The functions $U(z), W(z), P(z), \Xi_x(z), \Xi_z(z)$, and $\Upsilon(z)$ in the above equations are given as

$$U(z) = (C_1 - C_2\lambda z)e^{-\lambda z} + (C_3 - C_4\lambda z)e^{\lambda z} \tag{2.83}$$
$$+\lambda^2 C_5 e^{-\delta z} + \lambda^2 C_6 e^{\delta z}$$
$$W(z) = [C_1 - (1+\lambda z)C_2]e^{-\lambda z} - [C_3 + (1-\lambda z)C_4]e^{\lambda z} \tag{2.84}$$
$$+\lambda\delta(C_5 e^{-\lambda z} + C_6 e^{\lambda z})$$
$$P(z) = (1-2\nu)(C_2 e^{-\lambda z} - C_4 e^{\lambda z}) \tag{2.85}$$
$$+(1-\nu)(\delta^2 - \lambda^2)(C_5 e^{-\delta z} + C_6 e^{\delta z})$$
$$\Xi_x(z) = (C_1 - C_2\lambda z)e^{-\lambda z} + (C_3 - C_4\lambda z)e^{\lambda z} \tag{2.86}$$
$$+ \left[\lambda^2 - \frac{(\delta^2 - \lambda^2)\nu}{1-2\nu}\right](C_5 e^{-\delta z} + C_6 e^{\delta z})$$

$$\Xi_z(z) = (C_1 - C_2\lambda z)e^{-\lambda z} + (C_3 - C_4\lambda z)e^{\lambda z} \tag{2.87}$$
$$+\frac{1}{1-2\nu}[\delta^2(1-\nu) - \lambda^2\nu](C_5 e^{-\delta z} + C_6 e^{\delta z})$$
$$\Upsilon(z) = (C_1 - C_2\lambda z)e^{-\lambda z} - (C_3 - C_4\lambda z)e^{\lambda z} \tag{2.88}$$
$$+\lambda\delta(C_5 e^{-\delta z} - C_6 e^{\delta z})$$

in which δ is given by Eq. 2.73.

As seen, provided that the wave height in the present case is taken as $2H_i$ in which H_i is the wave height of each of the two progressive waves (i.e., the incident wave and the reflected wave) forming the standing wave,

1. the standing-wave solution for the *shear stress* (Eqs 2.81 and 2.88), not surprisingly, coincides with the progressive-wave solution (Eqs 2.66 and 2.72) at the nodes (e.g., at $x = L/4$); and

2. the standing-wave solution for the *pressure* (Eqs 2.78 and 2.85), again, not surprisingly, coincides with the progressive-wave solution (Eqs 2.63 and 2.69) at the antinodes (e.g., at $x = 0$).

In Chapter 7, we shall return to Hsu and Jeng's (1994) solution in conjunction with the shear stress and the pressure in the soil.

Finally, it may be noted that Hsu and Jeng (1994) also have developed analytical solutions for the case of short-crested waves with an unsaturated and anisotropic soil of finite depth.

2.3 References

1. Biot, M.A. (1941): General theory of three-dimensional consolidation. Journal of Applied Physics, vol. 12, No. 2, 155–164.

2. Cheng, H.-D. and Liu, P.L.-F. (1986): Seepage force on a pipeline buried in a poroelastic seabed under wave loading. Applied Ocean Research, vol. 8, No. 1, 22–32.

3. Cheng, L., Sumer, B.M. and Fredsøe, J. (2001): Solutions of pore pressure buildup due to progressive waves. International Journal of Numerical and Analytical Methods in Geomechanics, vol. 25, No. 9, 885–907.

4. Dalrymple, R.A. and Liu, P.L.-F. (1978): Waves over soft muds: A two-layer fluid model. Journal of Physical Oceanography, vol. 8, November 1978, 1121–1131.

5. Dawson, T.H. (1978): Wave propagation over a deformable sea floor. Ocean Engineering, vol. 5, No. 4, 227–234.

6. Dawson, T.H., Shuyada, J.N. and Coleman, J.M. (1981): Correlation of field measurements with elastic theory of seafloor response to surface waves. Proceedings of the Offshore Technology Conference, 201–210.

7. Dean, R.G. and Dalrymple, R.A. (1984): Water Wave Mechanics for Engineers and Scientists. Prentice-Hall, Inc., NJ.

8. Fredsøe, J. and Deigaard, R. (1992): Mechanics of Coastal Sediment Transport, World Scientific, Singapore.

9. Gade, H.G. (1958): Effects of non-rigid impermeable bottom on plane surface waves in shallow water. Journal of Marine Research, vol. 16, Issue 2, 61–82.

10. Hsiao, S.V. and Shemdin, O.H. (1980): Interaction of ocean waves with a soft bottom. Journal of Physical Oceanography, vol. 10, Issue 4, 605–610.

11. Hsu, J.R.S. and Jeng, D.S. (1994): Wave-induced soil response in an unsaturated anisotropic seabed of infinite thickness. International Journal for Numerical and Analytical Methods in Geomechanics, vol. 18, No. 11, 785–807.

12. Hunt, J.N. (1959): On the damping of gravity waves propagated over a permeable surface. Journal of Geophysical Research, vol. 64, No. 4, 437–442.

13. Liu, X. and Garcia, M.H. (2007): Numerical investigation of seabed response under waves with free-surface water flow. International Journal of Offshore and Polar Engineering, vol. 17, No. 2, 97–104.

14. MacPherson, H. (1980): The attenuation of water waves over a non-rigid bed. Journal of Fluid Mechanics, vol. 97, No. 4, 721–742.

15. Madsen, O.S. (1978): Wave-induced pore pressures and effective stresses in a porous bed. Géotechnique, vol. 28, No. 4, 377–393.

16. Mallard, W.W. and Dalrymple, R.A. (1977): Water waves propagating over a deformable bottom. Proceedings of the 9th Offshore Technology Conference, Houston, TX, vol. 3, 141–146.

17. McDougal, W.G. and Sollitt, C.K. (1984): Geotextile stabilization of seabed: theory. Engineering Structures, vol. 6, Issue 3, 211–216.

18. McDougal, W.G., Tsai, Y.T., Liu, P.L-F. and Clukey, E.C. (1989): Wave-induced pore water pressure accumulation in marine soils. Journal of Offshore Mechanics and Arctic Engineering, ASME, vol. 111, No. 1, 1–11.

19. Mei, C.C. and Foda, M.A. (1981): Wave-induced responses in a fluid filled poroelastic solid with a free surface, a boundary layer theory. Geophysical Journal of the Royal Astronomical Society, vol. 66, Issue 3, 597–631.

20. Moshagen, H. and Tørum, A. (1975): Wave induced pressures in permeable seabeds. Journal of Waterways, Harbors and Coastal Engineering Division, ASCE, vol. 101, No. WW1, 49–57.

21. Murray, J.D. (1965): Viscous damping of gravity waves over a permeable bed. Journal of Geophysical Research, vol. 70, No. 10, 2325–2331.

22. Putnam, J.A. (1949): Loss of wave energy due to percolation in a permeable sea bottom. Transactions, American Geophysical Union, vol. 30, No. 3, 349–356.

23. Reid, R.D. and Kajiura, K. (1957): On the damping of gravity waves over a permeable seabed. Transactions, American Geophysical Union, vol. 38, Issue 5, 662–666.

24. Sleath, J.F.A. (1970): Wave-induced pressures in beds of sand. Journal of Hydraulics Division, ASCE, vol. 96, No. 2, 367–378.

25. Terzaghi, K. (1948): Theoretical Soil Mechanics. London: Chapman and Hall, John Wiley and Sons, Inc., NY.

26. Tzang, S.Y. (1998): Unfluidized soil responses of a silty seabed to monochromatic waves. Coastal Engineering, vol. 35, No. 4, 283–301.

27. Ulker, M.B.C. and Rahman, M.S. (2009): Response of saturated and nearly saturated porous media: Different formulations and their applicability. International Journal for Numerical and Analytical Methods in Geomechanics, vol. 33, No. 5, 633–664.

28. Ulker, M.B.C., Rahman, M.S. and Jeng, D.S. (2009): Wave-induced response of seabed: Various formulations and their applicability. Applied Ocean Research, vol. 31, No. 1, 12–24.

29. Verruijt, A. (1969): Elastic storage of aquifers. In: Flow Through Porous Media (ed. R.J.M. De Wiest), Chapter 8, Academic Press, New York.

30. Yamamoto, T. (1977): Wave-induced instability in seabed. Proceedings of the ASCE Special Conference, Coastal Sediments '77, Charleston, SC, 898–913.

31. Yamamoto, T. (1978): Seabed instability from waves. Proceedings of the 10th Offshore Technology Conference, Houston, TX, vol. 3, 1819–1828.

32. Yamamoto, T. (1981a): Wave-induced pore pressures and effective stresses in homogenous seabed foundations. Ocean Engineering, vol. 8, No. 1, 1–16.

33. Yamamoto (1981b): Ocean waves spectrum transformations due to seabed interactions. Proceedings of the 13th Offshore Technology Conference, Houston, TX, vol. 1, 249–258.

34. Yamamoto, T., Koning, H.L., Sellmeijer, H. and van Hijum, E. (1978): On the response of a poro-elastic bed to water waves. Journal of Fluid Mechanics, vol. 87, No. 1, 193–206.

35. Yamamoto, T. and Suzuki, T. (1980): Stability analysis of seafloor foundations. Proceedings of the Coastal Engineering Conference, 1799–1818.

Chapter 3

Residual Liquefaction

As seen in Chapter 1, when a loose soil is exposed to waves, and when the soil is undrained (fine sand or silt), it may be susceptible to liquefaction under large waves due to the buildup of pore-water pressure. This lique-faction is called residual liquefaction. This chapter is concerned with this type of liquefaction. The so-called momentary liquefaction will be treated in Chapter 4.

The chapter begins with the sequence of liquefaction process, including the buildup of pore-water pressure in a soil subjected to a progressive wave, the onset of liquefaction, the liquefaction stage itself, and eventually the dissipation of the accumulated pressure, and compaction.

The next section deals with the mathematical modelling of the residual liquefaction in which the differential equation governing the buildup of pore-water pressure will be derived for a progressive wave, based on the Biot equations. Solutions to the differential equation will be developed for two cases, (1) the case of infinitely large soil depth; and (2) that of finite soil depth. The section also includes the results of a model validation exercise, and a numerical example where it is illustrated how to use the mathematical model in a typical liquefaction assessment study in the field.

The following section concerns the centrifuge (physical) modelling of residual liquefaction. The centrifuge technique is widely used in geotechnical engineering for physical modelling studies. On the same principles, centrifuge wave testing has been developed relatively recently to study soil liquefaction under waves. The section actually aims at relating physical model studies carried out in standard wave flumes in a $1 - g$ environment to those made in centrifuge wave testing in an $N - g$ environment.

In the next section, attention is concentrated on mathematical modelling of compaction where a simple mathematical model is developed for the compaction stage of the liquefaction–compaction sequence. With this simple model, one is able to estimate the time scale of compaction in a field situation, an important quantity required in a liquefaction assessment exercise.

The chapter ends with three sections where the influence of three important effects on residual liquefaction has been discussed, namely the influence of clay content, the influence of cover stones (or surcharge), and the influence of current (e.g., tidal current).

3.1   Sequence of Liquefaction Process

Observations show that when a "fresh", loose granular soil with no history of liquefaction is exposed to a progressive wave (Fig. 2.4), and when it is undrained (as in the case of fine sand or silt), it goes through a number of different stages: with the introduction of the wave, first the pore pressure begins to build up. When the "accumulated" pore pressure reaches a critical level, the soil is liquefied. Liquefaction first occurs at the surface of the seabed, the mudline, and progresses downwards. This stage is followed by a compaction process. The latter process first begins at the impermeable base and gradually progresses in the upward direction as the pore water is drained out of the soil.

The sequence of the liquefaction process has been studied by Miyamoto, Sassa and Sekiguchi (2004) and Sumer, Hatipoglu, Fredsøe and Sumer (2004, 2006 a), two studies conducted independently from each other. Miyamoto *et al.* (2004), a theoretical and experimental investigation of liquefaction and solidification (of liquefied sand) during wave loading, did their experiments in a very small wave tank in a centrifuge (see Section 3.3.2) while Sumer *et al.* (2004, 2006 a) did their experiments in a standard wave flume. It was shown in Sumer *et al.* (2006 a) that their findings appear to be in agreement with the sequence of sediment behaviour reported in Miyamoto *et al.* (2004). The following description is mainly based on the work of Sumer *et al.* (2006 a).

This paragraph describes the test conditions in Sumer *et al.*'s experiment (2006 a). The soil in Sumer *et al.*'s (2006 a) experiment was placed in a box, 17.5 cm deep and 90 cm long, and the box was placed in the flume so that the soil surface was flush with the false bottom of the flume. The test set-up is shown in Fig. 3.1. The wave heights in the tests were in the

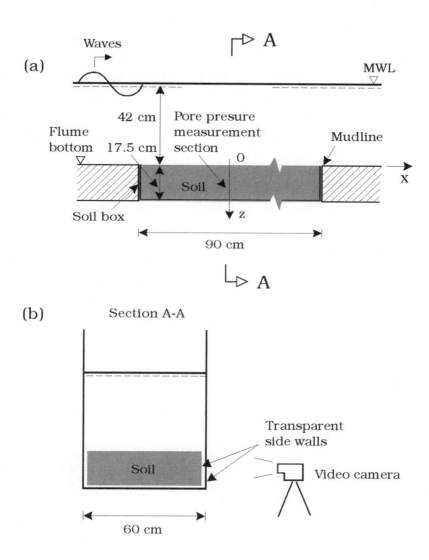

Figure 3.1: Test setup in Sumer *et al.*'s (2006 a) study.

range $H = 9 - 17$ cm, and the wave period was $T = 1.6$ s with the water depth $h = 42$ cm. The soil was silt with $d_{50} = 0.060$ mm and the geometric standard deviation $\sigma_g (=\sqrt{d_{84}/d_{16}}) = 1.8$. The quantity d_{50} is the grain size where 50% of soil is finer. Likewise, d_{84} is that where 84% is finer while d_{16} is that where 16% is finer. Other properties of the soil were as follows: the specific gravity of soil grains $s(=\gamma_s/\gamma) = 2.65$ (γ being the specific weight of water, and γ_s the specific weight of soil grains); the coefficient of lateral earth pressure $k_0 = 0.41$; the porosity $n = 0.42$; the total specific weight of the soil $\gamma_t = 19.25$ kN/m^3 (and therefore the submerged specific weight of the soil $\gamma' = \gamma_t - \gamma = 19.25 - 9.81 = 9.44$ kN/m^3); and the relative density $D_r = 0.38$ in which

$$D_r = \frac{e_{\max} - e}{e_{\max} - e_{\min}} \tag{3.1}$$

with e being the void ratio, and $e_{\max}$ and $e_{\min}$ the maximum and minimum void ratios, respectively. We note that, although the term density index is also used in the literature for the latter quantity, the term relative density is used here. However, this term should not be confused with its synonym "specific gravity". The values given for n, γ_t, and D_r above are all before-the-test values.

We now consider each stage of the liquefaction sequence (observed in Sumer et al.'s experiment) individually.

3.1.1   Buildup of pore pressure

Fig. 3.2 presents the pore-water pressure time series measured at different depths z (measured from the mudline, see Fig. 3.1) where p is the pore-water pressure in excess of the static pore-water pressure at these depths (also called excess pore pressure). The water surface elevation is plotted in the figure as $\eta/8$ for convenience, η being the water surface elevation; the reference level here is the still water level.

The figure shows that, with the introduction of waves, the excess pore water pressure begins to build up. As described in Section 1.2.1, the waves generate cyclic shear strains in the soil. If the grains are initially loosely packed, the cyclic shear strains in the soil will gradually rearrange the soil grains at the expense of the pore volume (Fig. 1.6). The latter effect will "pressurize" the water in the pores, and presumably lead to a buildup of

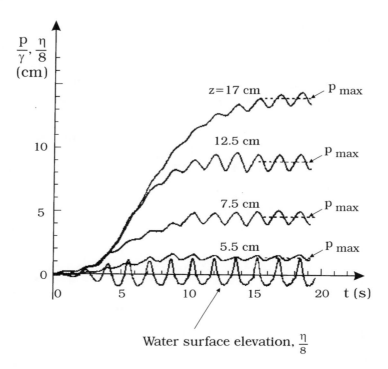

Figure 3.2: Time series of excess pore pressure in Sumer *et al.*'s (2006 a) Test 4 with $H = 16\,\mathrm{cm}$ and $T = 1.6\,\mathrm{s}$.

excess pore pressure in the case of an undrained soil. As the wave action continues, the excess pore pressure will continue to accumulate, as revealed by the measured time series in Fig. 3.2. See Video 2 on the CD-ROM accompanying the present book. When the accumulated pore pressure reaches a critical level, the soil will be liquefied. We also note that, from Fig. 3.2, the accumulated period-averaged pore water pressure, namely

$$\bar{p} = \frac{1}{T} \int_{t}^{t+T} p\, dt \qquad (3.2)$$

eventually reaches a constant value, $p_{\max}$.

Now, Fig. 3.3 illustrates the time development of the liquefaction depth in the same test. There are two sets of data: one (square symbols) obtained

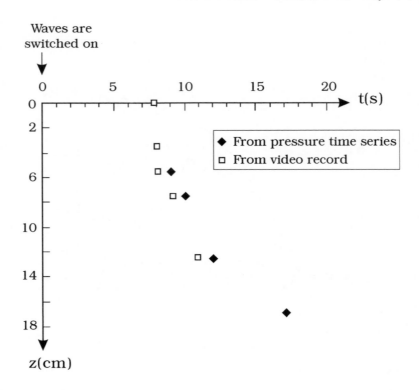

Figure 3.3:   Time development of liquefaction depth in Sumer *et al.*'s (2006 a) Test 4 with H = 16 cm and T = 1.6 s.

from the video record, and the other (diamond symbols) obtained from the pressure time series. In the former, the borderline between the liquefied soil and the rest of the soil was tracked as it evolved with time while the data points obtained from the pressure time series in Fig. 3.3 correspond to the time instants when the excess pore pressure reaches its maximum value p_{max}.

The data with square symbols clearly shows that the liquefaction first emerges at the mudline and subsequently spreads downwards. This observation agrees with that reported in Sassa and Sekiguchi (1999), Sassa, Sekiguchi and Miyamoto (2001), Teh, Palmer and Damgaard (2003) and Miyamoto *et al.* (2004).

Regarding the data with diamond symbols, although this data is associated with liquefaction, it does not correspond to the onset of liquefaction, contrary to what was perceived in Sumer *et al.* (2006 a). We shall return to this point later, in Section 3.1.2.

We also note that, in the experiments of Sumer *et al.* (2006 a), the liquefaction front travelled down to the impermeable base in all tests, and therefore the soil was liquefied across the entire sediment depth. When the soil depth is large, however, no significant shear deformations will be generated beyond a certain depth, and therefore the soil will be liquefied only to an intermediate depth (see, e.g., Teh *et al.*, 2003, and Sumer *et al.*, 2011, 2012).

Finally, we note that the Shields parameter, θ, in Sumer *et al.*'s (2006 a) experiments was significantly larger than the critical value, θ_{cr}, indicating that the sediment on the surface of the bed was set into motion before the onset of liquefaction. This was revealed visually in the experiments. Here, θ is defined by

$$\theta = \frac{U_f^2}{g(s-1)d_{50}} \tag{3.3}$$

in which U_f is the maximum value of the friction velocity at the bed

$$U_f = \sqrt{\frac{\tau_0}{\rho}} \tag{3.4}$$

with τ_0 being the maximum value of the bed shear stress on the surface of the bed induced by the wave boundary-layer flow, ρ is the water density, and g is the acceleration due to gravity, and θ_{cr} is the critical value of the Shields parameter corresponding to the initiation of sediment motion. The Shields parameter, a widely used parameter in hydraulic and coastal engineering, is a key parameter, characterizing the mobility of the sediment; the larger the value of the Shields parameter, the larger the mobility of the sediment under the shear stress on the bed generated by the boundary-layer flow. See, e.g., Fredsøe and Deigaard (1992) for general concepts of sediment transport in steady currents and waves.

3.1.2 Onset of liquefaction

This subsection addresses the issue of the onset of liquefaction. Sumer, Kirca and Fredsøe (2011, also published in 2012) studied the onset of liquefaction with the help of synchronized video recordings of the soil behaviour, and the pore-water pressure in a series of controlled liquefaction experiments. The procedure in Sumer *et al.*'s experiments was as follows:

(1) From the video recording, identify the time of the onset of liquefaction at a given depth.

(2) From the pressure time series at this depth, pick up the value of the period-averaged pore water pressure corresponding to the time of the onset of liquefaction. We designate this pressure by the symbol $\bar{p}_{cr}$ in which overbar indicates the period averaged value.

(3) Repeat this exercise for all the other depths where the synchronized pore pressure measurements were conducted.

It is important to elaborate on how Sumer *et al.* (2012) identified the onset of liquefaction that occurred on the video recording. The procedure was as follows: stop the video after an appreciable soil displacement due to liquefaction, taken arbitrarily like $O(10\,\text{mm})$; then measure this displacement, Δs, visually on the screen; subsequently, rewind the video by some 10 frames ($\sim 0.3\,\text{s}$), and stop it; measure Δs corresponding to this time, then repeat this exercise by rewinding the video by a small time increment each time, until Δs is no longer measurable; plot the obtained displacements Δs versus time, t, and; find the t-intercept of the curve $\Delta s = f(t)$ by extrapolating it linearly towards the t-axis. The time instant t-intercept is assumed to correspond to the inception (or onset) of liquefaction.

Now, Sumer *et al.*'s (2012) analysis of $\bar{p}_{cr}$ showed that $\bar{p}_{cr}$ is evidently equal to the initial mean normal effective stress, σ'_0,

$$\bar{p}_{cr} = \sigma'_0 \tag{3.5}$$

(see Fig. 3.4, reproduced from Sumer *et al.*, 2012). Here, σ'_0 is

$$\sigma'_0 = \frac{1}{3}(\gamma' z + k_0 \gamma' z + k_0 \gamma' z) = \gamma' z \frac{1 + 2k_0}{3} \tag{3.6}$$

in which γ' is the submerged specific weight of the soil (Appendix B),

$$\gamma' = \gamma_t - \gamma \tag{3.7}$$

and the quantity k_0 is the coefficient of lateral earth pressure (or the lateral stress ratio) at rest, and defined as the ratio of horizontal to vertical effective stress, $k_0 = \sigma'_h / \sigma'_v$ at rest (Lambe and Whitman, 1969, p. 127).

(The magnitude of k_0 depends on the amount of frictional resistance mobilized at contact points between grains. This is reflected in the following expression, the Jaky equation,

$$k_0 = 1 - \sin \varphi \tag{3.8}$$

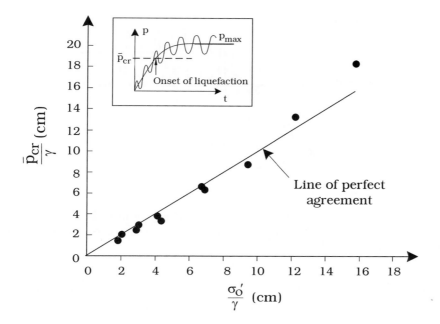

Figure 3.4: Critical values of the period-averaged excess pore pressure corresponding to the onset of liquefaction, plotted against the initial mean normal effective stress values at the same depth. Sumer *et al.* (2012).

in which φ is the friction angle (Lambe and Whitman, 1969, p. 127 and pp. 139–141). We note that, in Sumer *et al.*'s (2012) experiments, the friction angle was determined as the angle of repose in a simple test where the oven-dried soil was gently poured onto a horizontal plate in air, and the angle of repose was measured.)

From above considerations, the criterion for the onset of liquefaction can be given as:

$$\text{liquefaction occurs when } \bar{p} > \sigma_0' \qquad (3.9)$$

or, alternatively, in terms of the pore-pressure ratio

$$\text{liquefaction occurs when } \frac{\bar{p}}{\sigma_0'} > 1. \qquad (3.10)$$

It may be noted that, as seen, liquefaction occurs not at the time when the accumulated pore water pressure reaches p_{max} (as was perceived by Sumer

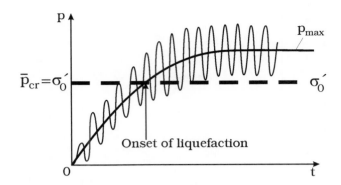

Figure 3.5:   Schematic description of buildup of pore pressure, the initial mean normal effective stress, σ_0', and p_{max}. Liquefaction occurs when the excess pore pressure reaches σ_0', which is distinctly smaller than p_{max}.

et al., 2006 a) but somewhat earlier, at the time when $\bar{p}$ reaches the value $\bar{p}_{cr} = \sigma_0'$, which is distinctly smaller than p_{max}; see the sketch in Fig. 3.5. See Video 2 on the CD-ROM accompanying the present book.

Fig. 3.6 displays the time series of pore pressure from a laboratory test (Sumer et al., 2010) with the values of σ_0' and p_{max} marked in the figure. The soil was silt with $d_{50} = 0.098\,\mathrm{mm}$. The wave height in the test was $H = 17\,\mathrm{cm}$, the period $T = 1.6\,\mathrm{s}$, with the water depth $h = 40\,\mathrm{cm}$. The pressure time series was obtained at a depth of $z = 16\,\mathrm{cm}$. Fig. 3.6 clearly shows that there is room for further buildup of pore pressure after the pressure reaches σ_0', and therefore, although marginal, there is still some effective stress as well as shear strength present (see also the analysis in Sassa and Sekiguchi, 2001, Figs 15–18). By contrast, when the complete liquefaction state is reached, there will be no room for further buildup of pore pressure, and therefore there will be no effective stress and shear strength present, and thus the soil becomes a fluid.

The liquefaction criterion had been an issue of much debate (see, e.g., Sumer et al., 2006 a). The question was whether the critical value of the period-averaged pore pressure, $\bar{p}_{cr}$, should be set equal to the initial effective stress (also termed the initial overburden pressure, or initial vertical effective stress, e.g. Teh et al., 2003), σ_{v0}',

$$\sigma_{v0}' = \gamma' z \tag{3.11}$$

or it should be set equal to the initial mean normal effective stress, σ_0', Eq. 3.6.

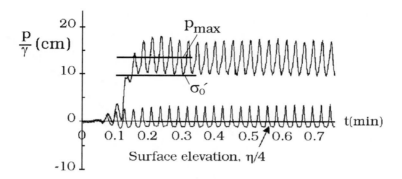

Figure 3.6:   Time series of pore pressure from a laboratory test with the values of σ_0' and $p_{\max}$ marked. Sumer *et al.* (2010).

Sumer *et al.*'s (2012) work did indicate (as discussed in the preceding paragraphs) that $\overline{p}_{cr} = \sigma_0'$, linked arguably to the fact that the latter compares two
isotropic quantities ($\overline{p}_{cr}$ and σ_0'), whereas the former compares one isotropic
quantity ($\overline{p}_{cr}$) with an anisotropic one (σ_{v0}'). The liquefaction criterion given
in Eq. 3.9 has been adopted by several studies in the past, e.g., Ishihara and
Yamazaki (1984, Eq. 36), McDougal, Tsai, Liu and Clukey (1989), Sumer,
Fredsøe, Christensen and Lind (1999), Jeng and Seymour (2007), and Jeng,
Seymour and Li (2007).

The above discussion is focused around the onset of liquefaction and the
initial mean normal effective stress, σ_0'. We now discuss the maximum pressure attained by the period-averaged pore pressure, for large times, $p_{\max}$.

Fig. 3.7 displays $p_{\max}$ values plotted against

$$(\gamma_{liq} - \gamma)z = \gamma_{liq}z - \gamma z \tag{3.12}$$

in which γ_{liq} is the specific weight of the liquefied soil, and the first term
on the right-hand side of the above equation, $\gamma_{liq}z$, is the actual pressure
at depth z in the liquefied soil, and the second term, γz, is the hydrostatic
pressure at the same depth, and therefore the quantity $(\gamma_{liq} - \gamma)z$ can be
interpreted as the pressure in excess of the hydrostatic pressure experienced
in the liquefaction stage, which we will designate with the symbol p_{liq}:

$$p_{liq} = (\gamma_{liq} - \gamma)z. \tag{3.13}$$

The p_{liq} values plotted in Fig. 3.7 are calculated from the above equation
in which γ_{liq} is calculated in two ways:

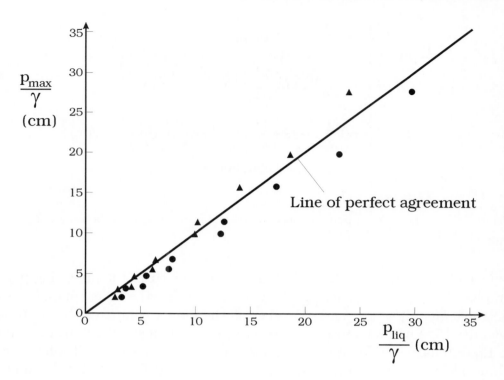

Figure 3.7:   Measured maximum excess pore pressure plotted against the excess pore pressure calculated from $p_{liq} = (\gamma_{liq} - \gamma)z$, at the same depth, in which γ_{liq} is calculated in two ways: (1) from $\gamma_{liq} = \gamma(s + e_{max})/(1 + e_{max})$, triangles; and (2) from the mathematical model of Sumer *et al.* (2006 b) (see Section 5.4), circles.

(1) It is assumed that the void ratio in the liquefaction stage is approximately equal to the maximum void ratio, and therefore Eq. 5.7 is used to calculate γ_{liq} (see discussion in Chapter 5, Section 5.3), $\gamma_{liq} = \gamma(s + e_{max})/(1 + e_{max})$. The triangular symbols in Fig. 3.7 correspond to this approximation.

(2) γ_{liq} is calculated from the mathematical model of Sumer *et al.* (2006 b) (see Chapter 5, Section 5.4). Circular symbols in Fig. 3.7 correspond to this approach.

It may be noted that the γ_{liq} value obtained from (1) is $\gamma_{liq}/\gamma = 1.76$ whereas that obtained from (2) is $\gamma_{liq}/\gamma = 1.94$.

Although not reported in Sumer *et al.* (2012), the data in Fig. 3.7 is actually from the same experiments as those discussed in conjunction with Fig. 3.4.

The solid line in Fig. 3.7 represents the line of perfect agreement. The figure implies that, not unexpectedly, p_{max}, the maximum pressure in excess of the hydrostatic pressure at depth z during the liquefaction stage, can, to a first approximation, be approximated to

$$p_{max} = (\gamma_{liq} - \gamma)z \tag{3.14}$$

i.e., the pressure in excess of the hydrostatic pressure experienced in the liquefaction stage.

Finally we note that, when inspected closely, Fig. 3.3 indicates that, at a given depth z, the maximum pressure p_{max} is reached later than the onset of liquefaction, in agreement with the description given in the preceding paragraphs.

3.1.3 Liquefaction stage

When the soil is liquefied, the effective stresses between the individual grains in the bed vanish, and therefore the water–soil mixture (as a whole) acts like a liquid.

In Sumer *et al.*'s (2006 a) experiments, one test was carried out with a wave height not large enough to cause liquefaction. p_{max} values measured in this test were substantially smaller than the corresponding values of the initial mean normal effective stress (e.g., $p_{max}/\gamma = 0.8$ cm versus $\sigma_0'/\gamma = 9.6$ cm at $z = 17$ cm), indicating that the soil was in the no-liquefaction state in this test. The result of this test and those from the liquefaction-regime tests are plotted in Fig. 3.8 for $z = 12.5$ cm. The initial mean normal effective stress σ_0' (Eq. 3.6) is also plotted in Fig. 3.8 as a reference line (dashed line). In the figure, Sumer *et al.*'s earlier data (circles, Sumer *et al.*, 1999) are also included along with the data from Sumer *et al.* (2006 a) (crosses). As seen, the transition between the no-liquefaction regime and the liquefaction regime occurs with a sudden jump. This "discontinuous" transition may be linked to the change in the state of the soil; the sediment is in the solid state in the no-liquefaction regime whereas, it is in the liquid state in the liquefaction regime.

With the soil liquefied, the water column and the liquefied soil will form a two-layered system of liquids of different density. The interface between the layers of this system will experience an internal wave (Fig. 3.9).

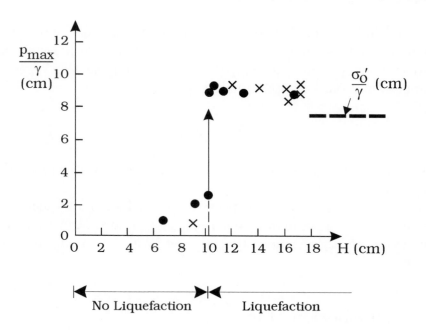

Figure 3.8:   Maximum excess pore pressure as function of wave height. Crosses: Sumer
et al. (2006 a) ($z = 12.5$ cm). Circles: Sumer *et al.* (1999) ($z = 12.1$ cm). σ_0' is initial mean
normal effective stress at $z = 12.5$ cm for Sumer *et al.* (2006 a) data.

For a two-layered system extending to infinity in the horizontal direction (Fig. 3.9), the ratio of amplitudes of the interfacial elevation and water
surface elevation (Fig. 3.9) is given by

$$\frac{a_2}{a_1} = \frac{1}{\omega^2} \cosh(\lambda h_1)(\omega^2 - g\lambda \tanh(\lambda h_1)) \tag{3.15}$$

with the dispersion relation

$$\left(1 - \left(\frac{g\lambda}{\omega^2}\right)^2\right) \tanh(\lambda h_1) \tanh(\lambda h_2)$$

$$= \frac{\rho_2}{\rho_1}\left(1 - \frac{g\lambda}{\omega^2}\tanh(\lambda h_1)\right)\left(1 + \frac{g\lambda}{\omega^2}\tanh(\lambda h_2)\right). \tag{3.16}$$

Here, ρ_1 and ρ_2 are the densities of water and liquefied soil, respectively, g
is the acceleration due to gravity, λ is the wave number, $\lambda = 2\pi/L$ (L being
the wave length) and ω is the angular frequency, $\omega = 2\pi/T$ (T being the
wave period). The preceding equations are those of Sassa *et al.* (2001) who
implemented Lamb's (1945) theory for wave propagation in a two-layer fluid
for the present two-layered system.

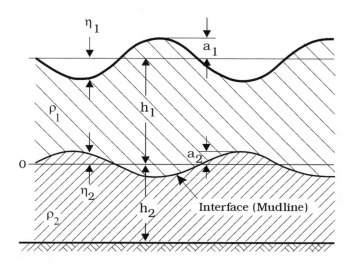

Figure 3.9: The liquefied soil at the interface of the two-layered system of liquids (water and liquefied soil) experiences internal waves.

Sumer *et al.* (2006 a) found that the experimental value of a_2/a_1, 0.125, was not radically different from that calculated from Eqs 3.15 and 3.16, namely 0.15. Sumer *et al.* (2006 a) also presented data related to the horizontal amplitudes of the orbital motion across the liquefied soil depth, including the amplitudes corresponding to the water particles just above the mudline. This data show that the motion in the liquefied soil is rather small (15–25% of that of water particles at the mudline) and rapidly decreases with the depth below the mudline.

3.1.4 Dissipation of accumulated pressure and compaction

Time series of pore pressure for large times

Recall the pore pressure time series given in Fig. 3.2, which displays the behaviour of the pore pressure for the initial 20 s of the process. Now, Fig. 3.10 displays the behaviour of the pore pressure for the entire length of the test (21.4 minutes). Sumer *et al.* (2006 a) present several similar pore pressure time series obtained for other wave heights as well.

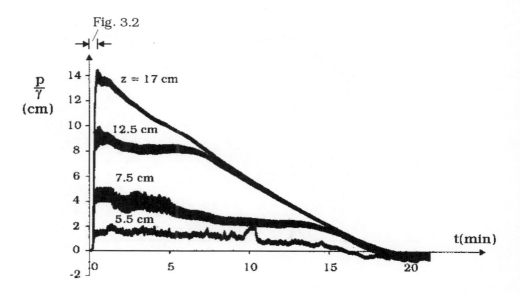

Figure 3.10:   Time series of excess pore pressure in Sumer *et al.*'s (2006) Test 4 with H = 16 cm, T = 1.6 s for the entire test duration.

The way in which the excess pore pressure varies with time (Fig. 3.10) over the entire length of the test is shown schematically in Fig. 3.11 in which $\bar{p}$ is the period-averaged pore-water pressure (Eq. 3.2).

Compaction process and dissipation of pore pressure

With the introduction of waves, first the excess pore pressure begins to build up (A in Fig. 3.11), as detailed in the preceding paragraphs. With the buildup of excess pore pressure, an upward-directed pressure gradient is generated; the accumulated pressure is largest at the impermeable base and smallest at the mudline (Fig. 3.10), generating an upward-directed pressure gradient.

This pressure gradient drives the water in the liquefied soil upwards while the soil grains settle through the water until they begin to come into contact with each other. The latter process (where grains come into contact) first begins at the impermeable base and gradually progresses in the upward direction. The process somewhat resembles self-weight consolidation of hydraulic fill (Powrie, 2004, p. 224) although, in the present case, waves continue during the process, whereas no waves, or any external pressure-generation mechanisms, exist in the case of the self-weight consolidation process. (We shall

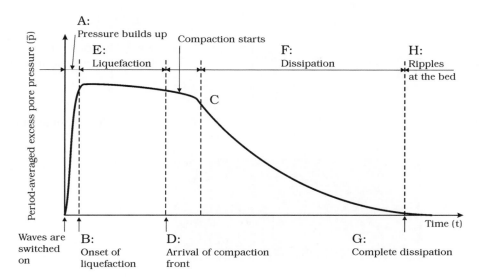

Figure 3.11:   Time series of period-averaged excess pore pressure at depth z. Schematic description.

return to the similarity between the present compaction process and the self-weight consolidation of hydraulic fill later in the section.)

This wave-induced compaction process has been studied independently by (1) Sumer *et al.* (2006 a) (first reported in Sumer *et al.*, 2004), using an ordinary wave flume, and (2) Miyamoto *et al.* (2004), utilizing a very small wave tank placed in a centrifuge. Miyamoto *et al.* (2004) termed the process "solidification" while Sumer *et al.* (2006 a) called it "compaction". The latter term is used in Sumer *et al.* (2006 a), on grounds that it refers to compaction of sand by waves, an expression frequently used in hydraulic/coastal engineering practice.

Fig. 3.12 depicts a sketch, a "snapshot", illustrating the soil with two layers, the top layer being in the liquefied state and the bottom layer in the solid state with the interface between the two layers (the compaction front) moving gradually in the upward direction. See Video 3 on the CD-ROM accompanying the present book. The visual observations (Sumer *et al.*, 2006 a) indicate that, at any time, there are two distinct layers of soil; one with a distinct "orbital motion" of soil particles (the top layer), and the other with no orbital motion at all (the bottom layer). Therefore the behaviour of the bed changes from essentially liquid in the upper layer to essentially solid

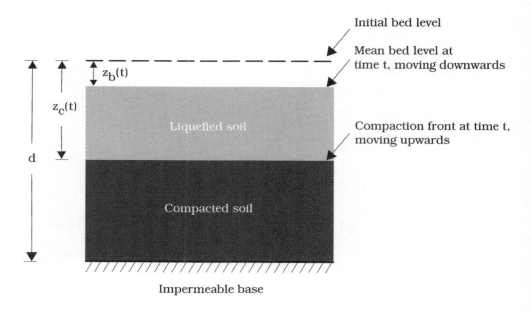

Figure 3.12:   Snapshot at time t after compaction starts at the impermeable base following the liquefaction of the soil across the entire depth.

in the lower layer. Miyamoto *et al.* (2004), too, observed such a distinct interface, the solidification front, in their centrifuge experiment.

The time at which the compaction front arrives at a certain depth z (obtained from video recordings) is marked on the pore pressure time series in Fig. 3.13, reproduced from Fig. 3.10 (vertical arrows: time of arrival of the compaction front for each time series except $z = 17$ cm where there were no compaction-front data available.) Now, going back to Fig. 3.11, liquefaction starts at B in Fig. 3.11 and in a short while the pore pressure reaches its maximum value, and then there will be practically no change in the pore pressure until sedimentation stops and compaction starts, i.e., the sediment grains come into contact. However, for some further time after the start of compaction, there will be practically no change in excess pore pressure. Thus, true compaction will start somewhere between D and C in Fig. 3.11. It may be noted that although it has been stated that there will be practically no change in the excess pore pressure, there is in fact a slight decrease in the excess pore pressure from A to C (the point where there is a substantial

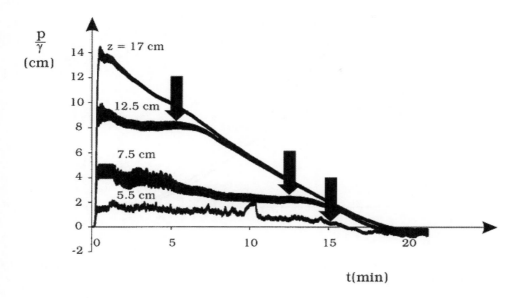

Figure 3.13: Time series of excess pore pressure in Sumer *et al.*'s (2006 a) Test 4 with H = 16 cm and T = 1.6 s. Vertical arrows: the time at which the compaction front arrives at a certain depth z.

change in the slope of the pressure time series) in Fig. 3.11. This is linked to the fact that the effective weight of the sediment above a z plane decreases gradually with time because of the downward transport of sediment grains across that plane (Miyamoto *et al.*, 2004).

Point C in Fig. 3.11 is actually the time where the pressure time series of the two neighbouring measurement points appear close together. When inspected closely, it is seen that the pressure time series come quite close to each other, but they do not "collapse" (Fig. 3.10). Consider, for example, the pressure time series for $z = 17$ cm and $z = 12.5$ cm in Fig. 3.13. When these two time series approach each other near C (Fig. 3.11), the excess pore pressure at $z = 12.5$ cm will become quite close (but not equal) to that at $z = 17$ cm (Fig. 3.14), implying that the upward-directed pressure gradient between these two points will still exist (although reduced quite considerably), and therefore the upward seepage flow will still continue at this depth (at $z = 12.5$ cm). This implies that the soil below this depth

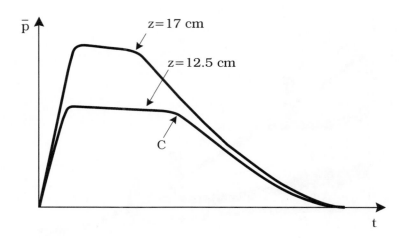

Figure 3.14:   Schematic description of the time series of excess pore pressure at two neigh-bouring measurement points.

will still experience compaction. As the water is drained out of the soil, the excess pore pressure will continue to be dissipated (F in Fig. 3.11 and Fig. 3.14).

As a result of the compaction process, the mudline will continuously move downwards (z_b, Fig. 3.12). One important point here is the following. As is well known from the wave boundary layer theory, there exists a sediment transport in the onshore direction because of the onshore-directed steady streaming present very near the bed under a progressive wave (see e.g. Fredsøe and Deigaard, 1992, p. 44). The question is: What is the contribution of this sediment transport in Sumer *et al.*'s (2006 a) experiment to the downward displacement of the mudline. (1) Sumer *et al.* (2006 a) calculated this contribution from the measured downward movement of the mudline and the before-the-test and after-the-test values of the bed porosity. (2) They also measured it by collecting the sediment transported onshore after the whole liquefaction/compaction sequence was completed, and determining the volume of the collected sediment. Their calculations and measurements showed that the contribution of this sediment transport was only $O(20\%)$. Therefore, the downward movement of the mudline was, in Sumer *et al.*'s experiment, mainly due to the compaction process, similar to that due to self-weight consolidation of hydraulic fill.

The time at which the excess pore pressure is completely dissipated (G in Fig. 3.11) coincides with the time at which the compaction front arrives at the mudline where the compaction process is all but over (i.e., no further settlement takes place and the excess pore pressure have substantially dissipated).

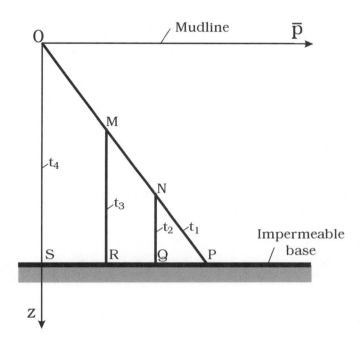

Figure 3.15: Period-averaged excess pore pressure distributions at different times: OP at time t_1, ONQ at time t_2, OMR at time t_3, and OS at time t_4.

Finally, Fig. 3.15 displays another representation (schematic) of the period-averaged pore water pressure as a function of the depth and time. Time t_1 corresponds to the time where the compaction process starts at the impermeable base while time t_4 corresponds to that where the compaction front arrives at the mudline, and therefore the compaction process ends. Times t_2 and t_3 are two arbitrary times between t_1 and t_4. One can easily construct this diagram from the time series given in, e.g., Fig. 3.13. As seen, the initial accumulated pore pressure distribution represented by the line OP gradually changes to ONQ at time t_2, to OMR at time t_3 and finally to OS (complete dissipation) at time t_4.

Degree of compaction

The quantity D_r, the relative density (Eq. 3.1), may be used to illustrate the degree of compaction of the soil after the liquefaction/compaction sequence. This quantity was determined before the test and after the test (i.e., after the soil underwent the liquefaction/compaction sequence) in Sumer *et al.*'s (2006 a) test. D_r was $D_r = 0.38$ before the test, as indicated earlier, while it was $D_r = 0.71$ for the wave height $H = 11$ cm, $D_r = 0.73$ for $H = 14$ cm, and $D_r = 0.78$ for $H = 17$ cm, corresponding to "dense" soil category, Table B.3, Appendix B. As seen, the compaction by the action of waves is immense. It may be noted that there is a trend that the compaction increases (D_r increases) with increasing wave height. This is expected because the larger the wave height, the larger the "shaking" of the sub-liquefied soil, and therefore the larger the density of the compacted sediment.

Similarity between wave-induced compaction and self-weight consolidation

In the preceding paragraphs, we have pointed out that the process of compaction somewhat resembles the self-weight consolidation of hydraulic fill (Powrie, 2004, p. 224). We have also noted, however, that the present compaction process differs from the self-weight consolidation process in that waves continue during the compaction process whereas no waves, or any external pressure-generation mechanism, exists in the case of the self-weight consolidation process.

In order to compare the compaction process with self-weight consolidation of hydraulic fill, Miyamoto *et al.* (2004) conducted the following two tests in their centrifuge wave testing: (1) a regular liquefaction test, similar to that presented in Fig. 3.13; and (2) a liquefaction test where the sediment was liquefied and immediately after that (but prior to the point where compaction starts), the waves were switched off and therefore the sediment began to consolidate, simulating the process of self-weight consolidation of hydraulic fill. They measured excess pore pressure in these two tests and compared the time development of the excess pore pressure with each other (Miyamoto *et al.*, 2004, Fig. 13). This comparison indicated that, in the second test, the excess pore pressure dissipated much more rapidly than in the first test. The slower dissipation of the pore pressure in the case when the waves continue can be explained by the fact that the waves and therefore the wave-induced shear strain in the sediment will generate additional excess pore pressure, and this effect will delay the dissipation of the excess pore pressure.

One final note is that it is expected that the compaction following a typical liquefaction/compaction sequence is arguably the most effective way to compact the sediment, and would therefore lead to, probably, relatively more dense soil, as the upward-directed pressure gradient generated during the process of the pressure buildup and liquefaction drives the water upward (or drains the water out), and therefore helps compact the soil even more effectively, in addition to the effect where the soil grains settle through the water.

Dissipation of pore pressure in the no-liquefaction regime

Fig. 3.16 displays the time series of the excess pore pressure in the case of the no-liquefaction regime from Sumer *et al.*'s (2006 a) work. p_{max} values are substantially smaller than the corresponding values of the initial mean normal effective stress, e.g., $\sigma_0' = 9.6$ cm at $z = 17$ cm, indicating that the soil is in the no-liquefaction state.

As seen, the way in which the excess pore pressure is dissipated is qualitatively much the same as in the case of the liquefaction regime, cf. Fig. 3.13. However, there are significant quantitative differences, as expected, such as, the dissipation time is much shorter; the upward-directed pressure gradient generated during the buildup of pore pressure is much smaller, and therefore it is much less effective in draining out the pore water, and consequently the compaction is much less than in the case of the liquefaction regime.

Bed forms upon completion of liquefaction/compaction process

Although not directly related to the subject matter, it is interesting to mention the following observation of Sumer *et al.* (2006 a). When the compaction process is completed (with the arrival of the compaction front at the mudline), ripples begin to emerge on the bed. As the waves continue, these initial ripples eventually grow in size and attain an equilibrium state where the bed is covered with fully-developed ripples (stage H in Fig. 3.11). Sumer *et al.*'s data (2006 a) showed that the ripple steepness (normalized with the angle of repose) for sediment with liquefaction history is the same as that in sediment with no liquefaction history. See Sumer *et al.* (2006 a) for further details.

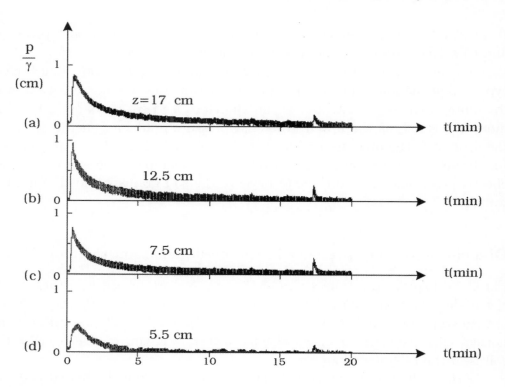

Figure 3.16:   Time series of excess pore pressure in the case of no liquefaction in Sumer *et al.*'s (2006 a) test 1 with H = 9 cm and T = 1.6 s.

This aspect of the problem will not be pursued further as it is beyond the scope of this book.

A great many works have been devoted to the experimental investigation of buildup of pore pressure induced by waves.

Clukey, Kulhawy and Liu (1983, 1985) report the results of an experimental investigation of the buildup of pore pressure induced by progressive waves in a laboratory flume. Tzang, Hunt and Foda (1992), Tzang (1998) and Sumer *et al.* (1999) present similar laboratory results. The sediment in the latter investigations was silt.

de Wit and Kranenburg (1992) report the results of an experimental study where the soil was a cohesive material; particular attention was concentrated on the liquefaction and erosion of China clay due to waves and current. Their results have shown that similar buildup of pore pressure is experienced in this kind of soil. Further results of the research undertaken by the same group have been reported in the publications by de Wit, Kranenburg and Battjes (1994), de Wit (1995), van Kessel, Kranenburg and Battjes (1996), de Wit and Kranenburg (1992), and van Kessel and Kranenburg (1998). Liquefaction of cohesive soils will be discussed further in Section 3.5.

Foda and Tzang (1994) give an account of what they call the resonant fluidization, a process where massive liquefaction failure due to buildup of pore pressure occurs by waves. The authors link this process to a strong channeling of the seepage flow within the silt bed. (See also the review article by Foda, 1995).

Tzang (1998) summarizes the results of a study on the oscillating component of the pore pressure in silt where there is a buildup of pore pressure. Tzang's (1998, p. 292) results show that the oscillating component behaves like that of a poro-elastic sandy soil. However, it may be noted that the accumulated pressure in Tzang's experiments was relatively small, mostly $O(0.1)$ times the overburden pressure (or even less) (Tzang, 1998, Table 4).

Finally, the Kyoto University group has studied buildup of pore pressure, liquefaction and compaction in a centrifuge wave tank facility; see among other publications, Sassa and Sekiguchi (1999) and Miyamoto et $al.$ (2004). A detailed account of their work will be given in Section 3.3.

Example 1. *Effect of irregular waves.*

Sumer et $al.$'s (1999) experiments showed that the process of buildup of pressure in irregular waves occurs in much the same way as in the case of the regular waves.

These authors addressed the question of how to define the wave parameters in the case of irregular waves with regard to the process of pore pressure accumulation. Several combinations of the wave heights (H_s, $H_s/\sqrt{2}$, etc.) and the wave periods (T_p, T_z, etc.) were tested to see which combination would give the best comparison with the regular wave results, when the results were plotted in terms of the wave height versus the number of waves to cause liquefaction. Here, H_s = the significant wave height, T_p = the peak period, and T_z = the mean zero upcrossing period. The results showed that the combination $H_s/\sqrt{2}$ and T_z gave the best agreement. Incidentally, the quantity $H_s/\sqrt{2}$ can be interpreted as the equivalent wave height of the irregular waves, since $H_s/\sqrt{2} \simeq (4\sigma_\eta)/\sqrt{2} = 2(\sqrt{2}\sigma_\eta) = 2\,a_\eta$, in which σ_η = the standard deviation of the surface elevation η,

$$\sigma_\eta = (\overline{\eta^2})^{1/2} \tag{3.17}$$

whereas, a_η = the amplitude of the surface elevation η when the waves are sinusoidal. Here, η is the surface elevation from the mean water level. Basic knowledge related to statistical treatment of irregular waves can be found in e.g. Sumer and Fredsøe (1997, pp. 297–317).

Residual liquefaction under irregular waves has also been investigated by Chen, Tzang and Ou (2008), and Dong and Xu (2010) and Xu and Dong (2011). Chen *et al.* (2008) found that what they call stepped pore pressure buildups (where the pore pressure buildup takes longer) are more likely to happen under irregular waves than under monochromatic waves, which is linked to the fact that, owing to non-uniform wave loading, the pore pressures cannot continuously and rapidly accumulate. Dong and Xu (2010), in a theoretical study, concluded that the liquefaction depth could be much deeper than that of regular waves, cautioning that the "current design practice, based on the regular wave theory, may underestimate the liquefaction depth and lead to unsafe design".

We note that, although not directly related, residual liquefaction under regular wave groups has been investigated by Tzang and Ou (2004).

Example 2. *Effect of history of wave exposure on the pressure buildup.*

To observe the effect of the history of wave exposure, Sumer *et al.* (1999) carried out the following systematic test where the soil was exposed more than once to a progressive wave with $H = 16.6$ cm, and $T = 1.6$ s. The following procedure was used:

1. Expose the soil to the waves for 20 minutes. Monitor the pore pressure.

2. Stop the waves. Wait for 10 minutes.

3. Switch on the waves again, and expose the soil to the same waves for another 20 minutes. Monitor the pore pressure.

4. Repeat steps 1–3 a number of times.

From these experiments it was found that $p_{\max}$, the maximum accumulated pore pressure (Fig. 3.5), was reduced tremendously when the soil was exposed to the waves for the second time. While p_{max}/γ at $z = 16.5\,\text{cm}$ was about $11\,\text{cm}$ for the first exposure, it was only $0.6\,\text{cm}$ when the soil was exposed to the same waves for the second time, an order of magnitude reduction. The pressure p_{max} was virtually nil, when the soil was exposed to the same waves for the third time. This behaviour is explained in the following.

The soil grains (when they are exposed to the waves for the first time) rearrange through the process of the buildup of pore pressure, liquefaction and dissipation (e.g., Fig. 3.11). When the soil is exposed to the waves for the second time, there will not be too much "room" for the grains to rearrange (the grains now being much more densely packed after the first "round" of buildup and dissipation of the pore pressure), and hence the pore pressure will accumulate only slightly.

Centrifuge wave-tank experiments of the Kyoto group (Miyamoto, Sassa and Sekiguchi, 2003) indicated similar results. These authors called this pre-shearing effect. They also developed an analytical model to reproduce the observed effect, and predicted the behaviour of a sand bed under a complex wave loading history.

Bjerrum (1973) illustrated the effect of preshearing on the generation of pore-water pressure, and thereby liquefaction potential, describing the test results of undrained simple shear with cyclic loading on fine sand samples prepared with relative densities of 80%. With a set of test results, Bjerrum (1973) observed that the preshearing in the test has apparently reduced the generation of pore pressures, by a factor ranging from 20 to 50.

The work by van Kessel and Kranenburg (1998) can also be consulted for a somewhat detailed discussion of this issue in relation to the seabed liquefaction.

From the preceding paragraphs it may be concluded that, *in a real-life situation, the soil will not be liquefied due to the wave action because of its long-term history of wave exposure.* However, important exceptions to this are (1) when the soil is "softened" or "loosened" (e.g., when it is used as a backfill material), or (2) when there is slow sedimentation, or (3) when the seabed is active, i.e., when there is continuous sediment transport in areas covered with sand waves in which case the top $O(1\,\text{m})$ (or more) of the soil is constantly being reworked by sediment transport. In these cases, the seabed sediment will be loose, presumably susceptible to residual liquefaction due to waves (when the sediment is fine, i.e., fine sand, silt, etc.). Loose seabed sediment is occasionally revealed by field surveys, indicating a loose, or even

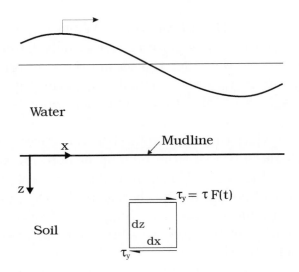

Figure 3.17:   The shear stress in the soil. F(t) is a periodic function of time.

very loose, sediment (see Appendix B, Table B.3, for the definition of various soil categories).

3.2   Mathematical Modelling

3.2.1   Peacock and Seed's (1968) experiment

As described in the previous section, under a progressive wave, a soil element of the seabed will undergo a cyclic shear stress variation (Fig. 3.17).

A similar process also occurs in the case of an earthquake, as mentioned previously (Chapter 1). A soil element below a horizontal ground surface during the earthquake undergoes a cyclic shear stress variation, as sketched in Fig. 3.18. This cyclic shear stress variation will lead to a progressive buildup of pore pressure, in precisely the same fashion as in waves, which may lead to the liquefaction of the soil (Section 1.3).

In order to simulate the pore-pressure accumulation during an earthquake, Peacock and Seed (1968) conducted laboratory experiments under cyclic stress conditions causing liquefaction of saturated sand in *undrained* simple shear tests. The following paragraphs will summarize the highlights of this important work. As will be seen in the next sub-section, the end result of

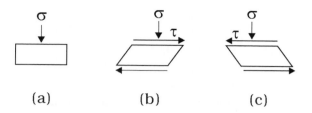

Figure 3.18: Idealized stress condition for element of soil below ground surface during an earthquake. Peacock and Seed (1968).

Peacock and Seed's work will be one of the key elements of the theory to be developed to describe the buildup of pore pressure under a progressive wave.

The equipment used for these tests essentially consisted of a simple shear box and an arrangement for applying a horizontal, cyclic, shear-stress load to the soil (Fig. 3.19).

The soil sample was *consolidated* under an initial confining pressure. This initial confining pressure is obviously apportioned by the soil and the water. The soil portion of the initial confining pressure, i.e., the effective stress, was $\sigma_0' = 5\,\mathrm{kg/cm^2}$, while the water portion, i.e., the initial pore pressure, was $p_0 = 1\,\mathrm{kg/cm^2}$ (Fig. 3.20) in Peacock and Seed's (1968) experiment.

The vertical load remained constant during the application of the cyclic shear stress in the experiments.

Fig. 3.20 displays the results of a typical test, reproduced from Peacock and Seed's paper. The top diagram shows the time series of the pore pressure, the middle diagram the time series of the shear strain, and the bottom diagram the time series of the applied shear stress.

As seen clearly from Fig. 3.20, the application of the cyclic shear stress on the soil sample generates an excess pore pressure, p, in the soil, and this pressure progressively builds up, as the cyclic loading continues. The action of the cyclic shear stress on the soil can be explained in the same way as in the case of the waves (Sections 1.2.1 and 3.1.1).

The process of the buildup of pore pressure will come to an end when the accumulated pore pressure reaches the level of the initial effective stress, σ_0'. When this point is reached, i.e., when

$$p = \sigma_0' \tag{3.18}$$

the total load will be carried by the water alone, and the effective stress will become zero, and therefore the soil will fail, the *soil liquefaction*. As seen, in

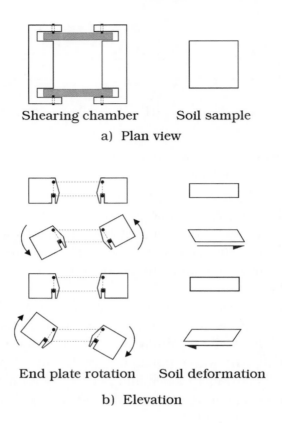

Shearing chamber   Soil sample

a) **Plan view**

End plate rotation   Soil deformation

b) **Elevation**

Figure 3.19:   Idealized stress condition for element of soil below ground surface during an earthquake, simulated in Peacocknd Seed's (1968) experiment.

the test presented in Fig. 3.20, this point is reached after 24 cycles; we shall return to this point later in the section.

(Incidentally, Fig. 3.20b clearly shows that the failure sets in precisely at the same instant as the liquefaction occurs, Fig. 3.20a. The fact that practically no significant shear strain/deformation occurs until this moment is reached, Fig. 3.20b, indicates that the soil failure is due only to the liquefaction alone, but not due to a combination of liquefaction and shear failure.)

The description in the preceding paragraphs implies that, for the generation of the buildup of pressure, the drainage of the pore water from the soil must be zero or very small. If the water can "escape" from the soil relatively

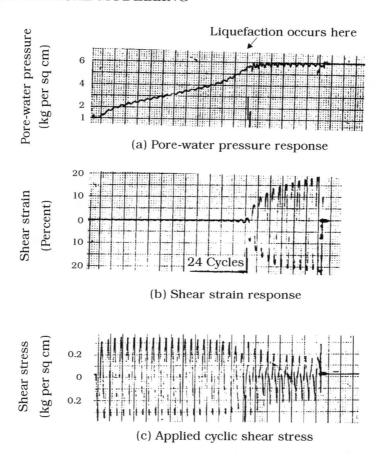

(a) Pore-water pressure response

(b) Shear strain response

(c) Applied cyclic shear stress

Figure 3.20:   Time series of pore pressure, shear strain, and shear stress in Peacock and Seed's (1968) experiment.

quickly, the pore pressure will be relieved, therefore no significant buildup of pressure will develop. This implies that, in practice, the pore-pressure accumulation occurs normally in soils with low permeability (such as silt).

Likewise, the pore-pressure accumulation develops only when the frequency of the cyclic loading is sufficiently high. If the frequency is low, the accumulated pore pressure will dissipate as rapidly as it develops; therefore, no significant pore pressure accumulation will take place.

An important quantity in the analysis of liquefaction is the number of cycles to cause liquefaction, N_ℓ. This quantity is mainly dependent on the

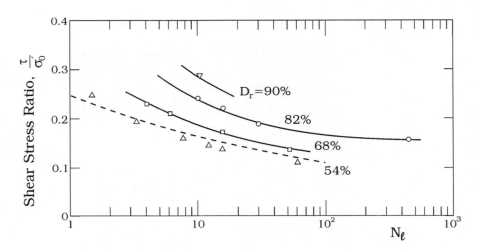

Figure 3.21:   Ratio of the amplitude of the shear stress in the soil to the initial effective stress versus the number of cycles to cause liquefaction. Alba *et al.* (1976).

following parameters

$$N_\ell = function \ (\tau, \ \sigma'_0, \ D_r)$$ (3.19)

in which τ is the amplitude of the oscillating shear stress acting on the soil, and D_r is the relative density of the soil defined by Eq. 3.1 with e = the void ratio, and $e_{\max}$ and $e_{\min}$ = the maximum void ratio (in the loosest condition) and minimum void ratio (in the densest condition), respectively, obtained in the way as described, for example, in Peacock and Seed (1968).

Peacock and Seed (1968) and Alba, Seed and Chan (1976) carried out undrained simple shear tests for different values of the relative density, and plotted the data N_ℓ versus the normalized shear stress τ/σ'_0. Fig. 3.21 displays this diagram (adapted from Alba *et al.*, 1976). It is seen that the number of cycles to cause liquefaction increases tremendously with decreasing shear stress, and with increasing relative density, as expected.

The variation in Fig. 3.21 can be represented by the following empirical equation:

$$\frac{\tau}{\sigma'_0} = \alpha N_\ell^\beta$$ (3.20)

or alternatively

$$N_\ell = \left(\frac{1}{\alpha} \frac{\tau}{\sigma'_0} \right)^{1/\beta} \tag{3.21}$$

in which α and β are two empirical coefficients. Clearly, Fig. 3.21 suggests that these coefficients are primarily a function of D_r, namely, $\alpha = \alpha(D_r)$ and $\beta = \beta(D_r)$. For $D_r = 0.54$, for example, $\alpha = 0.246$ and $\beta = -0.165$, the dashed line in Fig. 3.21 (McDougal *et al.*, 1989). We note that other effects also influence α and β, as will be seen later.

Example 3. *Discussion of the relation τ/σ'_0 versus N_ℓ.*

The tests leading to the plot in Fig. 3.21 were carried out by Alba *et al.* (1976), using a *large-scale* simple shear test facility with unidirectional cyclic stress applications on the assumption that the test results would provide an adequate basis for obtaining test data from which the field behaviour of sands might reasonably be determined. In this study, basically a bed of sand (Monterey No. 0 sand with a mean grain diameter of 0.36 mm and a uniformity coefficient of 1.5 – rather uniform soil; see Appendix B item 10 under subsection Ranges of Soil Properties), the size 2.3 m × 1.1 m × 0.1 m (depth), was constructed on a shaking table. The ends of the sample were tapered so that it was not in contact with the walls of the box and was free to undergo cyclic strains in response to the applied stresses, an important problem encountered in the previous small-scale simple shear tests. The bed was prepared by pluvial deposition of dry sand. The tests were carried out with saturated specimens under undrained conditions. The results obtained from the study may be summarized as follows:

1. There is clear evidence that the test results are significantly influenced by the length-to-depth ratio of the test samples. In most previous shaking-table studies where samples were in contact with the walls of the container, the test results were significantly influenced by the stiffness of the walls.

2. Generally good agreement was obtained between large-scale shaking-table tests and small-scale shear-box tests, and therefore carefully conducted small-scale simple shear tests provide data presumably representative of simple shear field conditions.

3. In the early stages of earthquake studies, the cyclic simple shear test equipment (the kind described in conjunction with Fig. 3.19) was not available, and therefore cyclic triaxial tests were used as a substitute (Seed, 1976). The cyclic shear stress ratio, $\tau_{\max}/\sigma'_{3c}$, required to cause liquefaction in the

case of the triaxial tests is not the same as τ/σ_0'. A conversion factor C_r defined by

$$\frac{\tau}{\sigma_0'} = C_r \frac{\tau_{\max}}{\sigma_{3c}'} \qquad (3.22)$$

ranges from 0.6 to 0.65, C_r increasing with decreasing N_ℓ.

Pointing out the fact that their study determined the shear stress ratio necessary to induce liquefaction in large samples of a normally consolidated clean medium sand, Alba *et al.* (1976) remarked that the experiments provided information on liquefaction potential in a relative density range for which little or no field data were available. They stressed, however, that before applying these results (Fig. 3.21) to the field, the influence of other factors must also be considered.

These factors were discussed in greater details in Seed (1976). First of all, the method of soil placement (soil structure) may affect the liquefaction resistance, e.g., the method of placement for constructed fills such as hydraulic fill of a pipeline trench. Secondly, the period under sustained load may have a significant influence on the behaviour of the soil; laboratory experiments show that characteristics of undisturbed samples with tens or hundreds of years of sustained pressure are different (larger) from those of freshly deposited samples of the same sand. Thirdly, the previous strain history is another influencing factor; experiments show that although the prior strain history causes no significant densification of the sand, it increases, however, the stress ratio required to cause liquefaction. Fourthly, the lateral earth pressure and overconsolidation is also an important factor; the larger the value of the coefficient of lateral earth pressure, the larger the stress ratio required to cause liquefaction. Although not mentioned in Seed (1976), the content of fines (clay) can also be an important factor (see Section 3.5).

3.2.2   Equation governing the buildup of pore pressure

Now, we return to the case where the soil is exposed to a progressive wave (Fig. 3.17).

As described earlier (Section 3.1.1), the pore pressure will begin to build up, as the waves progress, similar to Peacock and Seed's oscillating shear stress tests (Fig. 3.20a). Fig. 3.22 displays the time series of the pore pressure and the water surface elevation obtained in a wave flume with a silt

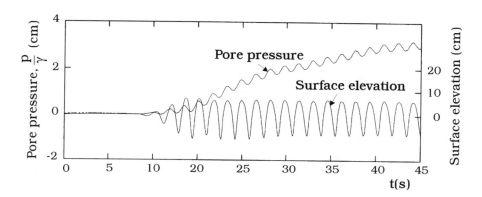

Figure 3.22:   Time series of excess pore pressure and surface elevation in Sumer *et al.*'s (1999) experiment. h = 42 cm, H = 10 cm, z = 16.5 cm, d = 17 cm, T = 1.6 s.

bottom, exposed to a progressive wave, from Sumer *et al.* (1999), illustrating that the pore pressure begins to accumulate upon the introduction of the waves. (It may be noted that the value of the relative density of the soil reported in Sumer *et al.*, 1999, is the after-the-test value; the before-the-test value was not measured in Sumer *et al.*, 1999.)

The purpose of this subsection is to derive the equation governing this process.

1) Period-averaged pore pressure

First, consider the pore pressure. This quantity is governed by the Biot consolidation equations (Eqs 2.30, 2.31, 2.32, and 2.38). Considering that the present process is a 2-D process (independent of the y-direction), and furthermore that the variations with respect to x are negligible (Fig. 3.17), Eq. 2.32 for the present case will be

$$G\frac{2 - 2\nu}{1 - 2\nu}\frac{\partial^2 w}{\partial z^2} = \frac{\partial p}{\partial z} \tag{3.23}$$

and Eq. 2.38

$$\frac{k}{\gamma}\frac{\partial^2 p}{\partial z^2} = \frac{n}{K'}\frac{\partial p}{\partial t} + \frac{\partial^2 w}{\partial z \partial t}. \tag{3.24}$$

Differentiating Eq. 3.23 with respect to t, and Eq. 3.24 with respect to z, one obtains

$$c_v\frac{\partial^3 p}{\partial z^3} = \frac{\partial^2 p}{\partial z \partial t} \tag{3.25}$$

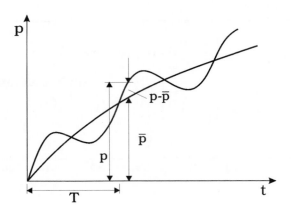

Figure 3.23:   Definition sketch. Two components of excess pore pressure: the residual component, $\bar{p}$, and the oscillating component, p-$\bar{p}$.

in which c_v is

$$c_v = \frac{Gk}{\gamma} \frac{2 - 2\nu}{(1 - 2\nu) + (2 - 2\nu)\frac{nG}{K'}} \qquad (3.26)$$

This quantity is termed the *coefficient of consolidation.*

Now, integrating Eq. 3.25 with respect to z gives

$$\frac{\partial p}{\partial t} = c_v \frac{\partial^2 p}{\partial z^2} + c \qquad (3.27)$$

in which c is an integration constant.

The above equation implies that the pore pressure satisfies the diffusion equation, with c_v, the coefficient of consolidation, acting as the familiar diffusion coefficient.

The actual physics of the coupling of the two processes, namely the poro-elastic deformation of the soil, and the Darcy-flow hydrodynamics, is presumably represented by the aforementioned diffusion mechanism. In other words, both the process of elastic deformation of the soil, and that of the Darcy-flow hydrodynamics are inherent in the diffusion representation in Eq. 3.27. This kind of treatment of the pore-water pressure will prove very useful not only in the present analysis but also in various other contexts, as will be seen later.

Now, we are not interested in the phase-resolved values of pressure, $p - \bar{p}$, but rather the period-averaged excess pressure, $\bar{p}$ (Fig. 3.23). "Moving" averaging therefore gives

$$\frac{\partial \overline{p}}{\partial t} = c_v \frac{\partial^2 \overline{p}}{\partial z^2} + f \qquad (3.28)$$

in which $\overline{p}$ is the period-averaged pore pressure, defined by Eq. 3.2.

In Eq. 3.28, apparently f is a source term, and represents the *total amount* of pore pressure generated per unit time and per unit volume of soil (including the pores). Although the above analysis implies that f is a function of t only, following McDougal *et al.* (1989), we will, for generality, consider it to be a function of both t and z, in conformity with the model developed (and validated) in the present section.

To summarize at this point, the period-averaged pore pressure satisfies the diffusion equation (Eq. 3.28). This is the *governing equation for the buildup of pore pressure*. Eq. 3.28 implies that the pore pressure is generated through the source term f, and it spreads out in the soil according to a diffusion process where c_v, the coefficient of consolidation, plays the role of the familiar diffusion coefficient.

Next, we shall study the source term, f.

2) The source term f

From Section 3.2.1, the generation of the pore pressure is achieved through the action of the cyclic shear stress in the soil. This will, at each point in the soil, lead to a pressure generation similar to that in Fig. 3.20a. This pressure can, in its simplest form, be represented by a linear variation with time (Fig. 3.24). (This is obviously valid up to the point of liquefaction). From Fig. 3.24:

$$\text{pressure generated} = \sigma_0' \frac{N}{N_\ell} \qquad (3.29)$$

in which N is the number of cycles. This is the pressure generated over the time period of NT. From Eq. 3.29, the pressure generated per unit time and volume (i.e., the quantity f) will then be

$$f = \frac{\sigma_0' \frac{N}{N_\ell}}{NT} \qquad (3.30)$$

or

$$f = \frac{\sigma_0'}{N_\ell T}. \qquad (3.31)$$

Here, σ_0' is the initial mean normal effective stress, defined by Eq. 3.6.

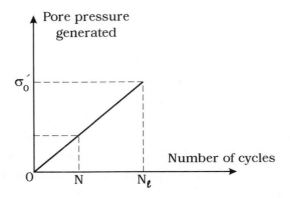

Figure 3.24:   The accumulated excess pore pressure increases linearly with the number of cycles. The simple, linear model.

The quantity N_ℓ in Eq. 3.29 is the number of cycles to cause liquefaction, and may be taken as that obtained from Peacock and Seed's (1968) and Alba *et al.*'s (1976) cyclic shear-stress tests discussed in Section 3.2.1 (Eq. 3.21), namely

$$N_\ell = \left(\frac{1}{\alpha}\frac{\tau}{\sigma_0'}\right)^{1/\beta} \tag{3.32}$$

where τ is, in the present case, the amplitude of the shear stress in the soil under the "forcing" progressive wave.

Clearly, Eq. 3.32 (with the coefficients α and β determined from the data in Fig. 3.21) was obtained in Alba *et al.*'s (1976) undrained simple shear tests. Therefore, the use of Eq. 3.32 is justified only if the soil is nearly undrained (silt and fine sand). However, the "drainage" effect present in the process is taken care of by the diffusion term $c_v \partial^2 p/\partial z^2$ in Eq. 3.28.

It may be noted that Eq. 3.31 represents essentially a linear mechanism. A nonlinear mechanism of pore pressure generation has been suggested by various authors, see, e.g., Seed, Martin and Lysmer (1976), Seed and Booker (1976), Alba *et al.* (1976), Rahman, Seed and Booker (1977), Seed and Rahman (1978), and Sekiguchi, Kita and Okamoto (1995), and recently Jeng *et al.* (2007). (See Example 5 for a detailed discussion of the latter study.)

Seed and Rahman (1978) were the first to adopt the model in Eq. 3.28 to describe the buildup of pore pressure under a progressive wave. Spierenburg (1987) and McDougal *et al.* (1989) have subsequently adopted similar approaches. The works by Barends and Calle (1985) and de Groot, Lindenberg and Meijers (1991) have considered similar theoretical descriptions of the process of pressure accumulation in the soil. Sekiguchi *et al.*'s (1995) study focused on the generation of the pore pressure (characterized in the present description by the term f, Eq. 3.28). Their poro-elastoplastic formulation enabled the researchers to obtain closed-form solutions for the accumulated pore pressure under cycling loading.

3.2.3 Solution to the equation of buildup of pore pressure. Infinitely large soil depth

The definition sketch is given in Fig. 3.17, with the soil depth $d \to \infty$.

As seen from Eqs 3.31 and 3.32, to be able to calculate the source term f in Eq. 3.28, we need N_ℓ; and for the latter we need τ, the amplitude of the cyclic shear stress in the soil (Fig. 3.17). The stresses in the soil with an infinitely large depth, exposed to a progressive wave, have been studied in Section 2.2.1, and closed solutions have been presented. From the latter, the amplitude of the shear stress, τ, is obtained as (Eq. 2.55)

$$\tau = p_b \lambda z \exp(-\lambda z) \tag{3.33}$$

in which p_b is the maximum value (the amplitude) of the pressure exerted on the bed, given in Eq. 2.46.

(Note that the shear stress given in Eq. 2.55 is obtained in the case of a soil in which no buildup of pressure takes place, and therefore it may not be entirely correct for the present case where there is a pore-pressure accumulation. No study is yet available in a closed-form solution, investigating, in a systematic manner, the soil stresses in the presence of pore pressure accumulation. However, we note that the soil-stress changes involving the vertical and horizontal effective stresses as well as the shear stresses during the process of pore pressure accumulation were *numerically* investigated by Sassa and Sekiguchi (2001, Figs 14–18). We also note that Tzang's (1998, p. 292) experimental investigation, although limited, suggests a poro-elastic response during the process of pore-pressure buildup; see discussion of Tzang's study in Section 3.1.4.)

Now, the solution to Eq. 3.28 (with f given in Eq. 3.31, σ_0' in Eq. 3.6, N_ℓ in Eq. 3.32, and τ in Eq. 3.33) is to be sought under the following initial and boundary conditions.

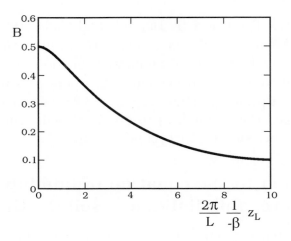

Figure 3.25:   Maximum liquefied depth, z_L, and the parameter B. Adapted from Jeng and Seymour (2007).

1. At the initial instant, there is no accumulated pore pressure:

$$t = 0 : \qquad \overline{p} = 0. \tag{3.34}$$

2. At the mudline, the pore pressure continuously dissipates:

$$z = 0 : \qquad \overline{p} = 0. \tag{3.35}$$

The solution to Eq. 3.28 can be obtained by the method of Fourier sine transform (Sneddon, 1957, p. 128 and p. 302):

$$\overline{p}(z,t) = \frac{2}{\pi} \int\limits_{t'=0}^{t} dt' \int\limits_{\xi=0}^{\infty} \exp\left[-c_v \xi^2 (t - t')\right] \sin(\xi z) \left[\int\limits_{z'=0}^{\infty} f(z') \sin(\xi z') dz' \right] d\xi \tag{3.36}$$

where $f(z')$ is to be calculated from Eqs 3.31–3.33.

Incidentally, Jeng and Seymour (2007) have developed an analytical solution for the governing equation, in a form different from Eq. 3.36, which was shown by the authors to be identical to that in Eq. 3.36. Jeng and Seymour (2007) have implemented the solution to determine the maximum liquefied depth, z_L. Their result is reproduced in Fig. 3.25 in which the liquefaction depth z_L appears on the horizontal axis. The quantity B on the vertical axis is defined by

$$B = \frac{(1 + 2k_0)\gamma' c_v \left(\frac{\lambda}{(-\beta)}\right)^2}{6A} \tag{3.37}$$

with A defined by

$$A = \frac{\gamma'(1 + 2k_0)}{3T} \left(\frac{3p_b\lambda}{\alpha(1 + 2k_0)\gamma'}\right)^{1/(-\beta)}. \tag{3.38}$$

Jeng and Seymour (2007) remark that, although the solution is only valid for infinitely large soil depth, it provides a reasonable estimate for deep foundations ($d/L > 0.5$ in which d is the soil depth).

3.2.4 Solution to the equation of buildup of pore pressure. Finite soil depth

In this case (Fig. 2.4), τ, the amplitude of the shear stress in the soil, is obtained from Eqs 2.66 and 2.72 as

$$\tau = |\tau_y| = p_b\{(C_1 - C_2\lambda z)e^{-\lambda z} - (C_3 - C_4\lambda z)e^{\lambda z} + \lambda\delta(C_5 e^{-\delta z} - C_6 e^{\delta z})\} \tag{3.39}$$

in which the coefficients C_i are depicted in Appendix D.

(Similar to the previous case, the shear stress given in Eq. 2.66 is obtained in the case of a soil in which no buildup of pressure takes place, and therefore it may not be entirely correct for the present case where there is a pore-pressure accumulation. As pointed out in the previous sub-section, no study is yet available in a closed-form solution, investigating in a systematic manner the soil stresses in the presence of pore pressure accumulation. However, see the note given in Section 3.2.3 in this regard.)

Regarding the initial and boundary conditions, those employed for the case of the infinite soil depth are also valid for this case (Eqs. 3.34 and 3.35), plus the pressure "flux" at the impermeable base should be zero:

$$z = d: \qquad \frac{\partial \overline{p}}{\partial z} = 0. \tag{3.40}$$

The solution to the governing equation (namely Eq. 3.28, with f given in Eq. 3.31, σ_0' in Eq. 3.6, N_ℓ in Eq. 3.32, and τ in Eq. 3.39) under these initial and boundary conditions is given as (Sumer and Cheng, 1999):

$$\bar{p}(z,t) = \frac{2}{\pi} \sum_{m=1}^{\infty} \sin\left[\left(m - \frac{1}{2}\right)\frac{\pi}{d}z\right] \tag{3.41}$$

$$\times \int_{t'=0}^{t c_v \left(\frac{\pi}{d}\right)^2} dt'$$

$$\times \int_{\xi=0}^{\pi} \exp\left[-\frac{1}{4}(2m-1)^2 \left[c_v \left(\frac{\pi}{d}\right)^2 t - t'\right]\right] \sin\left[\left(m - \frac{1}{2}\right)\xi\right] g(\xi)d\xi$$

in which $g(\xi)$ is

$$g(\xi) = \frac{1}{c_v}\left(\frac{d}{\pi}\right)^2 \frac{1}{T}\gamma'\frac{1+2k_0}{3}\xi\frac{d}{\pi}\left[\frac{1}{\alpha}\frac{3}{1+2k_0}\frac{\tau}{\gamma'\xi(d/\pi)}\right]^{-1/\beta} \tag{3.42}$$

with ξ defined by

$$\xi = \frac{z}{d/\pi}. \tag{3.43}$$

Recall that τ in Eq. 3.42, the amplitude of the shear stress, is a function of the vertical distance z (or alternatively, the nondimensional distance ξ, Eq. 3.43) and given in Eq. 3.39.

It may be noted that Cheng *et al.* (2001) solved Eq. 3.28 numerically (where the analytical expression given by Hsu and Jeng (1994), Eqs 2.66 and 2.72, was used). They obtained a good agreement between the numerical results and the analytical solution given in Eq. 3.41. They emphasized that a small error in the soil shear stress can lead to a large error in the accumulated pore pressure.

Validation of the mathematical model

The mathematical model given in the preceding paragraphs, namely Eqs 3.28, 3.32, and 3.39 with the solution in Eq. 3.41, has been validated by Sumer *et al.* (2011, also published in 2012) against a series of controlled experiments conducted by the latter authors.

Sumer *et al.* (2012) carried out their experiments in a wave flume. The soil was placed in a pit, 0.4 m deep and 0.78 m long, rigidly fixed to the flume. The wave heights in the tests were in the range $H = 7.7 - 18$ cm and the wave period was $T = 1.6$ s with the water depth $h = 55$ cm. The soil was silt with $d_{50} = 0.070$ mm. Other properties of the soil were as follows: the specific gravity of soil grains $s(=\gamma_s/\gamma) = 2.67$; the coefficient of lateral earth pressure $k_0 = 0.42$; the porosity $n = 0.51$; the total specific weight of the soil $\gamma_t = 17.95\,\mathrm{kN/m^3}$ (and therefore the submerged specific weight of the soil $\gamma' = \gamma_t - \gamma = 8.14\,\mathrm{kN/m^3}$); the coefficient of permeability $k = 0.0015\,\mathrm{cm/s}$; Poisson's ratio $\nu = 0.29$ (found from $\nu = k_0/(1 + k_0)$, Terzaghi, 1948 and Brinch–Hansen, 1957); and the relative density $D_r = 0.28$ with $e_{\max} = 1.2$ and $e_{\min} = 0.57$. The values given for n, γ_t, and D_r above are all before-the-test values.

Regarding the coefficients α and β in Eq. 3.21, these are essentially a function of the relative density, D_r, as suggested by Peacock and Seed's (1968) and Alba *et al.*'s (1976) experiments, already discussed in the preceding paragraphs. In Sumer *et al.* (2012), these coefficients are calculated from the empirical expressions,

$$\alpha = 0.34 D_r + 0.084, \qquad \beta = 0.37 D_r - 0.46 \qquad (3.44)$$

which were obtained from a curve-fit exercise to Alba *et al.*'s (1976) data, Fig. 3.21. Although not shown here for brevity, when plotted, the data associated with these coefficients indicate a clear trend that the above empirical equations can be extrapolated to the range $0 < D_r < 1$. As already pointed out (Section 3.2.1), Alba *et al.*'s (1976) study determined the shear stress ratio necessary to cause liquefaction of soil samples in large sizes of a normally consolidated clean medium sand. Alba *et al.* (1976) stressed, however, that before applying these results, the influence of other factors must also be considered (see also Seed, 1976), such as the method of soil placement, the period under sustained load, the previous strain history, and the lateral earth pressure and overconsolidation. Sumer *et al.* (2012) pointed out, however, that as the sediment bed conditions in their validation experiments are initially rather similar (normally consolidated clean sediment) to those in Alba *et al.*'s (1976) shear tests, the α and β coefficients can, to a first approximation, be determined from Eq. 3.44.

All the properties of the soil used in the Sumer *et al.* (2012) tests were determined through soil experiments in Sumer *et al.* (2012) except Young's modulus of elasticity, E. The latter quantity was tuned so that the model results for the accumulated pore water pressure match with the measured values. In this way, E was obtained as $5000\,\mathrm{kN/m^2}$. Sumer *et al.* (2012) discussed that the tuned value of Young's modulus of elasticity was consistent with values encountered in the literature with similar relative density.

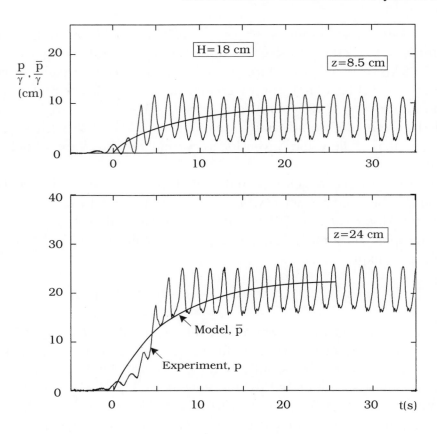

Figure 3.26:   Comparison of the mathematical model results with the experiment. Sumer *et al.* (2012).

Fig. 3.26 compares the time variation of the period-averaged pore water pressure, $\bar{p}$, calculated from the model, Eq. 3.41, with the time series of the pore water pressure, p, obtained in the experiments at two depths for a test where liquefaction occurred. Fig. 3.27 compares the model results for three different values of Young's modulus of elasticity, to illustrate the sensitivity of the results to the latter quantity. Fig. 3.28, on the other hand, compares the model results with the experiments illustrating the vertical distribution of $\bar{p}$ for different times. From Figs 3.26–3.28, it is seen that the model results are not radically different from the experiments. It should also be noted that the model can also capture, rather well, the buildup of pore water pressure in the case where no liquefaction was observed (see Sumer *et al.*, 2012, Fig. 8).

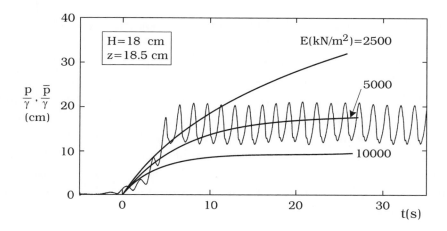

Figure 3.27: Sensitivity of the mathematical model results (solid lines) to Young's modulus of elasticity. Sumer *et al.* (2012).

As seen from Fig. 3.27, the model results change dramatically with changing the value of Young's modulus of elasticity. The buildup of pore water pressure decreases with increasing E. This is because, the larger the value of E, the smaller the cyclic elastic deformation caused by the wave, and therefore the smaller the pressure generated. Hence, the buildup of pore pressure should be smaller with increasing E. The latter can also be explained in terms of the dissipation of pore pressure. Namely, the coefficient of consolidation increases with increasing the modulus of elasticity (Eq. 3.26). On the other hand, the larger the dissipation, the smaller the buildup of pore pressure, and therefore the buildup of pore pressure should decrease with increasing E.

Using the criterion given in Eq. 3.10, Sumer *et al.* (2012) assessed the onset of liquefaction, utilizing the above mathematical model (Eq. 3.41). To this end, (1) the accumulated pore water pressure was predicted from the model, and subsequently the ratio $\bar{p}/\sigma_0'$ was calculated to observe whether or not liquefaction occurs according to Eq. 3.10; (2) furthermore, in the case of liquefaction, the number of cycles to cause liquefaction was also predicted, and the results were compared with the experiments. The results are summarized in Table 3.1.

As seen from Table 3.1, the agreement between the model prediction and the experiments is good except the test with $H = 12 \, \text{cm}$ where the experiments show that liquefaction occurs whereas the model predicts that it does not. This wave height is apparently rather close to the critical wave height for the onset of liquefaction (see Fig. 3.8 for the critical wave height for the onset of liquefaction). Sumer *et al.* (2012) report that no clear explanation was found for this discrepancy between the model prediction and the experiment. They add, however, the following: observations show that the transition from the no-liquefaction regime to the liquefaction regime occurs rather abruptly in terms of the wave height (similar to Fig. 3.8) whereas in the model this transition occurs gradually. Hence, in a narrow range of wave height around the critical regime, the model cannot capture the abrupt no-liquefaction-to-liquefaction transition.

Table 3.1.   Comparison of model with experiments. Number of waves to cause liquefaction. Sumer *et al.* (2012).

Wave height, H (cm)	Liquefaction Yes or No?		Number of waves to cause liquefaction at the bed surface	
	Model	Expt.	Model	Expt.
18	Yes	Yes	3	3.5
15	Yes	Yes	6	4
12	No	Yes	-	-
9.3	No	No[1]	-	-
8.4	No	No[1]	-	-
7.7	No	No	-	-

[1] Liquefaction did not occur initially. However, it emerged after a long transitional period. In addition, the liquefaction did not reach the impermeable base.

Sumer *et al.* (2012) also note that the response of the pore water pressure in the actual tests with $H = 8.4\,\text{cm}$ and $9.3\,\text{cm}$ was one with no liquefaction, as correctly predicted by the model. However, after a long "no-liquefaction" period in the experiments (11 waves, see Fig. 3.29), the response begins to change dramatically (Fig. 3.29) where the pore-water pressure builds up further, eventually exceeds the initial mean normal effective stress σ_0', and consequently the soil is liquefied. This kind of behaviour has been observed previously by Foda and Tzang (1994), Sumer *et al.* (1999), and Sassa *et al.*

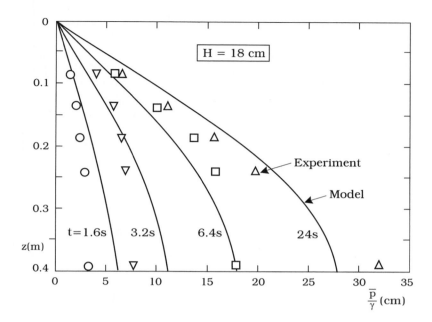

Figure 3.28: Comparison of the mathematical-model results with the experiment. Vertical distribution of $\bar{p}$ for different times. Sumer *et al.* (2012).

(2001). The physics behind this interesting transitional case have been not very extensively investigated. It may be that the soil being subject to a wave loading (initially too small to cause liquefaction) at that particular depth would "yield" after a substantial amount of exposure to wave loading, and, with this (and consequently with the large cyclic deformations), the soil grains begin to rearrange as in the case of large waves, and the soil is presumably liquefied. The present model does not include elements associated with the soil mechanics of this transient process, and therefore cannot capture the variation of the accumulated pore water pressure exhibited in Fig. 3.29.

Sassa *et al.* (2001, Figs. 12 c and d) present similar results from their centrifuge wave testing, and they consistently predicted and reproduced these results from their theoretical model, Sassa *et al.* (2001, Figs. 11 c and d). A notable feature of the behaviour exhibited in Sassa *et al.*'s (2001) experiments and in Fig. 3.29 is that the process of "re-buildup" of pore pressure takes place concurrently with an increase in the amplitude of the pore pressure. This feature has been captured by Sassa *et al.*'s (2001) theoretical model results.

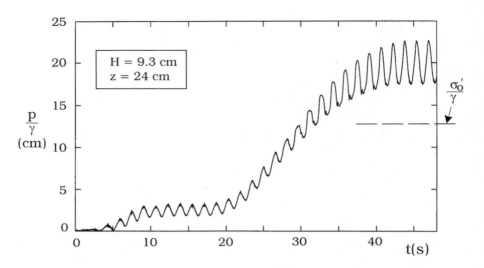

Figure 3.29:   Time series of p in Sumer *et al.*'s (2012) test 4 (see the wave height in the legend) with no-liquefaction-to-liquefaction transition.

Shinji Sassa (2012, personal communication) points out that the physics of the previously mentioned no-liquefaction-to-liquefaction transition is linked to the critical condition for the onset of liquefaction. Under the critical wave condition below which liquefaction does not occur, the deeper soil undergoes a low shear stress ratio at the beginning of the wave loading. However, with the liquefaction occurring at the uppermost layer, the liquefaction front will progress downwards, and the resulting two-layer fluid system will alter the shear stress in the sub-liquefied soil below the liquefaction front significantly (Sassa *et al.*, 2001, Eq. 13), and this will eventually lead to the re-buildup of the pore pressure. By contrast, in the case of large wave heights, the deeper soil will undergo a more severe shear stress ratio already at the beginning of the wave loading, and thus the previously mentioned transition pattern will not be experienced.

The model validation exercise in Sumer *et al.* (2012), the results of which are highlighted in the preceding paragraphs, provides confidence in the use of the mathematical model to make assessment of liquefaction potential.

In geotechnical engineering, one of the most popular methods for liquefaction potential assessment for earthquake-induced liquefaction is to use specially prepared charts where the quantity τ/σ_0', traditionally called Cyclic Stress Ratio (CSR), is plotted versus the so-called corrected standard penetration (or the corrected SPT blowcount), $N_{1(60)}$, cf., the bottom chart in Fig. 10.18 (see, e.g., Kramer, 1996, PIANC, 2001 and Idriss and Boulanger, 2008). The corrected standard penetration is obtained from Standard Penetration Tests (SPT), and characterizes, among others, the relative density of the soil (see Appendix C for a brief account of SPT); the larger the value of $N_{1(60)}$, the larger the relative density of the soil.

To get a sense of the CSR values experienced in the laboratory experiment described in the preceding paragraphs, the amplitude of the cyclic shear stress τ is calculated from Eq. 3.39 for the test of Sumer *et al.* (2012) where the wave height was $H = 18\,\text{cm}$. Recall that liquefaction occurred in this test (Table 3.1). The calculations show that the values of CSR apparently remain practically constant across the soil depth at $CSR = \tau/\sigma_0' = 0.13$.

Regarding the value of the corrected SPT blowcount, $N_{1(60)}$, clearly no such data existed for Sumer *et al.*'s (2012) test, as the SPT equipment cannot be applied in small-scale experiments (although there is always the option of converting CPT readings to SPT values (see Appendix C) with the CPT values obtained with the help of a laboratory-scale CPT equipment). Nevertheless, as an academic exercise, the corresponding SPT number $N_{1(60)}$ may be estimated from the knowledge that the soil in Sumer *et al.*'s (2012) test had a relative density of $D_r = 0.28$. From Table C.3, Appendix C, an estimate for $N_{1(60)}$ will be $N_{1(60)} = 6$.

Although purely an academic exercise, it will be seen that, not surprisingly, the pair $CSR = 0.13$ and $N_{1(60)} = 6$, corresponding to Sumer *et al.*'s (2012) wave-liquefaction test, lies in the "liquefaction" area, when it is plotted on the previously mentioned charts (e.g., Kramer, 1996, PIANC, 2001, Jefferies and Been, 2006, and Idriss and Boulanger, 2008).

Example 4. *Assessment of liquefaction potential. Numerical example.*

The soil properties are given as follows: the soil depth, $d = 1\,\text{m}$, the submerged specific weight, $\gamma' = 10.8\,\text{kN/m}^3$, the shear modulus, $G = 926\,\text{kN/m}^2$, the coefficient of permeability, $k = 1 \times 10^{-6}\,\text{m/s}$, the porosity, $n = 0.333$, the degree of saturation, $S_r = 1$, the coefficient of lateral earth pressure, $k_0 = 0.4$, Poisson's ratio, $\nu = 0.35$, and the empirical constants in the Seed equation (Eq. 3.21), $\alpha = 0.246$ and $\beta = -0.165$. (The latter values of α and β are actually taken from McDougal $et\ al.$ (1989, Fig. 3 and Table 1), and they represent the soil in their experiment with $D_r = 0.54$. These values are slightly different from those obtained through the empirical equations, Eq. 3.44, for the same relative density $D_r = 0.54$. We maintain the original values given in McDougal $et\ al.$ (1989), and reproduced in Sumer and Fredsøe, 2002, $\alpha = 0.246$ and $\beta = -0.165$, to avoid confusion.)

The water properties are: the specific weight of water, $\gamma = 9.81\,\mathrm{kN/m^3}$, and the bulk modulus of elasticity of water, $K = 1.9 \times 10^6\,\mathrm{kN/m^2}$.

The soil is exposed to a progressive wave with the following properties: the wave height, $H = 5\,\mathrm{m}$, the period $T = 13.7\,\mathrm{s}$, and the water depth, $h = 19\,\mathrm{m}$.

The question is whether the soil will be liquefied under the given wave climate.

Using Eq. 3.26, the coefficient of consolidation is found as $c_v = 4.1 \times 10^{-4}\,\mathrm{m^2/s}$. From the linear wave theory (Appendix A), the wave length is $L = 174\,\mathrm{m}$, and the wave number $\lambda = 2\pi/L = 0.036\,\mathrm{m^{-1}}$. The pore pressure is calculated from Eqs 3.41 and 3.42. The results are given in Figs 3.30 and 3.31.

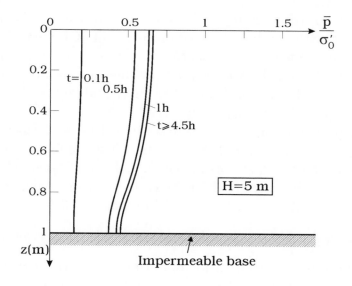

Figure 3.30:   Time development of the accumulated excess pore pressure. Liquefaction does not occur; $\bar{p}$ never exceeds σ_0'.

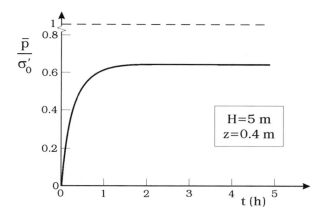

Figure 3.31: Time development of the accumulated excess pore pressure at $z = 0.4$ m. Liquefaction does not occur; $\bar{p}$ never exceeds σ'_0.

As seen from the figures, the pore pressure never reaches the initial mean normal effective stress σ'_0, i.e., $\bar{p}/\sigma'_0$ is always

$$\bar{p}/\sigma'_0 < 1.$$

Hence, liquefaction will not occur under this wave climate.

Now, Fig. 3.32 displays the corresponding pressure distributions when the wave height is $H = 6$ m. The pore pressure in this case reaches the initial mean normal effective stress σ'_0 within less than 15 minutes. Given the fact that 15 minutes is a reasonably short period of time for a given sea state (which might last as long as $O(4–5\,\text{hours})$ or even more), it may be concluded that there is a liquefaction potential for this second wave climate. Also note that liquefaction first starts at the surface of the soil and spreads downwards (Fig. 3.32), in agreement with the description given in conjunction with Fig. 3.3.

In-situ relative density, D_r

As seen from the preceding paragraphs, the relative density (or density index) of the soil, D_r, is one of the key quantities needed in a liquefaction assessment study. Therefore the in-situ value of the relative density needs to be determined in an accurate manner. The reader is referred to Appendix C for a detailed discussion on this issue.

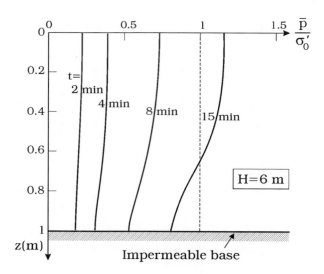

Figure 3.32:   Time development of the accumulated excess pore pressure. Liquefaction occurs; $\bar{p}$ exceeds σ_0' within less than 15 minutes.

Cyclic triaxial tests to determine α and β coefficients

This subsection focuses on the α and β coefficients of the mathematical model.

These coefficients can be determined from the empirical equations in Eq. 3.44, which were obtained from a curve-fit analysis to Alba *et al.*'s (1976) data. However, when it is deemed necessary, these coefficients should be determined from cyclic triaxial tests. The procedure may be described as follows:

(1) Collect soil samples from the site.

(2) Determine the *in-situ* relative density from SPT or CPT profiles (Appendix C).

(3) Reconstitute samples at the determined *in-situ* relative densities.

(4) Perform cyclic triaxial laboratory tests on these soil samples under the stress conditions that are typical for the field conditions. To this end, select some values (four or five, say) of the cyclic stress ratio ($CSR = \tau/\sigma_0'$) typical for the field, and apply these CSR values in the triaxial tests.

(5) Determine the number of cycles to cause liquefaction, N_ℓ, from the cyclic triaxial tests under the selected CSR stress conditions.

(6) Plot the CSR values versus the obtained values of N_ℓ (similar to Fig. 3.21), and get the values of α and β coefficients by fitting the expression in Eq. 3.20 to the plotted (CSR, N_ℓ) data.

(7) Implement the mathematical model (Example 4), using these values of α and β coefficients.

Example 5. *Mathematical modelling. Effect of nonlinear mechanism of pore pressure generation.*

Jeng *et al.* (2007) have studied the effect of nonlinear mechanisms of pore pressure generation on the buildup of pore pressure. They considered two kinds of pore pressure generation mechanisms: one is the linear mechanism (Eq. 3.29), and the other is the nonlinear mechanism given by Seed *et al.* (1976):

$$\text{pressure generated} = \sigma_0' \left\{ \frac{1}{2} + \frac{1}{\pi} \arcsin \left[2 \left(\frac{N}{N_\ell} \right)^{1/\theta} - 1 \right] \right\} \qquad (3.45)$$

in which θ is a shape factor taken as 0.7. This is the pressure generated over the time period of $t = NT$. The pressure generated per unit time and volume (i.e., the quantity f in Eq. 3.28) will then be

$$f = \frac{\partial}{\partial t} \left[\sigma_0' \left\{ \frac{1}{2} + \frac{1}{\pi} \arcsin \left[2 \left(\frac{N}{N_\ell} \right)^{1/\theta} - 1 \right] \right\} \right]. \qquad (3.46)$$

or

$$f = \frac{\partial}{\partial t} \left[\sigma_0' \left\{ \frac{1}{2} + \frac{1}{\pi} \arcsin \left[2 \left(\frac{t/T}{N_\ell} \right)^{1/\theta} - 1 \right] \right\} \right]. \qquad (3.47)$$

The latter equation reduces to Eq. 3.31 when the "pressure generated" is taken as in Eq. 3.29 with $N = t/T$.

Now, Jeng *et al.* (2007) solved Eq. 3.28 numerically with the source term predicted from Eq. 3.47 to get the pressure buildup for the nonlinear mechanism case, and compared the results with the linear-theory results for wave and soil conditions likely to be encountered in the field. The comparison indicated that, although the mechanism of pore pressure buildup and the nonlinear relation of pore pressure generation appear to be more important under a larger wave, longer wave period and shallower water depth, the linear and nonlinear results apparently practically coincide for the accumulated pore pressure values p/σ_0' larger than $O(0.1)$. These pressure values are the most important for practical applications.

Example 6. *Mathematical modelling. DIANA-SWANDYNE II code.*

Dunn, Vun, Chan and Damgaard (2006) adopted the code DIANA-SWANDYNE II (Dynamic Interaction and Nonlinear Analysis–Swansea Dynamic program version II) for waves, to study pore pressure variations (for both the phase resolved component and the period-averaged component). This code, developed for 2-D cases, uses the fully coupled Biot dynamic equation. The mathematical/numerical formulation of the code is described in detail by Chan (1988, 1995) and Zienkiewicz *et al.* (1990, 1999).

The model includes a constitutive model that can predict both residual and momentary liquefaction. However, the constitutive model requires a large number of material parameters to describe the loading and unloading behaviour of the soil, usually obtained from detailed laboratory testing such as triaxial tests.

Although the model has been validated for earthquake induced liquefaction, Dunn *et al.* (2006) implemented the model for wave loading; the model was, in the latter study, tested against the analytical solution of Hsu and Jeng (1994); and the laboratory experiments of Teh *et al.* (2003). The latter experiments involve pore-pressure buildup, and liquefaction.

The model, tested and validated, was subsequently implemented to study liquefaction around a buried pipeline (see the following example).

Example 7. *Liquefaction around buried pipelines.*

Sumer, Truelsen and Fredsøe (2006 c) studied buildup of pore-water pressure and the resulting liquefaction around pipelines buried in a soil exposed to a progressive wave. The buried model pipelines in the experiments were held stationary. The results indicated that both the buildup of pore pressure and the liquefaction are influenced by the presence of the pipe. The pore pressure builds up much more rapidly at the bottom of the pipe. By contrast, the pressure buildup at the top of the pipe is not influenced radically by the presence of the pipe.

Another interesting result obtained in Sumer *et al.*'s (2006 c) experiments is that the liquefaction first occurred at the bottom of the pipe and subsequently developed along the perimeter of the pipe upwards, evidently linked to the much more rapid development of the pressure buildup at the bottom of the pipe.

In Sumer *et al.*'s (2006 c) study, the influence of the "no-slip" (or "glue") condition at the pipe surface has also been investigated whereby the pipe was covered with a special Velcro material to roughen the pipe for "rough" pipe tests. These tests indicated that the influence of the no-slip condition at the pipe surface is very significant; the feature that was observed for smooth pipes, that the pore pressure builds up much more rapidly at the bottom of the pipe, apparently disappeared in the case of the rough pipe with the no-slip boundary condition. This is arguably linked to the constraint imposed by the rough surface where the soil adjacent to the pipe is not allowed to undergo much (shear) deformation.

As noted in the previous example, Dunn *et al.* (2006) applied the DIANA-SWANDYNE II code to liquefaction around buried pipelines. The buried pipeline simulations largely support the experimental findings of Sumer *et al.* (2006 c) and Teh *et al.* (2003). Most notably, Dunn *et al.* (2006) validated that the pore pressure builds up more quickly in the vicinity of the pipeline than in the far field; and a pipeline that is considered to be "fixed" causes the soil to be liquefied more quickly than in the case of an equivalent pipeline that is "free".

3.3 Centrifuge Modelling of Residual Liquefaction

3.3.1 Centrifuge testing. General

Geotechnical centrifuge testing is a technique widely used in geotechnical engineering for physical modelling studies. There are two kinds of geotechnical centrifuges: (1) arm centrifuges, and (2) drum centrifuges. The former is, by far, the most popular one. The book edited by Taylor (1995) gives a detailed account of various aspects of geotechnical centrifuge technology. Schofield (1980) also can be consulted for basic principles behind centrifuge modelling, including some applications.

The arm centrifuge essentially consists of a rotating arm, balanced by means of a counterweight. The arm itself contains a specimen box, placed at the end of the arm. Many geotechnical centrifuges have radii in the range 1.5–4 m (Powrie, 2004).

The principal idea behind the centrifuge testing is as follows. *In situ*, stresses change with depth, and it is known that the soil behaviour (stress–

strain relationship, friction angle) is a function of stress level. This implies that, in a regular physical model test, the soil behaviour will not be simulated correctly, because the stresses due to the soil self weight are too low.

In order to achieve the same stress level as in the prototype, soil (specimen) is placed at the end of the arm centrifuge, and the arm is rotated at a specified angular rotational speed, ω_c, so that the same stress level as in the prototype is achieved. Obviously, the larger the value of ω_c, the larger the stress level in the model. The angular rotational speed ω_c can be determined by a simple analysis. The following paragraphs will describe this.

The analysis will be developed for dry conditions.

Fig. 3.33a illustrates the soil specimen in the arm centrifuge model, simulating the soil in the prototype (Fig. 3.33b). The depth in the prototype is obviously measured vertically downwards (Fig. 3.33b) while the depth in the model is measured radially outwards (Fig. 3.33a), as indicated in Figs 3.33a and b.

The vertical stress in the prototype (Fig. 3.33b) is

$$\sigma_p = \rho_p g z_p \tag{3.48}$$

in which g is the acceleration due to gravity (Fig. 3.33b), z_p is the depth (Fig. 3.33b), ρ_p is the soil density, and the sub-index p indicates the prototype values.

Regarding the model, any element of the soil specimen in the model is subjected to an inertial (or centrifugal) acceleration

$$a_m = r\omega_c^2 \tag{3.49}$$

(Fig. 3.33a) in which r is the radius measured from the rotation axis to the element of the soil specimen.

Now, the stress in the model soil subjected to the acceleration $a_m (=r\omega_c^2)$ is

$$\sigma_m = \rho_m a_m z_m \tag{3.50}$$

in which σ_m is the stress in the radial (r) direction, z_m is the depth from the free surface of the soil specimen (Fig. 3.33a), ρ_m is the soil density in the model, and the sub-index m indicates the model values.

If the same soil is used in the model, and, furthermore, if the soil specimen is prepared such that it is subjected to a similar stress history (i.e., ensuring

(a) Model (arm centrifuge)

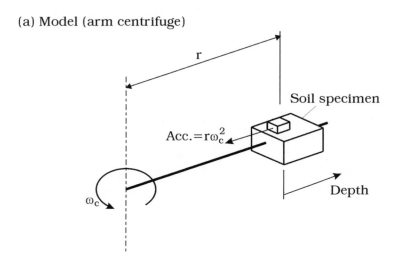

(b) Prototype

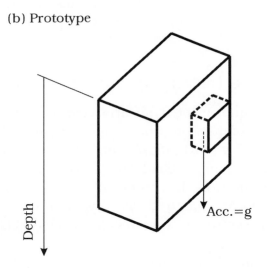

Figure 3.33: Centrifuge testing. (a): Model (arm centrifuge). (b): Prototype. Adapted from Schofield (1980).

that the packing of the soil particles is the same as in the prototype), then ρ_m will be the same as ρ_p. Then the requirement that we achieve the same stress level as in the prototype, $\sigma_m = \sigma_p$, reads

$$\rho_m a_m z_m = \rho_p g z_p \tag{3.51}$$

and solving a_m

$$a_m = \frac{z_p}{z_m}\, g \tag{3.52}$$

The ratio z_p/z_m,

$$N = \frac{z_p}{z_m} \tag{3.53}$$

is called the length (or model) scale. In order to achieve the same stress level as in the prototype, the centrifugal acceleration should, from Eq. 3.52, be selected as

$$a_m = N\, g \tag{3.54}$$

(For example, like $a_m = 50\ g$ or $100\ g$, etc.) Or alternatively, the angular rotational speed of the centrifuge should, from Eqs 3.49 and 3.52, be $\omega_c = (Ng/R_{eff})^{1/2}$ where R_{eff} is an effective centrifuge radius. Issues related to scaling laws and scaling errors in centrifuge modelling are discussed in greater details in Taylor (1995).

3.3.2   Centrifuge wave testing

Sassa and Sekiguchi (1999) implemented an arm centrifuge to study the sequence of liquefaction and compaction process under waves. The test set-up is shown in Fig. 3.34 with Fig. 3.34a illustrating the setup prior to spinning and Fig. 3.34b during spinning. (Incidentally, these authors reported the results of their subsequent work in a series of papers, among others, Sekiguchi, Sassa, Sugioka and Miyamoto, 2000, Miyamoto et al., 2003 and 2004. Earlier references for centrifuge wave testing include Sekiguchi and Phillips, 1991, and Sekiguchi et al., 1995 and 1998).

We will return to Sassa and Sekiguchi's (1999) study shortly. Now let us consider the principal equations when the pores of the soil contain fluid. In this case, the requirement that we achieve the same stress level as in the prototype of Eq. 3.51 applies to both the initial effective stress and the initial pore pressure, namely

$$(\rho_m - \rho_{fm})a_m z_m = (\rho_p - \rho_{fp})g z_p \tag{3.55}$$

$$\rho_{fm} a_m z_m = \rho_{fp} g z_p \tag{3.56}$$

in which ρ_{fm} is the model fluid density and ρ_{fp} the prototype fluid density. These equations lead to the same results as in the case of dry sediment, with Eqs 3.51 and 3.54, provided that the same soil is used in the model, and furthermore the soil specimen is prepared such that it is subjected to a similar stress history which ensures that the packing of the soil particles is the same as in the prototype.

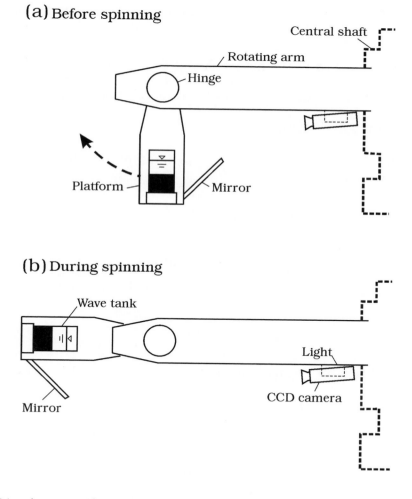

Figure 3.34: Arm centrifuge employed by Sassa and Sekiguchi (1999) for centrifuge wave testing.

Now, returning to Sassa and Sekiguchi's (1999) work, two issues addressed in this study are: (1) what is the wave frequency (or the wave period) in the model which ensures that the model wave is geometrically similar to the prototype wave; and (2) how can a model similarity be achieved for the process of the buildup of pore-water pressure?

Each issue is now considered individually.

Geometrical similarity of wave

To achieve a true geometrical similarity in the model, all lengths associated with the prototype wave should be scaled down according to the length scale N, namely

1. The water depth in the model should be selected as $h_m = h_p/N$,

2. The wave height in the model should be selected as $H_m = H_p/N$, and

3. The wave length in the model should be selected as $L_m = L_p/N$.

Now, the dispersion relation for a small-amplitude linear wave is, for prototype (Appendix A),

$$\omega_p^2 = g\lambda_p \tanh(\lambda_p h_p) \tag{3.57}$$

and that for model

$$\omega_m^2 = (Ng)\lambda_m \tanh(\lambda_m h_m) \tag{3.58}$$

in which λ_p and λ_m are, respectively, the prototype and model wave numbers:

$$\lambda_p = \frac{2\pi}{L_p}, \text{ and } \lambda_m = \frac{2\pi}{L_m} \tag{3.59}$$

with the condition for geometrical similarity (items 1 and 3 above):

$$h_m = h_p/N \text{ and } L_m = L_p/N. \tag{3.60}$$

From Eqs 3.57–3.60, one obtains

$$\omega_m = N\omega_p \tag{3.61}$$

meaning that, in order to achieve a geometrical similarity between the model and the prototype wave, the model wave angular frequency should be selected according to Eq. 3.61. For example, for a model scale of $N = 50$, the model wave angular frequency should be 50 times higher than the prototype angular wave frequency, or, from $\omega = 2\pi/T$, the model wave period T_m should be 50 times smaller than the prototype wave period T_p; e.g., for a prototype wave period of $T_p = 10\,\mathrm{s}$, and with, for example, $N = 50$, the model wave period should be $T_m = 0.2\,\mathrm{s}$.

Model similarity concerning the buildup of pore pressure

Sassa and Sekiguchi (1999) derived the differential equation governing the buildup of pore pressure in the following non dimensional form (cf. Eq. 3.28):

$$\frac{\partial \bar{p}}{\partial(\omega t)} = \left(\frac{K}{m_v \mu \omega} \lambda^2 \right) \frac{\partial^2 \bar{p}}{\partial(\lambda z)^2} + \frac{1}{m_v} \frac{\partial \epsilon_{vol}^{(2)}}{\partial(\omega t)} \tag{3.62}$$

in which K is what they termed the intrinsic permeability coefficient, the quantity m_v, is the coefficient of compressibility of the soil skeleton, $\epsilon_{vol}^{(2)}$ is the plastic component of the volumetric strain of the saturated soil, and μ is the fluid viscosity. Notice that the time is normalized with $1/\omega$ (proportional to the wave period), and the depth with $1/\lambda$ (proportional to the wave length), the characteristic time and length scales of the forcing, respectively. Now, for a complete similarity between the model and prototype,

$$\frac{K}{m_v \mu \omega} \lambda^2 = \text{constant} \tag{3.63}$$

in the model and in the prototype. Sassa and Sekiguchi (1999) argued that K should be constant if the same soil with the same initial state of packing is used in the model. They also argued that m_v is the same since this quantity depends essentially on the effective confining pressure. Therefore, for a model similarity

$$\frac{\lambda^2}{\mu \omega} = \text{constant}$$

or

$$\frac{\lambda_m^2}{\mu_m \omega_m} = \frac{\lambda_p^2}{\mu_p \omega_p} = \text{constant} \tag{3.64}$$

From Eqs 3.59, 3.60b, 3.61, and 3.64,

$$\mu_m = N \mu_p \tag{3.65}$$

The preceding equation implies that, for a model similarity, the model liquid should have a viscosity which is N times larger than the prototype liquid, normally sea water. This leads to a very viscous liquid. Sassa and Sekiguchi (1999), with $N = 50$, used silicone oil in their centrifuge experiments.

The above analysis is related to the "diffusion" component of the accumulated pore pressure process, the first term on the right-hand side of Eq. 3.62. However, for a complete model similarity, the generation of the pore pressure should also be simulated properly. To address this problem, Sassa and Sekiguchi (1999) adopted the expression for the amplitude of the shear stress in the soil, τ, under a progressive wave for a soil with an infinitely large depth, Eq. 2.55:

$$\tau = -p_b \lambda z \exp(-\lambda z). \tag{3.66}$$

Subsequently, they form the shear-stress ratio, τ/σ'_{v0}, in which σ'_{v0} is the initial vertical effective stress (Eq. 3.11):

$$\frac{\tau}{\sigma'_{v0}} = \frac{p_b \lambda}{\gamma'} \exp(-\lambda z) \tag{3.67}$$

and at the mudline, the preceding equation reads

$$\chi_0 = \left(\frac{\tau}{\sigma'_{v0}}\right)_{z=0} = \frac{p_b \lambda}{\gamma'}. \tag{3.68}$$

Sassa and Sekiguchi (1999) argued that the latter quantity can represent the severity of the wave loading even though, they added, the real soil is not an ideal elastic body. They remarked that, in this sense, the latter parameter may be better understood as a convenient index by which to assess the intensity of wave loading exerted on a given soil deposit. Sassa and Sekiguchi (1999) called χ_0 the wave severity. From the point of view of model similarity, the above quantity should also be maintained the same as in the prototype to achieve a complete model similarity.

Sassa and Sekiguchi's (1999) test conditions and corresponding field conditions

Fig. 3.35 gives a schematic description of the wave tank used by Sassa and Sekiguchi (1999). As mentioned previously, this wave tank was mounted on the platform at the end of the arm of the centrifuge (Fig. 3.34). PPTs in the figure are the pore pressure transducers to measure pore-water pressure.

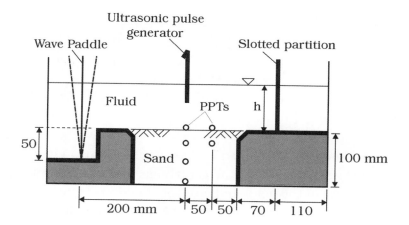

Figure 3.35: Schematic description of the wave flume in Sassa and Sekiguchi's (1999) centrifuge wave testing.

As mentioned earlier, silicone oil was used in the experiments, to fulfil the condition in Eq. 3.65. Two kinds of tests were conducted: progressive-wave tests, and standing-wave tests. In the context of the present section, we will consider the progressive-wave tests.

The test conditions were as follows.

Soil properties: sand with $d_{50} = 0.15$ mm; the specific gravity of grains, $s = 2.65$; the maximum and minimum void ratios, respectively, $e_{max} = 1.07$ and $e_{min} = 0.64$; the relative density, $D_r = 0.42$ (the relative density varied slightly over the tests carried out; D_r was 0.42 for two specific tests we will use in the following section); the void ratio, $e = 0.889$; the submerged specific weight of soil, $\gamma' = 428.5$ kN/m³; and the soil depth, $d = 100$ mm.

Wave properties: the water depth, $h = 90$ mm; the wave height, $H = 31$ mm (Test P4-1) and 33.4 mm (Test P5-1) (both wave heights calculated from the small amplitude linear wave theory, corresponding to the measured pore pressure values at the mudline); the wave period, $T = 0.0909$ s; and the wave length, $L = 515$ mm (calculated from the small amplitude linear wave theory).

Considering the model scale, $N = 50$, and the fact that the sediment properties of the model experiments remain the same as in the model, the above set of model conditions correspond to the following field conditions (see the above discussion under the subsection *Centrifuge testing. General*).

Soil properties: sand with $d_{50} = 0.15$ mm; the specific gravity of grains,

$s = 2.65$; the maximum and minimum void ratios, respectively, $e_{\max} = 1.07$ and $e_{\min} = 0.64$; the relative density, $D_r = 0.42$; the void ratio, $e = 0.889$; the submerged specific weight of soil, $\gamma' = 8.57\,\mathrm{kN/m^3}$ (calculated from $\gamma' = [(s - 1)/(1 + e)]\gamma$); and the soil depth, $d = 5\,\mathrm{m}$.

Wave properties: the water depth, $h = 4.5\,\mathrm{m}$; the wave height, $H = 1.55\,\mathrm{m}$ (Test P4-1) and $1.67\,\mathrm{m}$ (Test P5-1); the wave period, $T = 4.5\,\mathrm{s}$; and the wave length, $L = 25.7\,\mathrm{m}$.

In principle, the results obtained by Sassa and Sekiguchi can be viewed as field data since the stress level in the centrifuge experiments was maintained precisely the same as in the field, and also the liquid in the model was selected such that a complete similarity between the model and the prototype could be achieved. Therefore, in the following paragraphs, the Sassa and Sekiguchi (1999) results will be designated as field data (obtained through centrifuge experiments), and compared with similar results from standard wave-flume experiments of Sumer *et al.* (2006 a).

3.3.3   Comparison with standard wave-flume results

Governing nondimensional parameters

From dimensional considerations, the period-averaged accumulated pore pressure, $\bar{p}$, is described as a function of the following nondimensional quantities

$$\frac{\bar{p}}{p_{\max}} = f(\lambda z,\ \omega t,\ \chi_0,\ D_r,\ \lambda d,\ S) \tag{3.69}$$

in which $\bar{p}$ is, for convenience, normalized by $p_{\max}$, the maximum value attained by the time-averaged accumulated pore pressure for large times (Fig. 3.5). The relevance of the first three nondimensional quantities on the right-hand side of the above equation has been discussed in the previous subsection.

The relevance of the fourth nondimensional quantity D_r (the quantity characterizing the density of the sediment, loose or dense, etc.) is quite straightforward. The looser the sediment, the faster the buildup of pore pressure. By contrast, soils with a very large relative density may not even "permit" the pore pressures to build up, as there is not much room for grains to rearrange.

The fifth quantity, λd, represents the nondimensional depth of the sediment layer, and clearly the process of buildup of pore pressure should also be a function of this quantity.

The last parameter, S, in Eq. 3.69 is

$$S = \frac{(Gk/\gamma)T}{L^2} \quad (3.70)$$

and can be easily obtained by nondimensionalizing the differential equation that governs the buildup of pore pressure (Eq. 3.28). This is achieved in the same way as in Eq. 3.62. Note that, in this exercise, the coefficient of consolidation, c_v, in Eq. 3.28 is, from Eq. 3.26, taken as

$$c_v = \frac{Gk}{\gamma} \quad (3.71)$$

since

1. $G/K' \to 0$ (completely saturated soil), and

2. ν, the Poisson coefficient, is considered to be a constant with values from 0.3 to 0.4 during cycling loading (Lambe and Whitman, 1969, p. 160), and therefore can be left out for the present purpose.

Inserting Eq. 3.71 in Eq. 3.70, the parameter S can also be written as

$$S = \frac{c_v T}{L^2}. \quad (3.72)$$

The physical meaning of S can be explained as follows. Let

$$\ell^2 = c_v T \quad (3.73)$$

Considering that c_v acts as a diffusion coefficient (see the discussion in conjunction with Eq. 3.28 under Section 3.2.2), ℓ will represent the distance over which the accumulated pore pressure spreads over one wave period T. The quantity L (the wave length), on the other hand, is the length scale of the process. Hence, the parameter S, or alternatively $S^{1/2} = \ell/L$, will represent the distance over which the accumulated pore pressure spreads, normalized by the wave length. Obviously, the larger the value of S, the larger the spreading of the accumulated pressure. Eq. 3.72 can also be interpreted as the nondimensional time scale, analogous to the so-called time factor in the consolidation theory in soil mechanics (e.g., Lambe and Whitman, 1969, p. 408).

The above nondimensional framework, Eq. 3.69, will be the basis for comparison between the "field" data obtained through the centrifuge experiments of Sassa and Sekiguchi (1999) and the standard wave-flume data of Sumer *et al.* (2006 a).

Comparison

The "field" values of the soil and wave conditions of Sassa and Sekiguchi's (1999) centrifuge experiments along with the values of the nondimensional parameters χ_0, D_r, λd, and S are summarized in Table 3.2. Sumer et al.'s (2006 a) corresponding data from their standard wave-flume experiments also are summarized in Table 3.2.

Table 3.2. Field values corresponding to the centrifuge experiments of Sassa and Sekiguchi's (1999), and Sumer et al.'s (2006 a) standard wave-flume experiments.

Quantity	Field values corresponding to Sassa and Sekiguchi's (1999) centrifuge expt.		Standard wave-flume experiments from Sumer et al. (2006 a)	
	Test P4-1	Test P5-1	Test 1	Test 4
d_{50} (mm)	0.15	0.15	0.060	0.060
s	2.65	2.65	2.65	2.65
D_r	0.42	0.42	0.38	0.38
γ' (kN/m^3)	8.57	8.57	9.44	9.44
d (m)	5.0	5.0	0.175	0.175
λd	1.22	1.22	0.379	0.379
h (m)	4.5	4.5	0.42	0.42
H (m)	1.55	1.67	0.09	0.16
T (s)	4.5	4.5	1.6	1.6
ω (s^{-1})	1.396	1.396	3.925	3.925
L (m)	25.7	25.7	2.9	2.9
λ (m^{-1})	0.244	0.244	2.166	2.166
χ_0	0.13	0.14	0.07	0.125
S	$1.7{\times}10^{-5}-$ $1.7{\times}10^{-3}$	$1.7{\times}10^{-5}-$ $1.7{\times}10^{-3}$	$3.1{\times}10^{-5}$	$3.1{\times}10^{-5}$

The wave severity χ_0 in Table 3.2 is calculated from Eq. 3.68 in which p_b is calculated from the small amplitude linear wave theory (Eq. 2.46). The parameter S, on the other hand, is calculated from Eq. 3.70 in which G for both Sassa and Sekiguchi's "field" conditions and Sumer *et al.*'s (2006 a) laboratory conditions is taken as $G = 2500 \, \text{kN/m}^2$, corresponding to loose sand while the coefficient of permeability k is taken as $k = 10^{-5} - 10^{-3} \, \text{m/s}$ (Lambe and Whitman, 1969, p. 286) for the field conditions, and $k = 6.4 \times 10^{-7} \, \text{m/s}$, obtained from k (cm/s) $= 16(d_{10}(\text{cm}))^2$ (Lambe and Whitman, 1969, p. 290), for Sumer *et al.*'s wave-flume laboratory conditions. As the permeability varies, the value of S also varies; however, the true value for the considered sediment should be near the lower end of the range $S = 1.7 \times 10^{-5} - 1.7 \times 10^{-3}$, considering the sand size (fine sand).

Figs 3.36 and 3.37 present two comparisons between the Sassa and Sekiguchi centrifuge results and the Sumer *et al.* standard wave-flume results; in Fig. 3.36, the time series of the time-averaged pressure are compared in the no-liquefaction regime, and in Fig. 3.37, this is done for the liquefaction regime. Comparison is made in the nondimensional form implied by Eq. 3.69.

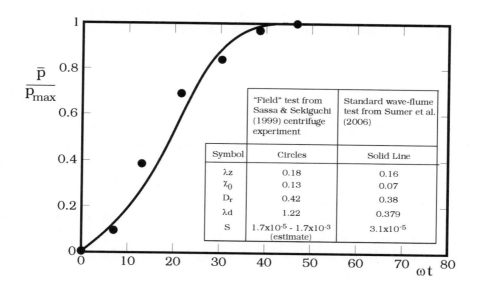

	"Field" test from Sassa & Sekiguchi (1999) centrifuge experiment	Standard wave-flume test from Sumer et al. (2006)
Symbol	Circles	Solid Line
λz	0.18	0.16
χ_0	0.13	0.07
D_r	0.42	0.38
λd	1.22	0.379
S	1.7×10^{-5} - 1.7×10^{-3} (estimate)	3.1×10^{-5}

Figure 3.36: Comparison between Sassa and Sekiguchi's (1999) 50 g centrifuge results and Sumer *et al.*'s (2006 a) 1 g standard wave-flume results. No-liquefaction regime. Test P4-1 with $z = 15 \, \text{mm}$ (0.75 m in the field) in Sassa and Sekiguchi (1999) and Test 1 with $z = 7.5 \, \text{cm}$ in Sumer *et al.* (2006 a).

Fig. 3.36 compares Test P4-1 measurement of Sassa and Sekiguchi at the depth $z = 0.75 \, \text{m}$ in the field (or $z = 15 \, \text{mm}$ in the centrifuge), with Test 1

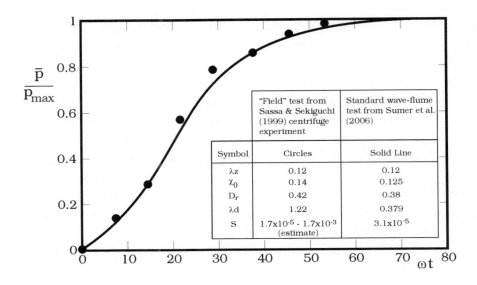

Figure 3.37:   Comparison between Sassa and Sekiguchi's (1999) 50 g centrifuge results and Sumer *et al.*'s (2006 a) 1 g standard wave-flume results. Liquefaction regime. Test P5-1 with $z = 10$ mm (0.5 m in the field) in Sassa and Sekiguchi (1999) and Test 4 with $z = 5.5$ cm in Sumer *et al.* (2006 a).

measurement of Sumer *et al.* at the depth $z = 7.5$ cm in the standard wave-flume test. The normalized depth values, namely $\lambda z = 0.18$ in Sassa and Sekiguchi and $\lambda z = 0.16$ in Sumer *et al.* are very close to each other, and therefore the results can be compared provided that they are plotted versus the nondimensional time ωt, as is done in the figure.

Fig. 3.37, on the other hand, compares Test P5-1 measurement of Sassa and Sekiguchi at the depth $z = 0.5$ m in the field (or $z = 10$ mm in the centrifuge), with Test 4 measurement of Sumer *et al.* at the depth $z = 5.5$ cm in the standard wave flume test. Notice that λz values of the two experiments are identical.

(It may be noted that, for the liquefaction-regime test, Test P5-1, there is one more pressure time series data (Sassa and Sekiguchi, 1999, Fig. 6c) with a λz value matching with a measurement depth of Sumer *et al.*'s (2006 a) Test 4 data. However, the way in which the pressure evolves at this depth in the Sassa and Sekiguchi test resembles the transitional pressure development displayed in Fig. 3.29, which is different from that observed in the standard liquefaction-regime wave-flume tests. Therefore, no comparison has been made for this data. See the discussion in conjunction with Fig. 3.29 for the transitional pressure development.)

Returning to Figs 3.36 and 3.37, although there are differences in the values of the nondimensional parameters χ_0 and D_r, and more significantly in the values of λd and S, the agreement between the standard wave-flume experiments and the centrifuge "field" experiments is striking. The latter has the following implication: standard wave-flume tests can also be used for a physical modelling study of the buildup of pore pressure and liquefaction provided that the results should be analysed and interpreted on the basis of the nondimensional representation described in Eq. 3.69.

Now, recall the soil depth in the above "field" experiment, namely $d = 5\,\text{m}$. In the corresponding centrifuge tests, the stress level associated with this depth was truly replicated, and therefore the soil behaviour (stress–strain relationship, friction angle) was reproduced as in the field. Since the buildup of pore pressure (both in the no-liquefaction regime and in the liquefaction regime) appear to be in very good agreement with that obtained in the standard wave-flume experiments, as revealed in Figs 3.36 and 3.37, it can be inferred that the change in the soil behaviour with depth is apparently not significant for such "shallow" soil depths. Since the wave-induced buildup of pore pressure and liquefaction is, for the most part, associated with shallow soil depths, the classic standard wave-flume tests appear to be a viable option for physical modelling studies of wave-induced liquefaction.

3.4 Mathematical Modelling of Compaction

As described in Section 3.1, the liquefaction is followed by the compaction where the upward-directed pressure gradient (generated by the liquefaction) drives the water in the liquefied soil upwards while the soil grains "settle", leading to a progressive compaction of the liquefied soil (Fig. 3.12). The compaction process first begins at the impermeable base and gradually progresses in the upward direction, as the pore water is drained out of the soil. The process is illustrated in Fig. 3.38. Sediment grains, as well as distances between the sediment grains in the liquefied-soil layer in Fig. 3.38, are grossly exaggerated for illustration purposes.

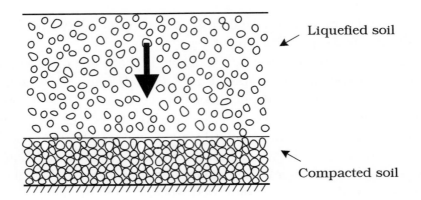

Figure 3.38:   Definition sketch. Snapshot of soil with the compacted-soil and the liquefied-soil layers. Distance between grains in the liquefied soil layer is grossly exaggerated for illustrations purposes.

Now, consider the compaction front in Fig. 3.39, at time t at the level $z + dz$. This compaction front will move to the level z during a small time interval of dt. As described in Section 3.1.4, this is actually due to the downward movement of sediment grains into the incremental volume dz during the time interval dt; the grains will move into this volume, and remain there to nearly "complete" the compaction of the soil layer between z and $z + dz$, according to the description of the process in Section 3.1.4. The amount of sediment that passes through the horizontal plane at the level z (per unit area) during the time interval dt is

$$V_s = (cw)(1 \times 1)dt \tag{3.74}$$

in which c is the volume concentration corresponding to the liquefied soil, and w the fall velocity of sediment grains.

Since the sediment settles in a very heavy "suspension", the fall velocity in the above equation is no longer equal to that given for a dilute suspension, w_0, but rather it is expressed as

$$w = (1 - c)^n w_0 \tag{3.75}$$

(the so-called hindered settlement). This subject is discussed in details in Chapter 5 in connection with a mathematical model developed for the density of liquefied soil (Section 5.4.1).

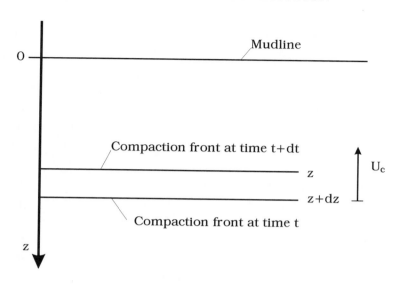

Figure 3.39: Compaction front moves upward from z+dz to z during the time increment (t, t + dt) with U_c being the velocity of the compaction front.

The power n (not to be confused with the porosity) in Eq. 3.75 can be taken as constant for applications to wave-induced liquefaction (see the discussion in Section 5.4.1). The value of n reported in the literature has been determined on the basis of sedimentation and fluidization experiments, and found to be 4.5 (see Section 5.4.2). However, Sumer *et al.* (2006 b), in connection with the mathematical model referred to in the preceding paragraphs (Section 5.4.1), argued that n should assume values smaller than 4.5 since the grains settle in an upward-directed pressure-gradient field in the case of liquefied soil. They found that the value of the density of the liquefied soil they predicted from their model matched with the measured values when $n = 2.7$ (Section 5.4.2), and recommended that n be taken as $n = 2.7$. We will adopt this value for the present analysis.

Now, the amount of sediment V_s in Eq. 3.74 is nothing but the solid part of the compacted soil volume in $dz \times 1 \times 1$ by the time $t + dt$. Therefore, the porosity of the compacted soil at the level z will be

$$n_1 = \frac{(dz \times 1 \times 1) - V_s}{dz \times 1 \times 1} \tag{3.76}$$

and can, to a first approximation, be taken as a constant.

Eqs 3.74–3.76 can be solved for $dz/dt \, (= U_c)$, the velocity of the compaction front:

$$U_c = \frac{c}{1 - n_1}(1 - c)^{2.7}w_0. \tag{3.77}$$

This equation enables us to predict U_c, and more importantly, the time scale of the compaction (or solidification) process from the time where the compaction starts at the impermeable base to the time where the compaction front reaches the mudline.

To this end, the concentration of the liquefied soil needs to be known (Eq. 3.77). The mathematical model of Sumer *et al.* (2006 b) mentioned in the preceding paragraphs (see Section 5.4.1) predicts this quantity as a function of the specific gravity of sediment grains γ_s/γ, the specific gravity of the soil γ_t/γ (before liquefaction), and the coefficient of lateral earth pressure k_0 (before liquefaction). We note that Sumer *et al.*'s (2006 b) mathematical model (Section 5.4.1) assumes that the concentration remains unchanged with the depth z, although this quantity is expected to be a very weak function of z, with c increasing with increasing z.

As a final note, one may argue that the water in the incremental volume dz moves upwards due to the upward movement of the compaction front, with a velocity of approximately $dz/dt \, (= U_c)$, and therefore it would be more correct to write the amount of sediment that passes through the horizontal plane at the level z (per unit area) during the time interval dt in Eq. 3.74 as

$$V_s = (c(w - U_c))(1 \times 1)dt. \tag{3.78}$$

However, this effect will, to a first approximation, be taken care of by employing Eq. 3.75. Therefore, the expression for V_s will, in the present analysis, be maintained as that given in Eq. 3.74.

Example 8. *The time scale of the compaction process.*

In a laboratory test reported by Sumer *et al.* (2006 b), a soil with a depth of $d = 0.175 \, \mathrm{m}$ was liquefied by $H = 0.17 \, \mathrm{m}$ high and $T = 1.6 \, \mathrm{s}$ waves (Test 5 in Sumer *et al.*, 2006 b, Fig. 8). The grain size in the test was $d_{50} = 0.078 \, \mathrm{mm}$, the coefficient of lateral earth pressure $k_0 = 0.4$, the total specific gravity of the soil before the test $\gamma_t/\gamma = 1.97$, and the porosity after the test (with the completion of compaction after the sequence of liquefaction and compaction processes), $n_1 = 0.354$.

1. Calculate the velocity of the compaction front U_c; and

2. Based on the latter, find the time scale for the completion of the compaction process.

From Sumer *et al.*'s (2006 b) mathematical model, the concentration c is found to be $c = 0.54$ for the values of the specific gravity of sediment grains $\gamma_s/\gamma = 2.65$, the specific gravity of the soil $\gamma_t/\gamma = 1.97$ and the coefficient of lateral earth pressure $k_0 = 0.40$. We note that Sumer *et al.* (2006 b) gives a "design chart" for c as a function of γ_t/γ, with two values of k_0, namely $k_0 = 0.4$ and 0.5, and for $\gamma_s/\gamma = 2.65$, reproduced in Figs 5.8 and 5.9. This chart can be used to pick up the corresponding value of the concentration for a given set of values of γ_t/γ, and k_0.

For $d_{50} = 0.078$ mm, the fall velocity in dilute concentration is $w_0 = 0.5$ cm/s at 20°C water temperature (Vanoni, 1975, p. 25). Inserting $c = 0.54$, $w_0 = 0.5$ cm/s and $n_1 = 0.354$, the velocity of the compaction front is, from Eq. 3.77, obtained $U_c = 0.051$ cm/s, or approximately 0.5 mm/s.

The time for the compaction front to move from the impermeable base to the surface of the soil can, to a first approximation, be estimated, using

$$T_c = \frac{d}{U_c} \tag{3.79}$$

in which d is the soil depth. This gives

$$T_c = \frac{d}{U_c} = \frac{17.5}{0.051} = 341 \text{ s or approximately 6 minutes.}$$

Note that the distance that the compaction front travels is taken as the initial soil depth, d, in Eq. 3.79. However, this depth, in reality, is decreased somewhat, due partly to compaction, and partly to the onshore sediment transport (see the discussion in Section 3.1.4).

Fig. 3.40 presents the pressure time series measured in the test (Test 5 in Sumer *et al.*, 2006 b, Fig. 8). The time from the instant where the compaction starts to that of the complete dissipation of excess pressure (the compaction time, T_c) is found to be $T_c = 7.1$ minutes (Fig. 3.40). As seen, the time scale calculated from the above simple model, $T_c \simeq 6$ minutes, is not radically different from the experimental value, $T_c = 7.1$ minutes.

The fact that the present model somewhat underpredicts the true time scale may be linked to the delay in the dissipation of the pore pressure.

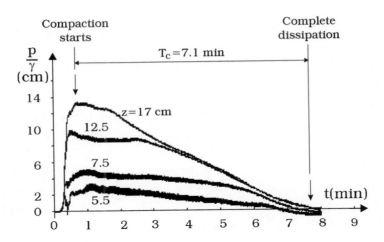

Figure 3.40:   Time series of excess pore pressure in Test 5 in Sumer *et al.*'s (2006 b) study. T_c is the compaction time for the compaction front to move from the impermeable base to the surface of the soil.

This delay, as discussed earlier in conjunction with compaction versus self-weight consolidation (Section 3.1.4), is caused by the fact that the waves and therefore wave-induced shear strains in the compacted sediment will generate additional excess pore pressure, and this effect will delay the dissipation of the excess pore pressure. The present model does not capture this aspect of the problem, and therefore the time scale is underpredicted (as a conservative value) relative to the true value. Nevertheless, this simple model (Eqs 3.77 and 3.79) can be used for a quick assessment of the time scale of the compaction process.

Example 9. *Assessment of time scale of compaction in a field situation. Numerical example.*

A soil with a depth of $d = 10\,\mathrm{m}$ is liquefied by $H = 3\,\mathrm{m}$ high and $T = 5.3\,\mathrm{s}$ waves at a water depth of $h = 17.7\,\mathrm{m}$. The soil class is given as silty sand. The grain size is $d_{50} = 0.1\,\mathrm{mm}$, the coefficient of lateral earth pressure $k_0 = 0.5$, the total specific gravity before the liquefaction $\gamma_t/\gamma = 1.92$, and the specific gravity of sediment grains $\gamma_s/\gamma = 2.65$. The post-liquefaction porosity n_1 is unknown.

Estimate the time scale of compaction with a sensitivity analysis in terms of n_1.

From Sumer *et al.*'s (2006 b) mathematical model, the concentration c is found to be $c = 0.52$ for the values of the specific gravity of sediment grains $\gamma_s/\gamma = 2.65$, the specific gravity of the soil $\gamma_t/\gamma = 1.92$ and the coefficient of lateral earth pressure $k_0 = 0.5$.

For $d_{50} = 0.1\,\mathrm{mm}$, the fall velocity in dilute concentration is $w_0 = 0.8$ cm/s at 20°C water temperature (Vanoni, 1975, p. 25). Inserting $c = 0.52$, $w_0 = 0.5\,\mathrm{cm/s}$ and taking, to a first approximation, n_1 as the minimum value of porosity, n_{min}, which may be considered to have values in the range $n_{min} \sim 0.2 - 0.3$ for the present sediment (Lambe and Whitman, 1969, p. 31), (1) the velocity of the compaction front, U_c, and (2) the time for the compaction front to move from the impermeable base to the mudline, T_c, are calculated from Eqs. 3.77 and Eq. 3.79, respectively, and tabulated in Table 3.3. (It may be noted that the values given in the latter reference for n_{min} are $n_{min} = 0.23$ for silty sand, 0.17 for fine to coarse sand, and 0.29 for clean uniform sand.)

Table 3.3. Estimated values for the time scale of compaction for different values of the post-liquefaction porosity.

n_1	U_c (cm/s)	T_c (hours)
0.2	0.045	6.2
0.25	0.048	5.8
0.3	0.051	5.4

From the results summarized in Table 3.3, it appears that the results are not too strongly sensitive to the value of the post-liquefaction porosity, and therefore it may be concluded that the time scale of the compaction is like $T_c = O(6\text{ hours})$.

3.5 Influence of Clay Content

The seabed in the previous sections is considered to contain one type of soil such as silt. However, it is not uncommon that the seabed may contain clay, and therefore the seabed soil in these cases acts as a composite soil such as, for example, clayey silt or clayey sand. Kirca *et al.* (2014) conducted liquefaction experiments to study the influence of clay content in such soils. It turns out that the influence of clay content is very significant. The following paragraphs will summarize the results of Kirca *et al.*'s study (2014).

Kirca *et al.* (2014) used in their experiments two kinds of sediment: silt with $d_{50} = 0.070$ mm, and sand with three different kinds of particle size distributions with $d_{50} = 0.17$ mm, 0.4 mm, and 0.9 mm. These sediments were used to obtain silt–clay, and sand–clay mixtures, respectively. The clay used in the experiments was a blue-clay. The clay fraction was obtained as 38%, silt 56% and fine sand 6%. The clay properties were as follows: plastic limit 17.8%, liquid limit 33.9%, plasticity index 16.3%, specific gravity 2.9, lime content 1%, organic content 1%, undrained shear strength 1.75×10^3 Pa, water content 33.6%, and median diameter, $d_{50} = 0.0041$ mm. The wave parameters on the other hand were: wave height $H = 7.6 - 18.3$ cm, wave period $T = 1.6$ s, and water depth $h = 55$ cm.

Silt–clay mixture

Fig. 3.41 compares two pore pressure responses obtained in Kirca *et al.*'s experiments; Fig. 3.41a is the pore water pressure time series in the case of silt alone, and Fig. 3.41b is that in the case of silt–clay mixture with a clay content of $CC = 8\%$. The experiments were carried out under exactly the same wave conditions ($H = 14.5$ cm and $T = 1.6$ s), and the time series shown in the figure corresponds to the measurement point at $z = 18.5$ cm, z being the distance from the mudline. The clay content CC here is defined by $CC = W_c/W_m$ in which W_c is the weight of the clay, and W_m is that of the mixture. Fig. 3.41 shows that even a clay content of as low as $CC = 8\%$ makes a dramatic change in the response of the excess pore pressure. The number of waves to reach liquefaction decreases approximately from 8.5 (silt alone) to 4.5 (silt–clay mixture), and similarly, the number of waves for excess pore pressure to reach $p_{\max}$ decreases from 13 (silt alone) to 6 (silt–clay mixture).

Fig. 3.42 displays the behaviour of the excess pore pressure for large times. As in Figs. 3.41a and b, the change in the behaviour of the excess pore pressure for large times is also dramatic. The liquefaction stage (from the time of the onset of liquefaction until approximately the time the excess pore pressure begins to fall off, marked C in Fig. 3.11) lasts considerably longer in the case of the silt–clay mixture.

Furthermore, Kirca *et al.*'s (2014) data indicated (not shown here) that the critical wave height for the onset of liquefaction (see the discussion in

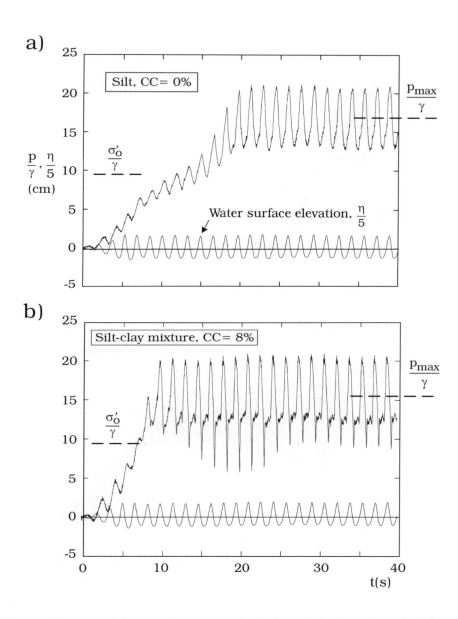

Figure 3.41: Excess pore pressure time series with silt and silt–clay mixture under exactly the same wave conditions. Kirca *et al.* (2014).

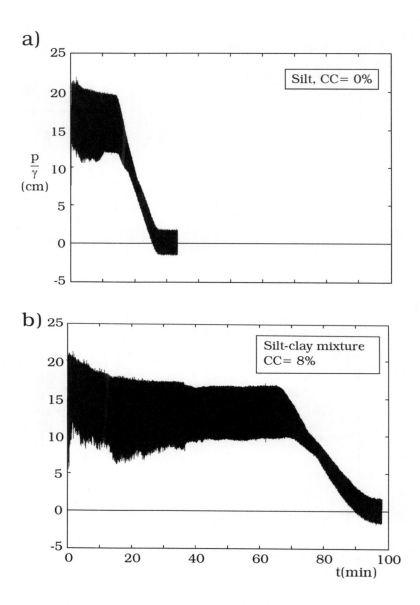

Figure 3.42:   Excess pore pressure time series with silt and silt–clay mixture for large times under exactly the same wave conditions. Kirca *et al.* (2014).

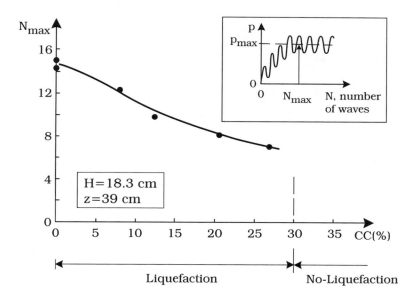

Figure 3.43: Number of waves for the excess pore pressure to reach p_{max} as function of clay content in Kirca *et al.*'s (2014) experiments.

conjunction with Fig. 3.8) is reduced from about 10.3 cm (silt alone) to about 8.9 cm (silt–clay mixture) for the same value of the clay content, i.e., $CC = 8\%$.

All this indicates that the soil becomes more prone to liquefaction when 8% clay is added. Kirca *et al.* (2014) report that this behaviour was observed, regardless of the depth, z, and the clay content, ranging from $CC = 8\%$ to approximately 30%.

Fig. 3.43 illustrates the number of waves, N_{max}, for the period-averaged pore-water pressure to attain its maximum value plotted versus the clay content for the wave height $H = 18.3$ cm and the measurement depth $z = 39$ cm. From the figure: (1) N_{max} decreases with the clay content. Kirca *et al.* (2014) note that their data showed the same behaviour, irrespective of the wave height H and the depth z. (2) No liquefaction is observed for the clay content larger than $CC \approx 30\%$.

Fig. 3.44 displays the duration of liquefaction, T_ℓ, as a function of the clay content; see the inset in Fig. 3.44 for the definition of T_ℓ. The quantity T_ℓ can be considered as a time scale characterizing the time duration over which the soil remains in the liquefied state (see Fig. 3.11) at a given depth

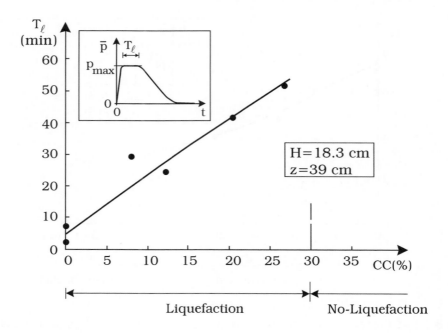

Figure 3.44:   Time scale characterizing the time period over which the soil remains in the liquefied state at depth $z = 39$ cm in Kirca *et al.*'s experiments (2014).

z. Fig. 3.44 shows that T_ℓ increases with increasing clay content CC. For large clay contents, CC larger than $CC \approx 30\%$, the silt–clay mixture is not liquefied, as already indicated in conjunction with the previous figure.

The results summarized in the preceding paragraphs imply that, for the given clay, the silt–clay mixture for CC values smaller than $CC \approx 30\%$ becomes more susceptible to liquefaction, with the liquefaction susceptibility increasing with increasing clay content. Kirca *et al.* (2014) linked this to the soil permeability.

The effect of the permeability (or alternatively the coarseness, and the grain-size distribution) on the buildup of excess pore pressure and liquefaction has been studied in the past in conjunction with laboratory simulations of seismic-induced liquefaction, see e.g. Prakash (1981, p. 280 and p. 310). The latter studies have shown that the smaller the permeability, the more susceptible the soil to liquefaction. Soil cannot be liquefied for large permeabilities; all pore pressures developed in the soil will dissipate as rapidly as they develop.

	a)	b)	c)
	Non-plastic	Low plasticity clayey silt/sand	High plasticity clayey silt/sand
Schematic representation of particle arrangement	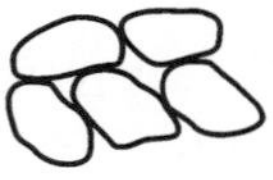		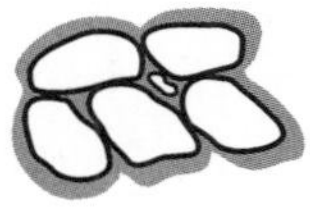
Special micro fabric features	Sand-to-sand or silt-to-silt contacts	Open microfabric, "clay bridges"	Clay matrix

Figure 3.45: Illustrations from Gratchev *et al.*'s (2006) scanning electron microscope study. Microfabric of various mixtures.

However, as already stressed, the above behaviour is for values of clay contents smaller than $CC \approx 30\%$. For clay contents larger than $CC \approx 30\%$, Kirca *et al.*'s experiments showed that the silt–clay mixture is not liquefied, a completely different behaviour. Kirca *et al.* (2014) explained this behaviour as follows.

Gratchev *et al.* (2006) studied the microfabric of the artificial mixtures by means of scanning electronic microscope (SEM). Fig. 3.45, adapted from Gratchev *et al.* (2006), essentially summarizes the results of their SEM study. Depending on the clay content, the silt–clay mixture may be a low-plasticity clayey silt (corresponding to low levels of clay content), or it may be a high-plasticity clayey silt (corresponding to high levels of clay content), or it may be in between. Gratchev *et al.* (2006) report that the low-plasticity clayey silt has a plasticity index of smaller than 4% while the high-plasticity clayey silt has a plasticity index of larger than 15%. Kirca *et al.* (2014) note that, except the clay-alone case, data regarding the plasticity index for their silt–clay mixtures are not available; they gave, however, the plasticity index of their clay as 16.3%.

Now, Kirca *et al.* (2014) indicate that their liquefaction results suggest that the microfabric of their silt–clay mixtures is much like that of low-plasticity clayey silt/sand of Gratchev *et al.* (2006) (Fig. 3.45b). This type of microfabric will allow the silt grains to rearrange under cyclic shear strains (a very important element in the buildup of excess pore pressure and eventual liquefaction). With the permeability decreasing with increasing clay content in this type of soil, this will obviously result in an increasing degree of susceptibility to liquefaction.

In the case when the clay content becomes greater than $CC \approx 30\%$, however, the soil acts like a high-plasticity clayey silt with its microfabric illustrated in Fig. 3.45c. The silt grains in this case are encapsulated completely with clay matrix, and therefore they cannot rearrange under cyclic shear strains due to the cohesive character of the mixture, presumably leading to resistance to liquefaction, as revealed by Kirca et al.'s experiments.

Recent work, notably by Guo and Prakash (1999) and Gratchev et al. (2006), unified two contradictory views (and observations) of the effect of clay content on liquefaction resistance of soils in connection with seismic induced liquefaction, namely the view that liquefaction resistance increases with increasing clay content (e.g., Seed et al., 1985), and the opposite, that liquefaction resistance decreases with increasing clay content (e.g., Troncoso, 1990). In that context, Guo and Prakash (1999) concluded that

> "the pore pressure buildup increases because the fine particles of clay reduce the hydraulic conductivity of the mixture, leading to higher pore pressures in the low plasticity range, while plasticity imparts some cohesive character to the mixture and therefore increased resistance to liquefaction in the high plasticity range", in complete accord with the above explanation.

At this juncture it may be noted that the limiting value beyond which the silt–clay mixture is not liquefied, i.e., $CC \approx 30\%$ in Kirca et al.'s (2014) studies, is expected to be dependent on the type of the clay, and therefore should not be considered to be a constant set of threshold value. Clearly this threshold value is to be determined from experiments.

The topic is further discussed at the end of this section for cases where the soil is not a composite soil like clayey silt or clayey sand but rather clay.

Sand–clay mixture

Although limited, Kirca et al.'s (2014) sand–clay mixture tests under waves showed that sand may become susceptible to liquefaction with clay content. For instance, sand with $d_{50} = 0.4\,\mathrm{mm}$ was liquefied by adding clay with $CC = 11\%$, while sand with $d_{50} = 0.17\,\mathrm{mm}$ was liquefied with a clay content of $CC = 11.5\%$ and partially liquefied with a clay content of as small as $CC = 2.9\%$. (It may be noted, however, that the sand with $d_{50} = 0.9\,\mathrm{mm}$ in Kirca et al.'s experiments with CC approximately 10% was not liquefied.)

These results are important in the sense that even a small amount of clay content in a coarse sediment (fine sand, or even medium sand) may be a serious threat to soil stability, contrary to the general perception that sand under waves is liquefaction resistant.

Remarks on practical applications for silt–clay and sand–clay mixtures

As seen, (1) the liquefaction susceptibility of silt (normally considered liquefaction prone) is increased with increasing clay content, and also (2) sand (normally considered liquefaction resistant) may be liquefied with some clay content. The question is: how can the effect of clay content be incorporated in liquefaction-potential assessments? The following paragraph will address this.

The data analysis of Kirca *et al.* (2014) has shown clearly that the liquefaction susceptibility of the soil is directly linked to the soil permeability. Therefore, whether or not a marine soil with a given clay content has potential for liquefaction in real-life situations may be assessed, using standard liquefaction assessment methods (e.g., using the mathematical model described in the preceding paragraphs in Sections 3.2.3, or 3.2.4; see also the numerical example given under Section 3.2.4, Example 4). To this end, the permeability of the mixture needs to be determined, along with the other soil parameters essential to conduct the assessment exercise. The latter parameters include the submerged specific weight, the porosity, the coefficient of lateral earth pressure, the relative density, the modulus of elasticity, Poisson's ratio, and the soil depth. These parameters are to be determined from soil tests with reconstituted soil samples where possible. Otherwise, sensitivity studies need to be carried out for the undetermined soil parameters.

Liquefaction of some clay and mud. Generation of fluid mud layers by waves

From the previous subsections it can be inferred that cohesive sediments such as clay are not liquefiable. However, laboratory observations suggest that waves can liquefy cohesive sediment, resulting in the formation of high-density mud layers close to the bed (see, e.g., Lindenberg, van Rijn and Winterwerp (1989) and de Wit and Kranenburg (1992) and de Wit (1995).

The thickness of mud layers *in the field* can be as much as $O(10\,\mathrm{cm})$ and may be even larger. A detailed account of fluid mud is given in Whitehouse,

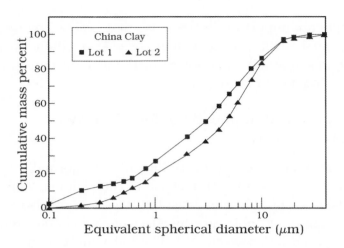

Figure 3.46:   Particle size distributions in de Wit's (1995) experiments.

Soulsby, Roberts and Mitchener (2000) where the prediction of the conditions
for fluid mud to be produced by waves and/or currents is among the topics
reviewed.

The question is: can cohesive sediment, for example clay, be liquefied by
waves?

de Wit experiments. In order to address the above question, de Wit
(1995) made an extensive series of experiments in a wave flume and in an
oscillating water tunnel, using various types of cohesive sediments.

Fig. 3.46 gives the particle size distribution of one of the sediment, namely
China Clay, de Wit used in the experiments, with $d_{50} = 0.003 - 0.005\,\mathrm{mm}$.

Fig. 3.47 presents the pore-water pressure time series obtained in one of
the de Wit tests with the sediment given in the previous figure (Fig. 3.46).
This test was conducted in the wave flume. The test set-up was qualitatively
similar to that in Fig. 3.1. The water depth was 30 cm while the soil depth
was 19 cm in this particular test. The sediment was placed in the sediment
box by allowing a homogeneous suspension to settle and consolidate in this
test (and indeed in all the flume tests). This type of placement is designated
as deposited bed in de Wit. The quantity z in the figure is the distance
measured from the surface of the bed downwards. σ', on the other hand,
is described as the effective weight of the overlying mud by de Wit (1995,
p. 124). No details are given in de Wit (1995) of how σ' was determined.

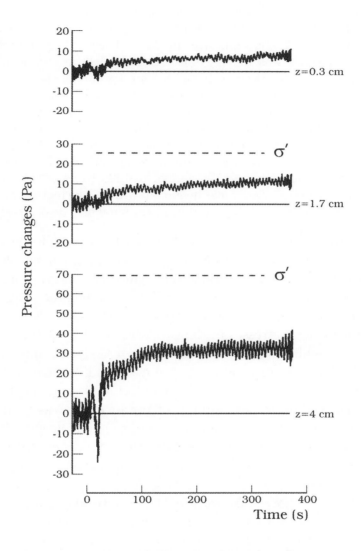

Figure 3.47:   Excess pore pressure time series in de Wit's (1995) experiments. Experiment III, China Clay, test 1. To convert Pa to cm water column, divide by 100.

de Wit (1995, p. 122) describes the process of the wave-and-sediment-bed interaction in the above mentioned test as follows:

> "movements were observed in the uppermost part of the bed as soon as the first waves with the wave height of about 2 cm travelled over the test section. A very thin layer of the bed seemed to be mobilized in which material was moving forward and backward over the bed. No movements were observed in the remaining part of the bed. Then, after about one minute small spots of fluid mud arose, apparently at random, at the bed surface. The sizes of these patches increased with time. The fluid mud was oscillating in patches and its movement seemed to be out of phase with the movement of the water just above it. After approximately 15 minutes the entire surface of the bed seemed to be liquefied." de Wit (1995, p. 122, Table 6.2.1) gives the consolidation period and wave height at which liquefaction occurred for this experiment as 6 days and 2 cm (or smaller), respectively.

The preceding description along with Fig. 3.47 (although the accumulated pore pressures do not reach σ' values) imply that the clay in the de Wit experiment was liquefied, a result contrary to the information given in conjunction with the influence of clay content on liquefaction in the preceding paragraphs. de Wit (1995) links this to the shear stresses in the soil caused by the wave; the wave-induced shear stresses in the soil (see, e.g. Fig. 2.2) are presumably larger than the yield stress of the clay used in the test, and therefore this latter effect breaks the microfabric of the clay, and consequently the cyclic shear stresses and their associated shear deformations in the soil rearrange the clay grains, leading to the buildup of pore-water pressure and eventual liquefaction, in exactly the same manner as in the noncohesive sediment. We note that the yield stress of clay depends on the solid concentration, the composition of the clay, and the electrolyte concentrations, among other things (de Wit, 1995, Chapter 2).

de Wit (1995, p. 128) substantiated the above explanation by comparing the estimated value of the shear stress induced by the wave with the yield stress of the consolidated China Clay, estimated from the rheological measurements. The shear stress induced by the wave was estimated, using Spierenburg's (1987) model (cf., Hsu and Jeng, 1994). It was found that the stresses induced by the wave in the soil may be of the same order of magnitude or larger than the yield stress of the consolidated clay.

de Wit (1995) offered the above description as the mechanism responsible for the generation of mud layers. A similar description is also given by Whitehouse *et al.* (2000, Chapter 6, p. 110).

Returning to Fig. 3.47, the fact that the accumulated pore-water pressures in Fig. 3.47 apparently do not reach the σ' values indicated in the figure is explained by de Wit (1995) as follows. When the wave was introduced in the test, the consolidation had not yet been completed, and there was some pore-water pressure in excess of the static pore-water pressure, and the difference between the accumulated pore-water pressure and the pressure corresponding to σ' is attributed to this effect.

Behaviour of mud samples from the field. de Wit's (1995) laboratory experiments demonstrated that cohesive sediment can be liquefied under waves. The question is: can mud (collected from the field) be liquefied? Lindenberg *et al.* (1989) addressed this issue by testing mud samples. They carried out two tests in a specially designed shaking device. In the first test, they used the samples without any treatment, and the test results were that no liquefaction occurred. In the second test, the mud sample was heated to remove the organic material, and saturated with de-aired water, and subsequently was subjected to the same shaking test in the same device. In this latter test, the top layer was liquefied. These results indicate the following: (1) the organic material makes the sediment more "cohesive", and therefore more liquefaction resistant. (2) When the sediment contains gas/air, the pore pressures will be dissipated as they build up, and therefore no accumulation of pore pressure will take place, and hence no liquefaction.

Lindenberg *et al.* (1989) point out that a consequence of their observations is that data concerning gas content or organic material should be measured when the liquefaction behaviour of natural muds has to be determined or explained.

Wave damping and generation of fluid mud layers in the field. Wave damping is an important aspect of mud layers. The current understanding of wave damping by mud layers is that the damping occurs due to viscous dissipation in fluid mud layers, the viscosity of a fluid mud being a few orders of magnitude larger than that of water (Maa and Mehta, 1990).

The issues such as the generation of mud layers, whether or not mud can be liquefied, and the wave damping by fluid mud are still a matter of large debate in the coastal engineering community (Sahin, Sheremet and Safak (2012) and Sahin, Safak, Sheremet and Mehta (2012)), Winterwerp, de Boer, Greeuw and van Maren (2012). The mechanism of the generation of mud layers in the real life is still not well understood. No data of pore pressure in fluid mud is yet available. Clearly, this is arguably the only way to check for the occurrence of liquefaction of mud in the field, and therefore these kinds of pore pressure measurements are much needed.

Winterwerp *et al.* (2012), in a recent study, report an interesting experiment. They tested mud samples (from the Wadden Sea coast in the Netherlands) in triaxial equipment. The purpose of the experiment was to assess whether the samples would liquefy under the cyclic stresses to be expected under storm conditions. None of the samples liquefied, even at cyclic stresses well beyond those to be expected under storm conditions, an inconsistent result contrary to (1) the observations that waves are significantly damped along the Wadden Sea coast during storm conditions, and (2) the results from Winterwerp *et al.*'s field work, suggesting mud deposits with strength around the liquid limit, two results suggesting the occurrence of fluid mud layers during storm conditions.

Winterwerp *et al.* (2012, pp. 108–109) proposed the following explanation for this apparent inconsistency. First of all, from their triaxial tests, they concluded that the mud deposits along the Wadden Sea coast cannot liquefy, neither fresh, nor consolidated, under the wave action, consistent with the results of Lindenberg *et al.*'s study (1989) referred to above. They remarked that the mud was just too stiff, and the wave-induced stresses were too low to cause liquefaction. However, they hypothesized that *the mud layer formed not from the wave-induced liquefaction but rather from deposition upon exceeding the flow carrying capacity.*

It appears that our understanding of mud processes is still far from being complete, and therefore this aspect of the problem would be worth exploring further.

3.6   Influence of Cover Stones/Surcharge

Surface protection by cover stones over a liquefiable soil (e.g., backfill soil, silt or fine sand, in a trench) is a method to protect the soil against scouring. Scouring may be caused by effects such as current, combined wave and current, and wave-induced steady streaming near the bed. While a fairly substantial amount of knowledge has been gained on the behaviour of cover stones/riprap on a liquefaction-resistant sediment bed in the past decade or so (see, e.g., Sumer and Fredsøe, 2002), relatively little study has been carried out on the behaviour of cover stones/riprap on a liquefiable sediment bed (Sekiguchi *et al.*, 2000 and Sumer *et al.*, 2010).

In Sekiguchi *et al.*'s (2000) study, the bed (sand sized $d_{50} = 0.15\,\text{mm}$) was covered completely or partially with gravel (the size $D_{50} = 3\,\text{mm}$). Centrifuge wave testing was used. A steady upward seepage flow was maintained during the tests. Even with the presence of this upward seepage flow, the soil was not liquefied in the tests when the soil surface was covered with gravel completely. Although limited to four tests (two tests with gravel covering the entire soil surface and two tests with that covering the soil surface only partially), Sekiguchi *et al.*'s experiments indicated for the first time that cover gravel/stones/riprap could be an option to protect soils (hydraulic fill or naturally deposited) against liquefaction. The latter is discussed in Section 11.1, Chapter 11.

The questions arising from Sekiguchi *et al.*'s study were (1) can a liquefaction-prone soil underneath such a protection system be liquefied even if it is fully covered? (2) What is the behaviour of the cover stones if the soil underneath is liquefied (issues involving sinking of the stones and their penetration distance)? (3) What is the effect of a filter layer used between the cover stones and the soil? (4) Can cover stones be used as a counter measure against liquefaction?

Sumer *et al.* (2010) have undertaken a study basically to address these questions and to get an understanding of the processes involved in the interaction between cover stones and a liquefiable sediment bed. This section essentially addresses the question in item 1 above, and is based on Sumer *et al.*'s (2010) study. The question in item 2 is addressed in Chapter 6 while the questions in items 3 and 4 are addressed in Chapter 11.

Sumer *et al.* (2010) did their experiments in a wave flume. The soil was silt with $d_{50} = 0.098$ mm. Stones, the size of 4 cm, were used as cover material. The schematic description of the test set-up is illustrated in Fig. 3.48. One-, two- and three-layer cover stones were tested in the experiments.

Fig. 3.49 displays the time series of pore water pressure for four cases in Sumer *et al.*'s study (2010): (a) with no cover stones; (b) with cover stones, one-layer deep; (c) with cover stones, two-layers deep; and (d) with cover stones, three-layers deep. The wave height was $H = 17$ cm, the period $T = 1.6$ s, with the water depth $h = 40$ cm. The pressure time series shown in Fig. 3.49 were obtained at a depth of $z = 16$ cm. The initial mean normal effective stress values are also marked in Fig. 3.49. The latter values are calculated from

$$\sigma_0' = (\gamma'z + p_s)\frac{1 + 2k_0}{3} \tag{3.80}$$

in which p_s is the surface loading (or the surcharge) corresponding to the cover stones, which is actually the submerged weight of the N-layer stones

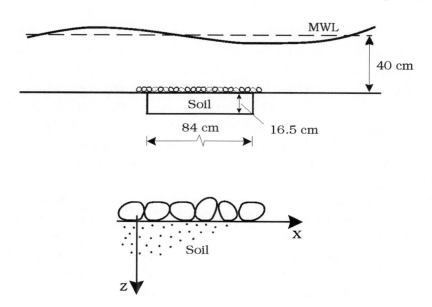

Figure 3.48: Stone cover layer over a liquefiable bed exposed to a progressive wave in Sumer *et al.*'s (2010) experiments.

per unit area of the bed, and given by

$$p_s = wN - (1 - n)\overline{D_z}N\gamma \tag{3.81}$$

in which w is the weight of stones per unit area of the bed per stone layer, N the number of layers, and $\overline{D_z}$ the mean stone height. Clearly when there is no surcharge, Eq. 3.80 will reduce to Eq. 3.6.

p_s given in Eq. 3.81 is for shallow soil depths. We note that p_s in Eq. 3.81 should be reduced according to the spreading of the loaded area with depth. This is important for not-too-shallow soil depths (see Powrie, 2004, Fig. 6.7). We shall return to this later.

Now, as seen from Fig. 3.49, the soil under the cover stones was liquefied in the case of the one- and two-layer cover stones (Figs 3.49b and c, respectively), while liquefaction did not occur in the case of the three-layer cover stones (Fig. 3.49d). This implies that, although covered by stones, the soil in the case of the one- and two-layer cover stones was subject to cyclic shear stresses, which were presumably sufficiently large to liquefy the soil.

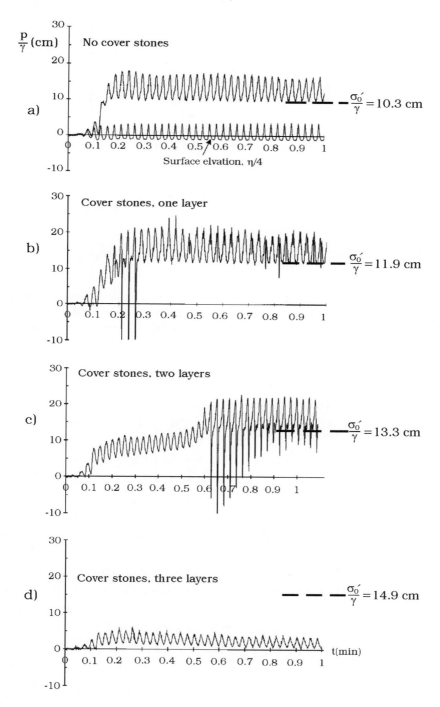

Figure 3.49:   Time series of excess pore pressure. Influence of cover stones. Sumer *et al.* (2010).

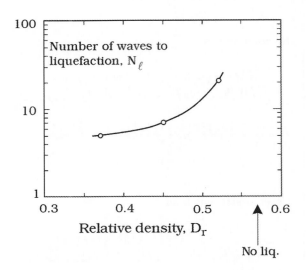

Figure 3.50: Liquefaction resistance as a function of soil relative density. The first data point to the left: no-stone case; the middle data point: one-layer stone cover case; the third data point: two-layer stone cover case; and "no-liq.": three-layer stone cover case.

(Sumer *et al.*, 2010, report that the measured pressure time series at the other depths, $z = 5, 10$, and $13 \, \text{cm}$, also revealed the same behaviour as in Figs 3.49b–d.)

The observed behaviour of the soil in Figs 3.49b–d is linked to the soil densification. Sumer *et al.* (2010) data show that the soil density, D_r, increases with the thickness of the cover layer. Fig. 3.50 displays the number of waves to cause liquefaction for the tests depicted in Fig. 3.49, plotted against D_r. The figure shows that while it takes only 5 waves for the soil to be liquefied in the case of no stones with the soil density $D_r = 0.37$ (see also Fig. 3.49a), it takes 7 waves in the case of the one-layer cover stones with $D_r = 0.45$ (Fig. 3.49b), and 21 waves in the case of the two-layer cover stones with $D_r = 0.52$ (Fig. 3.49c) to reach liquefaction, and liquefaction does not occur at all in the case of the three-layer cover stones with $D_r = 0.57$ (Fig. 3.49d).

In the three-layer case, the soil is densified so much that the resulting soil density is presumably too large, $D_r = 0.57$, for the liquefaction phenomenon to occur. To test this argument, Sumer *et al.* (2010) carried out a supplementary test in which the stones were first laid three-layers deep, and then they were removed, and subsequently the waves were switched on with the bed exposed directly to the action of the waves. This test showed that lique-faction did not occur, implying that D_r was too large to cause liquefaction even without the stones.

At this juncture, it is interesting to note that, in Sekiguchi *et al.*'s (2000) two experiments where liquefaction did not occur (cited in the preceding paragraphs), the coarse sand ($D_{50} = 0.3$ cm) simulating the gravel cover had a thickness of 1 cm in one test, and 2 cm in the other (corresponding to approximately three- and six-layer covers, respectively). Now, this thick-ness was apparently large enough to cause an increase in the soil density ($D_r = 0.5 - 0.6$, the range reported in Sekiguchi *et al.*) so that the soil was presumably not liquefied. Sekiguchi *et al.* (2000) report that the thicknesses of the gravel cover in their centrifuge experiments were designed such that, with a centrifugal acceleration of 50 g (the acceleration in their test), the cover layer thickness corresponded to a prototype thickness of 0.5 m and 1 m. Sumer *et al.*'s (2010) standard wave-flume experiments and Sekiguchi *et al.*'s (2000) centrifuge experiments (representing prototype conditions) are com-plementary, and both indicate the same mechanism regarding the response of soil beneath a cover layer exposed to a progressive wave.

In order to observe what happens in a test simulating dumping of stones in a real-life situation (Herbich *et al.*, 1984), Sumer *et al.* (2010) carried out a supplementary test where stones were released at a distance 15 cm from the bed. This test showed that no liquefaction occurred. The soil in this test was compacted by the falling stones over the entire soil surface, in a way similar to dynamic compaction used in practice for soil improvement, thereby increasing the relative density of the soil, which was, again, too large (also revealed by relative density measurements) for the soil to liquefy.

It may be mentioned that, although it is probably only of academic inter-est, Sumer *et al.* (2010) also did experiments with very densely packed stone covers. In these latter experiments the stones were arranged like a "jigsaw puzzle". In these tests, even with one-layer stone cover, the soil underneath was not liquefied. It is important to note that this was for the same wave conditions under which the soil was liquefied with even two-layer stone cover (Fig 3.49b and c). Sumer *et al.* (2010) linked this to the diminished expan-sion and contraction of the soil under such densely packed stone cover.

Returning back to the issue of the mean normal effective stress in the case of not-too-shallow soil, p_s in Eq. 3.80 should be replaced by $\alpha\, p_s$:

$$\sigma'_0 = (\gamma' z + \alpha\, p_s)\frac{1 + 2k_0}{3} \tag{3.82}$$

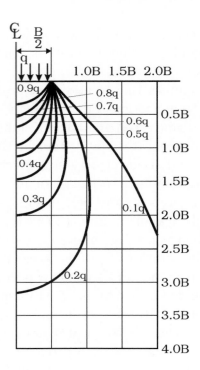

Figure 3.51:   Contours of increase in vertical stress below a strip footing (taken from Powrie, 2004). q is the load per unit area of the bed. The quantity B is the width of the strip footing.

in which α is a factor related to the spreading of the loaded area with the soil depth, and can be taken from Fig. 3.51 where the latter is given for a strip footing with a loading of q (in the present application q is equivalent to p_s given in Eq. 3.81). B in the figure is the width of the loaded area. Only the right half of the contour plot is shown due to symmetry. Clearly the figure shows that α can, to a first approximation, be taken as unity for shallow soil depths, and therefore Eq. 3.82 reduces to Eq. 3.80, as already pointed out.

3.7   Influence of Current

The analysis presented in this section and the end diagrams given in Figs 3.52–3.54 have been developed by the author's colleague Professor David R. Fuhrman of Technical University of Denmark, DTU Mekanik, Section for Fluid Mechanics, Coastal and Maritime Engineering.

It is known that when a wave encounters a current, the wave characteristics change, with the wave height and the wave length changing as a function of the current velocity, meaning that the bed pressure will also change with the current velocity. Clearly, in this case, a new wave loading will be generated, and therefore the soil will undergo cyclic shear stresses/strains induced by this new wave loading. This section will focus on the influence of the current on liquefaction potential under a given set of wave and current conditions. The subject will be developed in four steps:

1) No-current situation

The no-current situation (i.e., the wave-alone case) is characterized by three quantities: the water depth h_0, the wave height H_0, and the wave period T. The wave length L_0 is determined by the dispersion relation (Eq. A.8) from the small-amplitude linear-wave theory (Appendix A):

$$\omega^2 = g\lambda_0 \tanh(\lambda_0 h_0) \tag{3.83}$$

in which g is the acceleration due to gravity, ω is the angular frequency of the wave, $\omega = 2\pi/T$, and the quantity λ_0 is the wave number, $\lambda_0 = 2\pi/L_0$. In the present analysis the symbols with subindex 0 correspond to the no-current situation (waves alone).

The dispersion relation is, for convenience, written in the following form:

$$\Omega_0^2 = \lambda_0 h_0 \tanh(\lambda_0 h_0) \tag{3.84}$$

in which Ω_0 is a nondimensional quantity, the wave-frequency parameter, defined by

$$\Omega_0 = \omega \sqrt{\frac{h_0}{g}} \tag{3.85}$$

Based on conservation of energy flux, one can relate the wave height to the deep-water wave height by the relation (Svendsen, 2006, p. 136)

$$\frac{H_\infty^2}{H_0^2} = \tanh(\lambda_0 h_0) \, [1 + G_0] \tag{3.86}$$

in which H_∞ is the deep-water wave height, and the quantity G_0 is defined by

$$G_0 = \frac{2\lambda_0 h_0}{\sinh(2\lambda_0 h_0)}.$$ (3.87)

2) Combined wave–current situation

When the waves propagate on a current, the wave characteristics such as the wave height and the wave length will change; the wave height will decrease, and the wave length will increase if the wave propagates in the same direction as the current, and the opposite if the wave propagates in the direction opposite to the current. In this process, there will be a small change in the water depth while the wave period will remain unchanged.

In the current situation, the dispersion relation reads

$$(\Omega - \lambda h \ Fr)^2 = \lambda h \tanh(\lambda h)$$ (3.88)

in which Ω is the wave-frequency parameter for the combined-wave-and-current situation

$$\Omega = \omega \sqrt{\frac{h}{g}}$$ (3.89)

Fr is the Froude number

$$Fr = \frac{U}{\sqrt{gh}}$$ (3.90)

in which U is the current velocity with $U > 0$ corresponding to the current in the same direction as the wave propagation, and $U < 0$ the opposite. Formally, one can also define a Froude number corresponding to the no-current-case water depth, h_0, namely,

$$Fr_0 = \frac{U}{\sqrt{gh_0}}.$$ (3.91)

Similar to the no-current case, based on conservation of wave action, one can relate the wave height to the deep-water wave height by

$$\frac{H_\infty^2}{H^2} = \frac{\Omega^2[[1 + G]\sqrt{[\tanh(\lambda h)]/(\lambda h)} + 2 \ Fr]}{\Omega - \lambda h \ Fr}$$ (3.92)

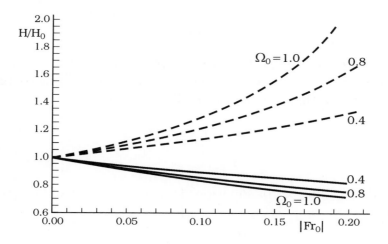

Figure 3.52: Wave height in the combined-wave-and-current situation. H_0 is the no-current wave height. Solid lines: following current. Dashed lines: opposing current.

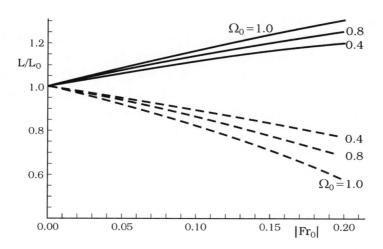

Figure 3.53: Wave length in the combined-wave-and-current situation. L_0 is the no-current wave length. Solid lines: following current. Dashed lines: opposing current.

in which G is defined by

$$G = \frac{2\lambda h}{\sinh(2\lambda h)} \tag{3.93}$$

(G should not be confused with the shear modulus.)

The water depth, h, in the current case can be related to that in the no-current case, h_0, through the Bernoulli equation

$$\frac{h}{h_0} + \frac{1}{2}Fr_0^2 = 1. \tag{3.94}$$

Now, the preceding set of equations (Eqs 3.84–3.94) can be solved for the quantities

$$\frac{H}{H_0}, \quad \frac{L}{L_0}, \quad \text{and} \quad \frac{h}{h_0} \tag{3.95}$$

in terms of the Froude number, Fr_0, and the wave-frequency parameter, Ω_0. The results are plotted in Figs 3.52–3.54 for the ranges of the latter parameters encountered in practice, the dashed lines illustrating the variations for the case where the wave propagates in the direction opposite to the current, and the solid lines the variations for the case where the wave propagates in the same direction as the current.

As seen, when the wave propagates in the opposite direction to the current, the wave height can be greatly increased, by as much as a factor of 2. Likewise, the wave length can be decreased significantly, by as much as 60%. (The effect is similar to that in the process of shoaling. The wave height increases and the wave length decreases as the wave propagates towards the smaller and smaller water depths.) These two effects can have significant implications for the buildup of pore-water pressure and liquefaction, as will be demonstrated in the following paragraphs.

3) Procedure to assess liquefaction potential in the presence of current

Given (a) the wave parameters in the case of the no-current, namely the wave height H_0, the wave period T, and the water depth h_0, and (b) the current U, we can calculate a new set of wave parameters, namely the wave height H, the wave length L, and the water depth h, corresponding to the combined-wave-and-current situation. The liquefaction assessment exercise

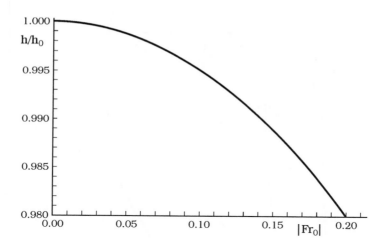

Figure 3.54: Water depth in the combined-wave-and-current situation. h_0 is the no-current water depth.

can then be worked out, based on this new set of wave parameters. The procedure may be summarized as follows.

(1) Calculate the wave length, L_0, for the no-current situation from the dispersion relation Eq. 3.83.

(2) Calculate the Froude number, Fr_0, from Eq. 3.91, and the wave-frequency parameter Ω_0 from Eq. 3.85.

(3) Pick up the values of $\frac{H}{H_0}$, $\frac{L}{L_0}$, and $\frac{h}{h_0}$ from Figs. 3.52–3.54, and from the latter, get the new wave characteristics H, L, and h.

(4) Subsequently, implement the mathematical model in Section 3.2.4, to make an assessment of liquefaction potential for this new set of wave parameters. The following numerical example will illustrate this.

4) Numerical example to assess liquefaction potential in the presence of current

We consider Example 4. The soil properties will be exactly the same as those given in Example 4. The wave properties will also be exactly the same as in Example 4, with the wave height $H_0 = 5\,\text{m}$, the period $T = 13.7$ s, and the water depth $h_0 = 19\,\text{m}$. The wave length from the dispersion relation Eq. 3.83 is $L_0 = 174\,\text{m}$. Using the mathematical model in Section 3.2.4, it was found in Example 4 that liquefaction does not occur as the pore

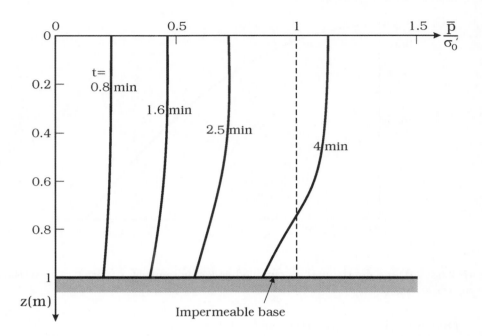

Figure 3.55:   Time development of the accumulated excess pore pressure. With the introduction of a 2 m/s opposing current, liquefaction occurs (see the text for details); $\overline{p}$ exceeds σ_0' within less than 4 minutes.

pressure never reaches the initial mean normal effective stress, i.e., $\overline{p}/\sigma_0' < 1$, see Fig. 3.30.

Suppose that the wave propagates against a current with a velocity of $U = -2$ m/s. The question is whether the soil is liquefied in this new combined-wave-and-current situation.

Now, the Froude number, Fr_0, and the wave-frequency parameter, Ω_0, from Eqs 3.91 and 3.85 are

$$Fr_0 = \frac{U}{\sqrt{gh_0}} = \frac{-2}{\sqrt{9.81 \times 19}} = -0.146 \qquad (3.96)$$

$$\Omega_0 = \omega\sqrt{\frac{h_0}{g}} = \frac{2\pi}{13.7}\sqrt{\frac{19}{9.81}} = 0.638. \qquad (3.97)$$

From Figs 3.52–3.54, for $|Fr_0| = 0.146$, and $\Omega_0 = 0.638$, using the dashed lines (opposing current), the quantities $\frac{H}{H_0}$, $\frac{L}{L_0}$, and $\frac{h}{h_0}$ are found to be as follows: $\frac{H}{H_0} = 1.28$, $\frac{L}{L_0} = 0.807$, and $\frac{h}{h_0} = 0.989$, and therefore the new wave parameters will be: the wave height $H = 6.39$ m, the period $T = 13.7$ s (the period will not change), the wave length $L = 140.7$ m, and the water depth $h = 18.8$ m.

With this new set of wave parameters, the corresponding pressure distributions are calculated from the mathematical model, using Eqs 3.41 and 3.42. The result is given in Fig. 3.55. The pore pressure in this case reaches the initial mean normal effective stress σ_0' within less than 4 minutes, and therefore it may be concluded that there is a liquefaction potential when the wave propagates against the given current.

3.8   References

1. Alba, P.D., Seed, H.B. and Chan, C.K. (1976): Sand liquefaction in large-scale simple shear tests. Journal of Geotechnical Engineering Division, ASCE, vol. 102, No. GT9, 909–927.

2. Barends, F.B.J. and Calle, E.O.F. (1985): A method to evaluate the geotechnical stability of off-shore structures founded on a loosely packed seabed sand in a wave loading environment. Behaviour of Offshore Structures, Elsevier Science Publishers B.V. Amsterdam, 643–652.

3. Bjerrum, L. (1973): Geotechnical problems involved in foundations of structures in the North Sea. Géotechnique, vol. 23, No. 3, 319–358.

4. Brinch-Hansen, J. (1957): Calculation of settlements by means of pore pressure coefficients. Acta Polytechnica, vol. 235, Civil Engineering and Building Construction Ser., vol. 4, No. 8, UDC 624.131.526.

5. Chan, A.H.C. (1988): A Unified Finite-element Solution to Static and Dynamic Geomechanics Problems. Ph.D. Thesis, University College of Swansea, Wales.

6. Chan, A.H.C. (1995): User Manual for DIANA-SWANDYNE II. Department of Civil Engineering, University of Birmingham.

7. Chen, Y.-L., Tzang, S.-Y. and Ou, S.-H. (2008): Application of the EEMD method to investigate pore pressure buildups in a wave-fluidized sandbed. Proceedings of the 31st International Conference on Coastal

Engineering, 31 August–5 September, 2008, Hamburg, Germany, vol. 2, 1614–1624.

8. Cheng, L., Sumer, B.M. and Fredsøe, J. (2001): Solutions of pore pressure buildup due to progressive waves. International Journal of Numerical and Analytical Methods in Geomechanics, vol. 25, issue 9, 885–907.

9. Clukey, E.C., Kulhawy, F.H. and Liu, P.L.-F. (1983): Laboratory and field investigation of sediment interaction. Cornell University, Itacha, New York, Geotechnical Engineering Report 83-9, and Joseph H. DeFrees Hydraulics Laboratory Report 83-1, October 1983.

10. Clukey, E.C., Kulhawy, F.H. and Liu, P.L.-F. (1985): Response of silts to wave loads: Experimental study. In: Strength Testing of marine sediments, Laboratory and In-Situ Measurements, ASTM STP 883, R.C. Chaney and K.R. Demars, Eds., American Society for testing and materials, Philadelphia, 381–396.

11. de Groot, M.B., Lindenberg, J. and Meijers, P. (1991): Liquefaction of sand used for soil improvement in breakwater foundations. Proceedings of the International Symposium Geo-coast '91, Yokohama, 555–560.

12. de Wit, J. (1995): Liquefaction of cohesive sediments caused by waves. Communications on Hydraulic Engineering, Report No. 95-2, Delft University of Technology, Faculty of Civil Engineering, June 1995. Report submitted as a Doctoral Thesis to Delft University of Technology, The Netherlands.

13. de Wit, J. and Kranenburg, C. (1992): Liquefaction and erosion of China Clay due to waves and current. Proceedings of the 23rd International Conference on Coastal Engineering, 4–9 October, 1992, Venice, Italy, 2937–2947.

14. de Wit, J., Kranenburg, C. and Battjes, J.A. (1994): Liquefaction and erosion of mud due to waves and current. Abstract Book of the 24th International Conference on Coastal Engineering, ICCE '94, 23–28. October 1994, Kobe. Japan, 278–279.

15. de Wit, J. and Kranenburg, C. (1997): The wave-induced liquefaction of cohesive sediment beds. Estuarine, Coastal and Shelf Science, vol. 45, No. 2, 261–271.

16. Dong, P. and Xu, H. (2010): An ensemble modelling for the assessment of random wave-induced liquefaction risks. Abstract Book of the 32nd International Conference on Coastal Engineering, ICCE 2010, June 30–July 5, 2010, Paper No. 214.

17. Dunn, S.L., Vun, P.L., Chan, A.H.C. and Damgaard, J.S. (2006): Numerical modeling of wave-induced liquefaction around pipelines. Journal of Waterway, Port, Coastal, and Ocean Engineering, ASCE, vol 132, No. 4, 276–288.

18. Foda, M.A. (1995): Sea Floor Dynamics. Advances in Coastal and Ocean Engineering, vol. 1, ed. P.L.-F. Liu, World Scientific, 77–123.

19. Foda, M.A. and Tzang, S.-Y. (1994): Resonant fluidization of silty soil by water waves. Journal of Geophysical Research, vol. 99, No. C10, 20,463–20,475.

20. Fredsøe, J. and Deigaard, R. (1992): Mechanics of Coastal Sediment Transport, World Scientific, Singapore.

21. Gratchev, I.B., Sasa, K., Osipov, V.I. and Sokolov V.N. (2006): The liquefaction of clayey soils under cyclic loading. Engineering Geology, vol. 86, No. 1, 70–84.

22. Guo, T. and Prakash, S. (1999): Liquefaction of silts and silt–clay mixtures. Journal of Geotechnical and Geoenvironmental Engineering, ASCE, vol. 125, No. 8, 706–710.

23. Herbich, J.B., Schiller, R.E., Dunlap, W.A. and Watanabe, R.K. (1984): Seafloor Scour, Design Guidelines for Ocean-Founded Structures, Marcel Dekker, Inc, New York and Basel.

24. Hsu, J.R.S. and Jeng, D.S. (1994): Wave-induced soil response in an unsaturated anisotropic seabed of finite thickness. International Journal for Numerical and Analytical Methods in Geomechanics, vol. 18, No. 11, 785–807.

25. Idriss, I.M. and Boulanger, R.W. (2008): Soil Liquefaction During Earthquakes. Earthquake Engineering Research Institute, Oakland, California, USA. Original Monograph Series MNO-12, 243 p.

26. Ishihara, K. and Yamazaki, A. (1984): Analysis of wave-induced liq-
    uefaction in seabed deposits of sand. Soils and Foundations, vol. 24,
    No. 3, 85–100.

27. Jefferies, M. and Been, K. (2006): Soil Liquefaction. A Critical Stage
    Approach. Taylor & Francis, London and New York, 479 p.

28. Jeng, D.S. and Seymour, B.R. (2007): Simplified analytical approxi-
    mation for pore-water pressure buildup in marine sediments. Journal
    of Waterway, Port, Coastal, and Ocean Engineering, ASCE, vol 133,
    No. 4, 309–312.

29. Jeng, D.S, Seymour, B.R. and Li, J. (2007): A new approximation
    for pore pressure accumulation in marine sediment due to water waves.
    International Journal for Numerical and Analytical Methods in Geome-
    chanics, vol. 31, No. 1, 53–69.

30. Kirca, V.S.O., Fredsøe, J. and Sumer, B.M. (2012): Wave liquefac-
    tion in soils with clay content. Proceedings of the 8th International
    Conference on Coastal and Port Engineering in Developing Countries.
    COPEDEC 2012, IIT Madras, Chennai, India, 20–24 February 2012,
    395–402.

31. Kirca, V.S.O., Sumer, B.M. and Fredsøe, J. (2014): Influence of clay
    content on wave-induced liquefaction. Journal of Waterway, Port,
    Coastal, and Ocean Engineering, ASCE. In print.

32. Kramer, S.L. (1996): Geotechnical Earthquake Engineering. Prentice
    Hall, New Jersey, USA, xviii+653 p.

33. Lamb, H. (1945): Hydrodynamics. 6th edition. Dover Publications,
    New York, NY.

34. Lambe T.W. and Whitman, R.V. (1969): Soil Mechanics. John Wiley
    and Sons, Inc., 553 p.

35. Lindenberg, J., van Rijn, L.C. and Winterwerp, J.C. (1989): Some
    experiments on wave-induced liquefaction of soft cohesive soils. Journal
    of Coastal Research, Special Issue, No. 5, 127–137.

36. Maa, P.-Y. and Mehta, A.J. (1990): Soft mud response to water waves. Journal of Waterways, Port, Coastal and Ocean Engineering, ASCE, vol. 116, No. 5, 634–650.

37. McDougal, W.G., Tsai, Y.T., Liu, P.L-F. and Clukey, E.C. (1989): Wave-induced pore water pressure accumulation in marine soils. Journal of Offshore Mechanics and Arctic Engineering, ASME, vol. 111, No. 1, 1–11.

38. Miyamoto, J., Sassa, S. and Sekiguchi, H. (2003): Preshearing effect on the wave-induced liquefaction in sand beds. Proceedings of Soil and Rock America, Boston, US, vol. 1, 1025–1032.

39. Miyamoto, J., Sassa, S. and Sekiguchi, H. (2004): Progressive solidification of a liquefied sand layer during continued wave loading. Géotechnique, vol. 54, No. 10, 617–629.

40. Peacock, W.H. and Seed, H.B. (1968): Sand liquefaction under cyclic loading simple shear conditions. Journal of Soil Mechanics and Foundations Engineering, ASCE, vol. 94, No. SM3, 689–708.

41. PIANC (2001): Seismic Design Guidelines for Port Structures. Working Group No. 34 of the maritime Navigation Commission, International Navigation Association (PIANC). A book published by A.A. Balkema Publishers, XV + 474 p, Lisse/Abingdon/Exton (PA)/Tokyo.

42. Powrie, W. (2004): Soil Mechanics. 2nd edition. Spon Press, Taylor and Francis Group, London, x+675p.

43. Prakash, S. (1981): Soil Dynamics. McGraw-Hill.

44. Rahman, M.S., Seed, H.B. and Booker, J.R. (1977): Pore pressure development under offshore gravity structures. Journal of Geotechnical Engineering Division, ASCE, vol. 103, No. GT12, 1419–1436.

45. Sahin, C., Safak, I., Sheremet, A. and Mehta A.J. (2012): Observations on cohesive bed reworking by waves: Atchafalaya Shelf, Louisiana. Accepted for publication in the Journal of Geophysical Research, Oceans.

46. Sahin, C., Sheremet, A. and Safak, I. (2012): Coupled wave-bed dynamics, Atchafalaya Shelf, Louisiana. Proceedings of the 22nd International Offshore and Polar Engineering Conference, Rhodes, Greece, June 17–22, 2012, 1420–1424.

47. Sassa, S. and Sekiguchi, H. (1999): Wave-induced liquefaction of beds of sand in a centrifuge. Géotechnique, vol. 49, No. 5, 621–638.

48. Sassa, S. and Sekiguchi, H. (2001): Analysis of wave-induced liquefaction of sand beds. Géotechnique, vol. 51, No. 2, 115–126.

49. Sassa, S., Sekiguchi, H. and Miyamoto, J. (2001): Analysis of progressive liquefaction as a moving-boundary problem. Géotechnique, vol. 51, No. 10, 847–857.

50. Schofield, A.N. (1980): Cambridge geotechnical centrifuge operations. Géotechnique, vol. 30, No. 3, 227–268.

51. Seed, H.B. (1976): Some aspects of sand liquefaction under cyclic loading. Behaviour of Offshore Structures (BOSS '76), Proceedings, The Norwegian Institute of Technology, Norway, 374–391.

52. Seed, H.B. and Booker, J.R. (1976): Stabilization of Potentially Liquefiable Sand Deposits Using Gravel Drain Systems. Report No. EERC 76–10, Earthquake Engineering Research Center, University of California, Berkeley, CA, April 1976.

53. Seed, H.B., Martin, P.P. and Lysmer, J. (1976): Pore-water pressure changes during soil liquefaction. Journal of Geotechnical Engineering Division, ASCE, vol. 102, No. GT4, 323–346.

54. Seed, H.B. and Rahman, M.S. (1978): Wave-induced pore pressure in relation to ocean floor stability of cohesionless soil. Marine Geotechnology, 3, No. 2, 123–150.

55. Seed, H.B., Tokimatsu, K., Harder, L.F. and Chung, R. (1985): Influence of SPT procedures in soil liquefaction resistance evaluations. Journal of Geotechnical Engineering, ASCE, vol. 111, No. 12, 861–878

56. Sekiguchi, H., Kita, K. and Okamoto, O. (1995): Response of poro-elastoplastic beds to standing waves. Soils and Foundations, vol. 35, No. 3, 31–42, Japanese Geotechnical Society.

57. Sekiguchi, H., Kita, K., Sassa, S. and Shimamura, T. (1998): Generation of progressive fluid waves in a geo-centrifuge. Geotechnical Testing Journal, vol. 21, No. 2, 95–101.

58. Sekiguchi, H. and Phillips, R. (1991): Generation of water waves in a drum centrifuge. Proceedings of the International Conference CENTRIFUGE 91, Boulder, USA, 343–350.

59. Sekiguchi, H., Sassa, S., Sugioka, K. and Miyamoto, J. (2000): Wave-induced liquefaction, flow deformation and particle transport in sand beds. Proceedings of the International Conference GeoEng2000, Melbourne, Australia, Paper No. EG-0121.

60. Sneddon, I. (1957): Elements of Partial Differential Equations. McGraw-Hill, New York.

61. Spierenburg, S.E.J. (1987): Seabed Response to Water Waves. Ph.D. dissertation, Delft University of Technology, The Netherlands.

62. Sumer, B.M. and Cheng, N.-S. (1999): A random-walk model for pore pressure accumulation in marine soils. Proceedings of the 9th International Offshore and Polar Engineering Conference, ISOPE-99, Brest, France, 30 May–4 June, 1999, vol. 1, 521–526.

63. Sumer, B.M., Dixen, F.H. and Fredsøe, J. (2010): Cover stones on liquefiable soil bed under waves. Coastal Engineering, vol. 57, No. 9, 864–873.

64. Sumer, B.M. and Fredsøe, J. (1997): Hydrodynamics Around Cylindrical Structures, World Scientific, Singapore, 530 p. Second edition 2006.

65. Sumer, B.M. and Fredsøe, J. (2002): The Mechanics of Scour in the Marine Environment. World Scientific, Singapore, 552 p.

66. Sumer, B.M., Fredsøe, J., Christensen, S. and Lind, M.T. (1999): Sinking/floatation of pipelines and other objects in liquefied soil under waves. Coastal Engineering, vol. 38, No. 2, 53–90.

67. Sumer, B.M., Hatipoglu, F. and Fredsøe, J. (2004): The cycle of soil behaviour during wave liquefaction, Book of Abstracts, Paper 171,

29th International Conference on Coastal Engineering, 19–24 September, 2004, National Civil Engineering Laboratory (LNEC), Lisbon, Portugal.

68. Sumer, B.M., Hatipoglu, F., Fredsøe, J. and Sumer, S.K. (2006 a): The sequence of soil behaviour during wave-induced liquefaction. Sedimentology, vol. 53, 611–629.

69. Sumer, B.M., Hatipoglu, F., Fredsøe, J. and Hansen, N.-E.O. (2006 b): Critical floatation density of pipelines in soils liquefied by waves and density of liquefied soils. Journal of Waterway, Port, Coastal and Ocean Engineering, ASCE, vol. 132, No. 4, 252–265.

70. Sumer, B.M., Truelsen, C. and Fredsøe, J. (2006 c): Liquefaction around pipelines under waves. Journal of Waterway, Port, Coastal and Ocean Engineering, ASCE, vol. 132, No. 4, 266–275.

71. Sumer, B.M., Kirca, V.S.O. and Fredsøe, J. (2011): Experimental validation of a mathematical model for seabed liquefaction in waves. Proceedings of the 21st International Offshore and Polar Engineering Conference, Maui, Hawai, USA, June 19–24, 2011, 1010–1018.

72. Sumer, B.M., Kirca, V.S.O. and Fredsøe, J. (2012): Experimental validation of a mathematical model for seabed liquefaction under waves. International Journal of Offshore and Polar Engineering, vol. 22, No. 2, June 2012, 133–141.

73. Svendsen, I.A. (2006): Introduction to Nearshore Hydrodynamics. World Scientific, Singapore, xxii+722 p.

74. Taylor, R.N. (1995): Centrifuge in modelling: principles and scaling effects. In: Geotechnical Centrifuge Technology. Blackie Academic & Professional, London, 19–33.

75. Teh, T.C., Palmer, A. and Damgaard, J. (2003): Experimental study of marine pipelines on unstable and liquefied seabed. Coastal Engineering, vol. 50, No. 1–2, 1–17.

76. Terzaghi, K. (1948): Theoretical Soil Mechanics. London: Chapman and Hall, John Wiley and Sons, Inc., NY, 510 p.

77. Troncoso, J.H. (1990): Failure risks of abandoned tailings dams. Proceedings of the International Symposium on Safety and Rehabilitation of Tailings Dams, ICOLD, Paris, 82–89.

78. Tzang, S.Y. (1998): Unfluidized soil responses of a silty seabed to monochromatic waves. Coastal Engineering, vol. 35, No. 4, 283–301.

79. Tzang, S.Y., Hunt, J.R. and Foda, M.A. (1992): Resuspension of seabed sediments by water waves. Abstract Book of the 23rd International Conference on Coastal Engineering, ICCE '92, 4–9 October, 1992, Venice, Italy, 69–70.

80. Tzang, S.Y. and Ou, S.H. (2004): Flume measurements of regular wave group-induced pore pressures in a fluidized sandy bed. Proceedings of the 29th International Conference on Coastal Engineering, ICCE 2004, September, Lisbon, Portugal, vol. 2, 1830–1840.

81. van Kessel, T. and Kranenburg, C. (1998): Wave-induced liquefaction and flow of subaqueous mud layers. Coastal Engineering, vol. 34, No. 1–2, 109–127.

82. van Kessel, T., Kranenburg, C. and Battjes, J.A. (1996): Transport of fluid mud generated by waves on inclined beds. Proceedings of the 25th International Conference on Coastal Engineering, ICCE '96, 2–6 September, 1996, Orlando, FL, USA, vol. 3, 3337–3348.

83. Vanoni, V. (1975): Sedimentation Engineering. ASCE-Manuals and Reports on Engineering Practice, No. 54, 745 p.

84. Whitehouse, R., Soulsby, R., Roberts, W. and Mitchener, H. (2000): Dynamics of Estuarine Muds. Thomas Telford Publishing, London, UK.

85. Winterwerp, J.C., de Boer, G.J., Greeuw, G. and van Maren D.S. (2012): Mud-induced wave damping and wave-induced liquefaction. Coastal Engineering, vol. 64, June, 102–112.

86. Xu, H. and Dong, P. (2011): A probabilistic analysis of random wave-induced liquefaction. Ocean Engineering, vol. 38, No. 7, 860–867.

87. Zienkiewicz, O.C., Chan, A.H.C., Pastor, M., Paul, D.K. and Shiomi, T. (1990): Static and dynamic behaviour of geomaterials — A rational approach to quantitative solutions. Part I: Fully saturated problems. Proceedings of the Royal Society of London, Series A, vol. 429, 285–309.

88. Zienkiewicz, O.C., Chan, A.H.C., Pastor, M., Schrefler, B.A. and Shiomi, T. (1999): Computational Geomechanics with Special Reference to Earthquake Engineering. Wiley, Chichester, U.K.

Chapter 4

Momentary Liquefaction

As mentioned in Section 1.2.2, momentary liquefaction occurs during the passage of the wave trough. Under the wave trough, the pore pressure (in excess of the hydrostatic pressure) has a negative sign (Fig. 1.5b), and the pressure distribution across the soil depth will look like in Fig. 1.7a in the case of a completely saturated soil, or in Fig. 1.7b in the case of an unsaturated soil. In the latter case, the soil contains some air/gas, and therefore the pore pressure is "dissipated" very quickly with the depth.

In the case of the completely saturated soil, the vertical pressure gradient is not tremendously large (Fig. 1.7a) while, in the case of the unsaturated soil, it can be very large (Fig. 1.7b), particularly at small values of z, meaning that a substantial amount of lift can be generated at the top layer of the soil under the wave trough (Fig. 4.1). If this lift exceeds the submerged weight of the soil, the soil will fail, and, as a result, it will be liquefied. As noted in Section 1.2.2, the liquefaction occurs over a short period of time during the passage of the wave trough; for the rest of the wave period, the soil will remain in the no-liquefaction regime.

The large pressure gradient in the case of the unsaturated soil (Fig. 1.7b) is caused by the air/gas content of the soil. It may be mentioned that only a very small amount of gas (less than 1%) would be enough to cause a very large dissipation. We shall return to this issue later in the chapter.

This chapter presents a theoretical framework for the analysis of the momentary liquefaction. The analysis includes a general description of the process, including the criterion for the occurrence of momentary liquefaction; momentary liquefaction in the case of the completely saturated soil; and in the case of unsaturated soil. It will be shown that the momentary

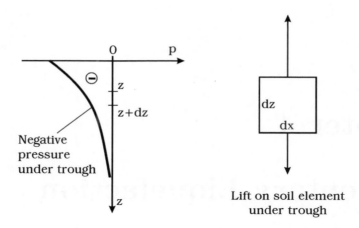

Figure 4.1:   Soil experiences a lift force during the passage of the wave trough.

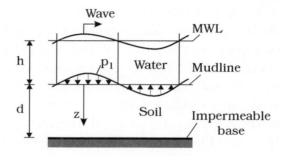

Figure 4.2:   Pressure in excess of the static pressure on the bed under a progressive wave.
Snapshot.

liquefaction develops only in the case of unsaturated soils. The chapter will
end with a detailed discussion of air/gas content of marine soils, one of the
key elements in the process of momentary liquefaction.

4.1   General Description

The soil is subject to a progressive wave (Fig. 4.2). The pressure caused by
this wave on the bed is

$$p_1 = p_b \exp[i(\lambda x + \omega t)] \qquad (4.1)$$

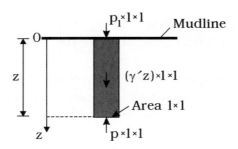

Figure 4.3:   Forces on a soil column, the size $1 \times 1 \times z$.

in which p_b is the maximum value (the amplitude) of the bed pressure given in Eq. 2.46; see also Appendix A.

Consider the soil column, the size $1 \times 1 \times z$ (Fig. 4.3). The forces on this soil column in the vertical direction at any time t are as follows:

1. The force on the top surface of the soil column in excess of the static situation: $p_1 \times 1 \times 1$.

2. The force on the bottom surface of the soil column in excess of the static situation: $p \times 1 \times 1$.

The difference between these two pressure forces will induce a lift force on the soil column in excess of the static situation. Clearly, when this wave-induced lift force, $(p - p_1) \times 1 \times 1$, exceeds the submerged weight of the soil, $(\gamma'z) \times 1 \times 1$, the soil will fail, the momentary liquefaction. Therefore lique-faction will occur when

$$p - p_1 \geq \gamma'z \qquad (4.2)$$

or

$$\gamma'z - p + p_1 \leq 0. \qquad (4.3)$$

(The friction forces are practically zero at the instant of failure.)

Alternatively, the liquefaction criterion can also be given in terms of the vertical effective stress. The vertical effective stress at depth z before the wave is introduced (i.e., the initial vertical effective stress) is

$$\sigma'_{v0} = \gamma'z. \qquad (4.4)$$

Now, let us introduce the wave. Let σ_z' be the vertical effective stress (in excess of the static situation) generated by the wave, at the same depth, z, and at time t.

With the introduction of the wave, the initial vertical effective stress will be reduced by an amount equal to σ_z'. Clearly, liquefaction will occur when this reduced vertical effective stress $\sigma_{v0}' - \sigma_z'$ becomes negative (or zero)

$$\sigma_{v0}' - \sigma_z' \leq 0 \qquad (4.5)$$

simply because the vertical effective stress in this case will be a tensile stress and therefore, as a result, the soil will fail. Hence, from Eqs 4.4 and 4.5, the liquefaction criterion will read: liquefaction will occur when

$$\gamma' z - \sigma_z' \leq 0. \qquad (4.6)$$

As will be seen later, the criterion based on the pore pressure (Eq. 4.3) and that based on the vertical effective stress (Eq. 4.6) lead to the same result, simply because the vertical effective stress σ_z' is practically the same as $p - p_1$, considering the force balance (i.e., the balance of forces in excess of the static situation) in the vertical direction.

The following two sections will discuss the momentary liquefaction in the following two cases: (1) saturated soil, and (2) unsaturated soil, utilizing the criterion in Eq. 4.3. The criterion in Eq. 4.6 will also be discussed.

4.2   The Case of Saturated Soil

In the case of a saturated soil with an infinitely large depth (Fig. 4.2 with $d \to \infty$), the pore pressure p and the vertical effective stress σ_z' have been obtained from the solution of the Biot equations in Chapter 2, Section 2.2.1, Eqs. 2.52 and 2.54, respectively, with p_b given in Eq. 2.46.

Inserting Eq. 2.52 (Yamamoto et al.'s, 1978, solution) and Eq. 4.1 in Eq. 4.3, the left-hand side of the inequality in Eq. 4.3 can be worked out to check whether or not the liquefaction condition is satisfied. Namely, liquefaction occurs if

$$\frac{\gamma'}{\gamma} \frac{z}{H} - \frac{0.5}{\cosh(\lambda h)} \left[1 - \exp\left(-\lambda H \frac{z}{H}\right)\right] \exp[i(\lambda x + \omega t)] \leq 0 \qquad (4.7)$$

or, in terms of wave steepness H/L, liquefaction occurs if

$$\frac{\gamma'}{\gamma}\frac{z}{H} - \frac{0.5}{\cosh(\lambda h)}\left[1 - \exp\left(-2\pi\frac{H}{L}\frac{z}{H}\right)\right]\exp[i(\lambda x + \omega t)] \leq 0. \qquad (4.8)$$

The soil will be most susceptible to liquefaction when the wave height is largest. The largest wave height for a given wave length may be given approximately as (Isaacson, 1979)

$$\left(\frac{H}{L}\right)_{max} = 0.14\tanh(\lambda h). \qquad (4.9)$$

(The waves higher than that given in the preceding equation will break.) Hence, inserting $(H/L)_{max}$ from Eq. 4.9 into Eq. 4.8, the left-hand side of the inequality in Eq. 4.8 is calculated. For $\gamma'/\gamma = 0.8$, a typical value, it can be readily seen that the liquefaction criterion, Eq. 4.8, will not be reached, i.e., the left-hand side of the inequality in Eq. 4.8 will never be negative. It may be noted that application of the criterion in Eq. 4.6 also gives the same result.

The above result means that *normally the momentary liquefaction will not occur in a completely saturated soil.*

In the case of the completely saturated soil with a finite depth, d (Fig. 4.2), a similar exercise using the analytical solution given by Hsu and Jeng (1994) for the pore-water pressure leads to the same conclusion.

4.3 The Case of Unsaturated Soil

4.3.1 Infinitely large soil depth

In the case of an unsaturated soil with an infinitely large depth ($d \rightarrow \infty$, Fig. 4.2), the pressure in the soil has been calculated by various investigators; e.g., Yamamoto *et al.* (1978) (see Section 2.2.1) and Mei and Foda (1981).

Sakai, Hatanaka and Mase (1992 a) were one of the first to address the problem of momentary liquefaction in the case of unsaturated soils. They used Mei and Foda's (1981) solution, and studied the momentary liquefaction for the special case of shallow-water waves.

As in Sakai *et al.* (1992 a), we adopt Mei and Foda's (1981) solution. However, the analysis we will carry out will not be limited with the shallow-water case.

Mei and Foda's (1981) solution for pore pressure under a progressive wave is

$$p = p_b \frac{1}{1+m} \exp(-\lambda z) \cos(\lambda x - \omega t) + p_b \frac{m}{1+m}$$

$$\times \exp\left(\frac{-z}{\sqrt{2}\delta}\right) \cos\left(\lambda x - \omega t + \frac{z}{\sqrt{2}\delta}\right) \tag{4.10}$$

in which

$$m = \frac{n}{(1-2\nu)}\frac{G}{K'} \tag{4.11}$$

$$\delta = \left[\frac{(k/\gamma)G}{\omega}\right]^{1/2} \left[n\frac{G}{K'} + \frac{1-2\nu}{2(1-\nu)}\right]^{-1/2} \tag{4.12}$$

in which K' is given by Eq. 2.36, namely

$$\frac{1}{K'} = \frac{1}{K} + \frac{1-S_r}{p_0}. \tag{4.13}$$

The quantity δ in Eq. 4.12 should not be confused with δ introduced in Eq. 2.73.

Mei and Foda (1981) obtained the above solution as the sum of two solutions representing the effects of (1) elastic response of an outer region, and (2) seepage-flow response of an inner region close to the surface of the seabed. Mei and Foda (1981) called this thin, inner region a "boundary layer". The first term in Eq. 4.10 is the outer solution while the second term is the boundary layer "correction", which includes a phase delay equal to $z/(\sqrt{2}\delta)$. The phase delay presumably takes care of the process of infiltration and exfiltration that takes place near the bed surface in what Mei and Foda (1981) called the boundary layer. When worked out from Eq. 4.10, the phase delay is seen to increase with z, experiences a maximum, and then begins to decrease and approaches to zero for large values of z within the boundary layer, as expected.

It should be noted that Mei and Foda's (1981) solution given above, Eqs 4.10–4.13, reduces to Yamamoto et al.'s (1978) solution, Eq. 2.52 (which is used in the previous section), when the degree of saturation is set equal to $S_r = 1$, in which case G/K' will be extremely small, $G/K' \to 0$, as discussed in Chapter 2, in conjunction with Yamamoto et al.'s (1978) work.

Whether or not liquefaction would occur for a given set of soil and wave parameters can be worked out using Eq. 4.3 with p to be calculated from Eq. 4.10. The following numerical example will illustrate this.

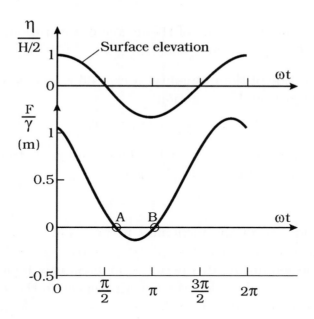

Figure 4.4:   Momentary liquefaction occurs when $F \leq 0$ in which $F = \gamma'z - p + p_1$, i.e., during the time period A–B. At $z = 0.5$ m. Infinitely large soil depth.

Example 10. *Check for momentary liquefaction. A numerical example.*

The soil properties are given as follows: the submerged specific weight, $\gamma' = 10.0\,\text{kN/m}^3$, the shear modulus, $G = 14220\,\text{kN/m}^2$, the coefficient of permeability, $k = 6 \times 10^{-5}\,\text{m/s}$, the porosity, $n = 0.333$, the degree of saturation, $S_r = 0.985$, and Poisson's ratio, $\nu = 0.35$.

The water properties are: the specific weight of water, $\gamma = 9.81\,\text{kN/m}^3$, and the bulk modulus of elasticity of water, $K = 1.9 \times 10^6\,\text{kN/m}^2$.

The soil is exposed to a progressive wave with the following properties: the wave height, $H = 25.6\,\text{m}$, the period $T = 12.3\,\text{s}$, and the water depth, $h = 65\,\text{m}$. The small amplitude wave theory (Appendix A) gives a wave length of $L = 224.5\,\text{m}$.

Questions:

1. Find whether the soil will undergo momentary liquefaction under the given wave climate at depth $z = 0.5\,\text{m}$.

2. What is the liquefaction depth, z_ℓ, defined by $z < z_\ell$ in which z indicates the depth at which the soil is liquefied?

3. What is the limiting value of the degree of saturation, S_r, that allows the soil to be liquefied at $z = 0.5\,\text{m}$?

The left-hand side of the inequality in Eq. 4.3 is calculated with $\lambda x = 0$, the bed pressure p_1 from the equation

$$p_1 = p_b \cos(\lambda x - \omega t) \tag{4.14}$$

and the pore pressure p from Eq. 4.10. The result is plotted in Fig. 4.4 in which the quantity F on the vertical axis is

$$F = \gamma' z - p + p_1 \tag{4.15}$$

i.e., the left-hand side of the inequality in Eq. 4.3. As seen, F becomes negative over the time duration A–B (during the passage of the wave trough), meaning that the soil during this period is momentarily liquefied. It may be noticed that there is a phase lead in F with respect to the surface elevation η. Recall that the pore pressure itself has a phase delay with respect to η.

Regarding the liquefaction depth, when the above exercise is repeated for various values of z, it will be found that the soil is liquefied for depths $z \leq 1.01\,\text{m}$, i.e., the liquefaction depth is $z_\ell = 1.01\,\text{m}$.

Regarding the limiting value of the degree of saturation, S_r, that allows the soil to be liquefied, this value is found to be $S_r = 0.989$; the soil will always be liquefied at $z = 0.5\,\text{m}$ if $S_r \leq 0.989$.

As seen, even a very small amount of gas or air in the soil (as small as approximately 1%) makes the soil prone to the momentary liquefaction (in this particular example, at depth $z = 0.5\,\text{m}$).

As mentioned previously, Eq. 4.6 can also be used to check for liquefaction. In this case we need the vertical effective stress σ_z'. Mei and Foda's (1981) solution gives this quantity as follows:

$$\sigma_z' = p_b \left(-\frac{m}{1+m} - \lambda z \right) \exp(-\lambda z) \cos(\lambda x - \omega t)$$

$$+ p_b \frac{m}{1+m} \exp\left(\frac{-z}{\sqrt{2\delta}} \right) \cos\left(\lambda x - \omega t + \frac{z}{\sqrt{2\delta}} \right). \tag{4.16}$$

Inserting the preceding expression in Eq. 4.6 and calculating the left-hand side of the inequality in Eq. 4.6 gives the same result as in Fig. 4.4 with F on the vertical axis replaced by

$$F = \gamma' z - \sigma_z'. \tag{4.17}$$

Momentary liquefaction for shallow water waves. Sakai *et al.*'s (1992 a) analysis

There are three kinds of waves: (1) shallow-water waves; (2) intermediate-depth waves; and (3) deep-water waves. Waves are called shallow-water waves if the water-depth-to-wave-length ratio is $h/L < 1/20$, or alternatively, in terms of wave number, if $\lambda h < \pi/10$. Waves are called intermediate-depth waves if $\pi/10 < \lambda h < \pi$, and deep-water waves if $\lambda h > \pi$ (see, e.g., Dean and Dalrymple, 1984).

Sakai *et al.* (1992 a) argue that since the main concern of their study is for surf-zone conditions, the shallow-water-wave approximation is justified. The theory of shallow-water waves imply that (Dean and Dalrymple, 1984)

$$\cosh(\lambda h) \simeq 1, \quad \frac{2\pi}{\lambda} \equiv L \simeq T\sqrt{gh}. \tag{4.18}$$

Using the above approximations, Sakai *et al.* (1992 a) examined whether or not Eq. 4.3 is satisfied, and worked out the liquefaction depth. Fig. 4.5 depicts Sakai *et al.*'s results (corrected by Law, 1993), giving the depth of momentary liquefaction.

As implied by Sakai *et al.*'s analysis, the momentary liquefaction does occur for the values of the parameters indicated in the legend in Fig. 4.5, and it can penetrate to depths as large as $0.5H$. Sakai *et al.* note that it is difficult to produce the momentary liquefaction in a small-size wave flume using a normal sand bed, because the value of $kG/(\rho g^2 Th)$ is two orders of magnitude larger than that in actual sandy beaches.

Gratiot and Mory (2000) extended the analysis of Sakai *et al.* (1992 a) to include the effect of the water-depth variations. In the analysis, the long wave approximation was not made. Basically, Gratiot and Mory's work is a sensitivity study of the depth of liquefaction with respect to the degree of saturation, the porosity, the permeability, the shear modulus, the Poisson ratio of the soil, the water depth, the wave height, and the wave period.

4.3.2 Finite soil depth

The above analysis is given for a soil with an infinitely large depth. For the case of a soil with a finite depth, d (Fig. 4.2), the pore pressure p can be obtained from the analytical solution of Hsu and Jeng (1994) for the case of unsaturated soil where the Biot equations are solved with S_r different from

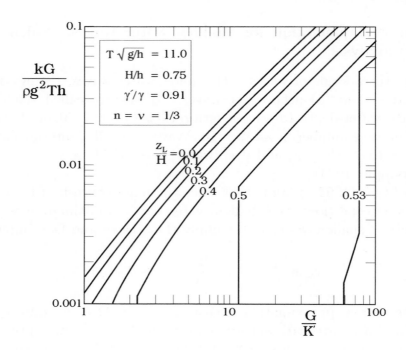

Figure 4.5:   Depth of momentary liquefaction. Sakai *et al.* (1992 a) and corrected by Law (1993).

unity. We note that the analytical solution of Hsu and Jeng (1994) given in Section 2.2.1 is for the case of saturated soil, and therefore can not be used in the present context, and therefore we need to resort to Hsu and Jeng's (1994) original publication.

The following is a summary of the procedure in implementing Hsu and Jeng's (1994) solution for the present, *unsaturated soil* case.

(1) Calculate the quantity δ in Eq. 2.73 for the unsaturated soil from

$$\delta^2 = \lambda^2 - \frac{i\omega\gamma}{k}\left(\frac{n}{K'} + \frac{1 - 2\nu}{2G(1 - \nu)}\right) \tag{4.19}$$

in which K' is the bulk modulus of elasticity related to the *true* bulk modulus of elasticity of water, K, by Eq. 2.36, as indicated in Chapter 2. The above equation reduces to Eq. 2.73 in the case of saturated soils, as $S_r = 1$, and therefore, from Eq. 2.36, $1/K' = 1/K$, with $\frac{n}{K} \ll \frac{1-2\nu}{2G(1-\nu)}$ since G/K is extremely small for most soils except for dense soil, as discussed in Section 2.2.1.

(2) Calculate a new parameter related to the saturation of the soil, λ_{H-J}, from

$$\lambda_{H-J} = \frac{\frac{n}{K'}}{\frac{n}{K'} + \frac{1-2\nu}{G}}(1 - 2\nu). \tag{4.20}$$

This parameter becomes nil for the case of saturated soil, and therefore is not involved in the Hsu and Jeng solution reproduced in Section 2.2.1. (Note that, in Hsu and Jeng's (1994) paper, the symbol λ is used for λ_{H-J}; the original symbol is replaced here with λ_{H-J} to avoid confusion, as the symbol λ in the present analysis is used for the wave number, $\lambda = 2\pi/L$.)

(3) Subsequently, calculate Hsu and Jeng's (1994) coefficients from Hsu and Jeng's (1994) original paper where the analytical expressions given for these coefficients include both δ and λ_{H-J}, the new coefficients.

(4) Note that, although the end results are not significantly affected, there are two typing errors in Hsu and Jeng's (1994) original publication: the product $\lambda_{H-J}(B_{11} + \lambda_{H-J}B_{12})$ in the expression for the coefficient C_{11} should read $\lambda_{H-J}(B_{11} + B_{12})$; and that $-2\delta k^2\lambda_{H-J}$ in the expression for the coefficient C_{32} should read $-2\delta^2 k\lambda_{H-J}$ in which k is the wave number used in the original publication. We emphasize that these are only typing errors; the authors' original code of computations is correct (Dong Jeng, 2012, personal communication).

(5) Calculate the pore pressure p from Eq. 42 in Hsu and Jeng's (1994) original paper. Note that the latter equation is different from Eq. 2.63 given in Chapter 2, as Eq. 2.63 is for fully saturated soils.

(6) Likewise, calculate the vertical effective stress σ'_z from Hsu and Jeng's (1994) Eq. 44. Similarly, the latter equation is different from Eq. 2.65 given in Chapter 2, as Eq. 2.65 is for fully saturated soils.

(7) Subsequently, insert p in Eq. 4.3, and check if the quantity $\gamma'z - p + p_1$ becomes negative (or zero), or alternatively insert σ'_z in Eq. 4.6, and check if the quantity $\gamma'z - \sigma'_z$ becomes negative (or zero), the momentary liquefaction.

The following numerical example will illustrate this exercise.

Example 11.   *Check for momentary liquefaction for a soil with a finite depth. A numerical example.*

The soil and water properties are given as in the previous example. The only difference between the present case and the case in the previous example is that the soil has now a finite depth of $d = 10\,\text{m}$, the soil layer being confined with an impermeable base below.

The soil is exposed to the same progressive wave.

As in the previous example, the question is whether the soil will undergo momentary liquefaction under the given wave climate at depth $z = 0.5\,\mathrm{m}$, and also, what is the liquefaction depth, z_ℓ?

The pore pressure and the vertical effective stress are calculated from Hsu and Jeng's (1994) analytical solution, following the procedure summarized in the preceding paragraphs, and the left-hand side of the inequality in Eq. 4.3 at $z = 0.5\,\mathrm{m}$ is worked out.

The result is shown in Fig. 4.6 in which the quantity F on the vertical axis is $F = \gamma'z - p + p_1$, the left-hand side of the inequality in Eq. 4.3. When compared with Fig. 4.4, Fig. 4.6 indicates that the amplitude of the pore pressure p in the case of the finite soil depth somewhat increases and, as a result, the soil is liquefied earlier (point A in Fig. 4.6), and the liquefaction lasts longer (the time interval between A and B in Fig. 4.6), cf. Fig. 4.4. This is linked to the lesser dissipation of the pore pressure because the soil is now confined with an impermeable base below.

Similar to the infinitely-large-soil-depth case, when the above exercise is repeated for various values of z, it will be found that the soil is liquefied for depths of approximately $z \leq 1.44\,\mathrm{m}$, i.e., the liquefaction depth is $z_\ell = 1.44\,\mathrm{m}$. The latter figure was $z_\ell = 1.01\,\mathrm{m}$ in the case of the infinitely large soil depth, consistent with the discussion in the preceding paragraph.

From the above results, it may be concluded that, as a first screening, Mei and Foda's (1981) simple expression for p given in Eq. 4.10 (with Eqs 4.11–4.18) can be used in Eq. 4.3 in the assessment of liquefaction potential. If it turns out that there is a liquefaction potential, then a detailed analysis may be carried out, utilizing the finite-depth solution of Hsu and Jeng (1994) for p along with Eq. 4.3.

As noted previously, Eq. 4.6 can also be used to check for liquefaction. In this case, the analytical expression given in Hsu and Jeng (1994) for the vertical effective stress σ'_z is picked up from the original publication (their Eq. 44), as was the case for p (their Eq. 42). Inserting σ'_z in Eq. 4.6 and calculating the left-hand side of the inequality in Eq. 4.6 gives the same result as in Fig. 4.6, with F on the vertical axis in the figure replaced by $F = \gamma'z - \sigma'_z$.

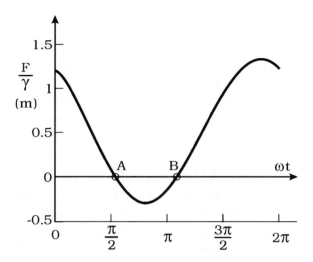

Figure 4.6: Momentary liquefaction occurs when $F \leq 0$ in which $F = \gamma' z - p + p_1$, i.e., during the time period A–B. At $z = 0.5$ m. Finite-soil-depth case with $d = 10$ m.

4.4   Miscellaneous

Momentary liquefaction and densification in seabed

Zen and Yamazaki (1990 a) developed an analysis similar to that in Section 4.1 for momentary liquefaction with the liquefaction criterion corresponding to Eq. 4.2. The authors also discussed densification in conjunction with the wave action. Now, recall the vertical force $(p - p_1) \times 1 \times 1$ exerted on the soil column in Fig. 4.3. This force is obviously directed downwards during the passage of the wave crest. Zen and Yamazaki (1990 a) remarked that when this force attains values larger than the submerged weight of the soil column, $(\gamma' z) \times 1 \times 1$, it will densify the soil.

Zen and Yamazaki (1990 a) tested their theory against laboratory experiments they conducted. The soil was placed in a test cylinder with 225 mm outer diameter and 200 mm inner diameter, and 2.1 m height, with a water column of 0.2 m depth at the top above the soil column. The water was pressurized in a cyclic manner with time, simulating the cyclic pressure variation at the interface between the soil column and the water, with the seabed-pressure amplitude in the range $p_b/\gamma \simeq 1 - 4$ m water column. The pore-water pressures and the settlements at various depths in the soil column were measured.

Zen and Yamazaki's (1990 a) measurements indicated that momentary liquefaction occurred. Of particular interest is the finding of the authors related to the soil densification. Very large settlements (up to 5 cm), as well as an increase in the relative density (up to $D_r = 1$) from the initial value $D_r = 0.5$ were measured for soil depths $z < 1$ m, and after 3,500 cycles, presumably as a result of the sequence of liquefaction and compaction, combined with the downward-directed force exerted on the soil column during the crest half period.

Zen and Yamazaki (1990 b), in a parallel study, developed a theoretical model for momentary liquefaction, and verified their model against their experiments conducted in the previously described test cylinder. The authors later extended their study to field observations and analysis of momentary liquefaction (Zen and Yamazaki, 1991). Their analysis of the field data of pore pressure revealed liquefaction periods in the top layer of sediment bed for $z < O(30 \, \text{cm})$ (Zen and Yamazaki, 1991, Fig. 16).

Numerical modelling, using OpenFOAM

Liu and Garcia (2007) have developed an iterative solution scheme for the Biot equations, using the finite volume method (FVM), and subsequently incorporated it in the so-called OpenFOAM, an open-source numerical code. The free surface was modelled by the volume of fluid (VOF) method, and the water wave was generated by the numerical wave maker boundary condition.

Two numerical tests were carried out. The first test validated their solution scheme for the Biot equations against the analytical solution of Jeng and Hsu (1996). Although they did not specifically study the momentary liquefaction for the case of progressive waves in the undisturbed case (i.e., in the absence of any structure), clearly their model can be used for check for momentary liquefaction.

In the second test, Liu and Garcia (2007) have studied the seabed response in the case when a finite-length, square-section cylinder (block) is placed in the seabed (half-buried). Here, the hydrodynamics, including the forces on the cylinder were computed, using the wave model described above. The coupling between the water wave and the seabed is through pressure and stress conditions on the seabed. Similar to the undisturbed case, momentary liquefaction can also be checked, using the model in the presence of the structure.

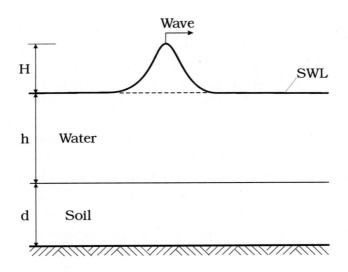

Figure 4.7: Solitary wave. Wave height is grossly exaggerated for illustration purposes.

Solitary wave

Merxhani and Liang (2012) developed a numerical model to calculate the seabed response to a solitary wave (see the definition sketch of a solitary wave in Fig. 4.7). The model solves the Biot equations, using the Finite Element Method. The authors tested and validated their model against the analytical solution of Hsu and Jeng (1994) for progressive waves (Chapter 2, Section 2.2.1). Having tested and validated the model, Merxhani and Liang (2012) implemented the model for a solitary wave, with the wave and soil conditions representative of the Ekofisk site of the North Sea. Parallel to the solitary-wave test, these authors also carried out a progressive-wave test for reference purposes. These numerical tests were carried out for two kinds of sediment, namely coarse sand and fine sand.

The wave characteristics were: the water depth, $h = 70\,\text{m}$; the wave height $H = 24\,\text{m}$; and the wave period for the progressive wave tests, $T = 15\,\text{s}$. (The wave period for solitary waves is $T = \infty$, and therefore a solitary wave is characterized by only two quantities, namely, the water depth, h, and the wave height, H, see, e.g., Svendsen, 2006, p. 406.)

The soil characteristics, on the other hand, were: the soil depth, $d = 25\,\text{m}$; the shear modulus, $G = 10^4\,\text{kN/m}^2$; the permeability, $k = 10^{-4}\,\text{m/s}$ for the fine sand, and $k = 10^{-2}$ m/s for the coarse sand; the porosity $n = 0.3$; Poisson's ratio, $\nu = 0.333$; and the degree of saturation, $S_r = 1.0$.

The results of Merxhani and Liang's (2012) study revealed that the pore pressure in excess of the static pressure is always positive. This is not unexpected, however, because there is no trough in the solitary wave. The fact that the pore pressures will never get negative values implies that no lift will be generated (cf. Fig. 4.1) and, as a result, the soil will always remain stable, i.e., no momentary liquefaction will be induced under a solitary wave.

This result can be obtained formally from Eq. 4.3. The left-hand side of the inequality in this equation can be worked out as follows.

For a small-amplitude solitary wave ($H/L \ll 1$ in which L being a length scale that characterizes the width of the surface-elevation variation, cf., Fig. 4.7), the pressure at the seabed in excess of the static pressure is

$$p_1 = \gamma \eta \tag{4.21}$$

in which η is the surface elevation measured from the Still Water Level (Svendsen, 2006, p. 419). Incidentally, for a small-amplitude solitary wave, the pressure in excess of the static situation is constant across the water depth, and equal to $\gamma \eta$.

Now, the pressure at any depth z in the soil is always smaller than the bed pressure, $p_1 = \gamma \eta$, but larger than nil, namely $0 < p \leq \gamma \eta$, and therefore the left-hand side of the inequality in Eq. 4.3, $\gamma' z - p + p_1$, will never become smaller than (or equal to) zero, and hence no liquefaction will occur.

It may be noted that the liquefaction criterion given in Eq. 4.6, when applied to the field case studied in Merxhani and Liang (2012), with the values of σ'_z taken from Merxhani and Liang's (2012) numerical solution, reveal that the 24 m solitary wave considered in the Merxhani and Liang work will not induce liquefaction, consistent with the above considerations (although one should note that the degree of saturation in the Merxhani and Liang work was taken as $S_r = 1$, a fully saturated soil).

To sum up, the work of Merxhani and Liang (2012) and the above analysis indicate that a solitary wave will not induce momentary liquefaction.

This is what occurs over a horizontal seabed. Recall, however, that a solitary wave can induce liquefaction on a coastal slope, mainly due to upward-directed pressure gradient in the soil generated during the rundown stage of the wave (Chapter 1, Section 1.4). The following subsection will focus on this.

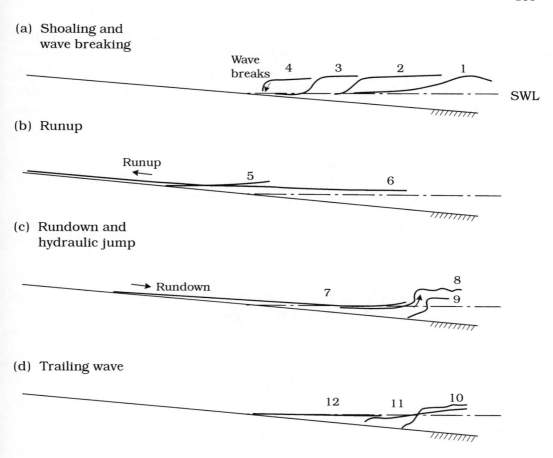

Figure 4.8: Sequence of a plunging solitary wave. Sumer *et al.* (2011).

Solitary wave on coastal slopes

Fig. 4.8, borrowed from Sumer *et al.* (2011), schematically illustrates the observed sequence of a plunging solitary wave on a coastal slope: (a) shoaling and wave breaking; (b) runup; (c) rundown and hydraulic jump; and (d) trailing wave. Numbers in the figure denote different stages of the breaking process. The breaking occurs with the wave curling over forward and impinging onto the water surface onshore with a small amount of air trapped inside the "tube" formed by the wave crest.

Sumer *et al.* (2011) conducted pore-water pressure measurements under a plunging solitary wave on a slope of 1:14, representative of a mild slope.

Fig. 4.9 from Sumer *et al.* (2011) shows three characteristic pore-pressure distributions obtained at a section slightly offshore the hydraulic-jump location (Fig. 4.8c), one at $t = 2.4$ s, the second one at $t = 4.9$ s, and the third one at $t = 7.5$ s. The time $t = 2.4$ s is associated with the runup stage (Fig. 4.8b) while $t = 4.9$ s and $t = 7.5$ s are associated with the rundown and hydraulic-jump stage (Fig. 4.8c).

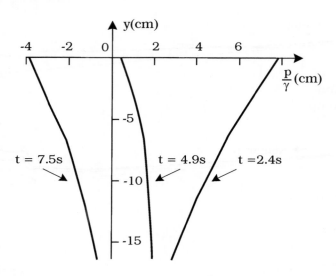

Figure 4.9:   Three characteristic excess pore pressure distributions at a section slightly offshore side of the hydraulic jump under a plunging solitary wave on a slope of 1:14. Location y=0 corresponds to the bed level (mudline). Laboratory experiment. Reproduced from Sumer *et al.* (2011).

As seen from Fig. 4.9, the pressure distribution at $t = 2.4$ s generates a downward-directed pressure gradient force, while the distributions at $t = 4.9$ s and 7.5 s generate upward-directed pressure gradient forces in the bed soil. As seen from the figure, the gradient of the pressure distribution in all three cases is largest near/at the bed surface, and therefore the previously mentioned downward- and upward-directed pressure gradient forces on the bed sediment become largest near/at the bed surface.

Both the downward-directed pressure gradient and the upward-directed pressure gradient are caused by the delay in the pore pressure in responding to the fluid loading, as can be readily seen from Sumer *et al.*'s (2011, Fig. 11) pore-pressure time series. The delay itself is linked largely to the time associated with infiltration and exfiltration processes.

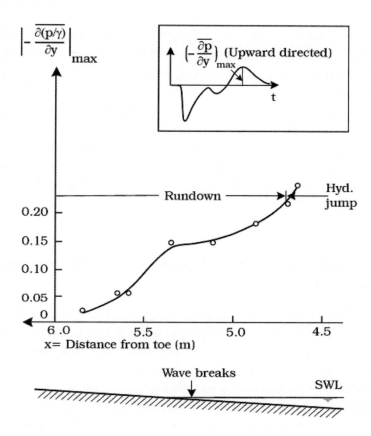

Figure 4.10: Maximum value of the upward-directed excess pore pressure gradient over a slope of 1:14, measured in Sumer *et al.* (2011). Location y=0 corresponds to the bed level (mudline), and y is measured from mudline upwards. Reproduced from Sumer *et al.* (2011).

Clearly the upward-directed pressure gradient force on the bed sediment is most interesting as it may cause momentary liquefaction in precisely the same way as in the momentary liquefaction process discussed in the previous sections. Fig. 4.10 presents the maximum values of the upward-directed pressure gradient measured in Sumer *et al.*'s experiment (2011) as a function of the offshore–onshore distance. The figure indicates that the sediment bed during this stage (rundown and hydraulic jump, Fig. 4.8c) is subject to an upward-directed force with values largest near the hydraulic jump. As seen, this force (which is, from Fig. 4.10, $(\frac{\partial(p/\gamma)}{\partial y})_{max} = O(0.05 - 0.25))$ can reach values as much as up to approximately 30% of the submerged weight of the sediment (which is like $(s-1)(1-n) \sim O(0.9)$) towards the end of rundown and hydraulic-jump stage in which s is the specific gravity of sediment grains and n the porosity.

The primary purpose of Sumer *et al.*'s (2011) work was to get an understanding of the flow and sediment transport induced by a breaking solitary wave over a slope, and therefore the subject was not pursued further. However, the topic was investigated from the point of view of liquefaction by Young *et al.* (2009). The following sub-section will highlight the latter study.

Liquefaction potential of coastal slopes under solitary waves. Young *et al.* (2009) work

Young *et al.* (2009) studied the liquefaction potential of coastal slopes under solitary waves. They developed a numerical model, consisting of two components. The first component (what the authors called the surface simulator) computes the wave, while the second component (the subsurface pore-water pressure simulator) computes the pore-water pressure.

The surface simulator solves the depth-averaged nonlinear shallow water equations and Boussinesq equations, using a finite volume method. Boussinesq equations are solved during the pre-breaking stage and the shallow water equations are solved for the post-breaking stages.

Regarding the subsurface pore-water pressure simulator, the pore-water pressure and deformation fields in the soil are solved simultaneously, using a finite element method. Two different soil constitutive models are considered, a linear elastic model, and a non-associative Mohr–Coulumb model (elastoplastic model).

Young *et al.* (2009) validated the numerical models against the results with analytical models, and against experimental measurements from a large-scale laboratory study of breaking solitary waves over a fine sand slope.

Having validated and tested the numerical model, Young *et al.* (2009) conducted numerical experiments for a full-scale simulation of a 10 m high solitary wave propagating over an initial water depth of 20 m, over two kinds of slopes, one with 1:15, and the other with 1:5, representing a mild and a steep slope, respectively. The depth of the soil to the impermeable base was 20 m. The sand was typical dense fine sand.

The model predicted significant amount of upward-directed pressure gradients. Young *et al.* commented that, during the rundown stage, the bed surface pressure drops to the atmospheric pressure as the water level drops to 0. They noted:

> "however, the pore-water pressure beneath the bed surface cannot dissipate as fast, which leads to upward pore pressure gradients that may cause liquefaction failure of the soil near the bed surface", in complete accord with the pressure-gradient generation mechanism, described in the previous subsection (from Sumer *et al.*, 2011).

It may be noted that the pore pressure responses predicted by the two models in Young *et al.*'s (2009) study appear to be essentially identical during the loading stage. However, significant differences are observed during the unloading stage; the elastoplastic model predicts significantly larger negative pressures, with much less dissipation with depth than the elastic model.

Young *et al.* (2009, Figs. 13 and 16) present a sequence of the pictures representing the time evolution of a liquefied layer beneath the bed surface. In the case of the 1:15 slope, the maximum depth of the liquefaction zone is predicted to be 2.8 m whereas, this figure is found to be 4.4 m in the case of the 1:5 slope.

Important results from Young *et al.* (2009) may be summarized as follows:

(1) The region immediately offshore from the shoreline is the most suscep- tible to liquefaction because the water level drops significantly below the still water level during the rundown (cf. Stages 8–10 in Fig. 4.8). Consequently, a seepage face is created between the initial shoreline and maximum rundown location due to the inability of the subsurface water table to respond to the rapid water surface elevation changes. In this region, the pore pressure can easily exceed the significantly reduced effective overburden pressure (due to drop in water level), which will cause liquefaction.

(2) The depth of the seepage face increases and the width of the seepage face decreases with increasing bed slope. The rate of bed surface loading and unloading due to wave runup and rundown, respectively, also increases with increasing bed slope. Therefore, the case with the steeper slope is more susceptible to liquefaction failure due to the higher hydraulic gradient.

(3) The results are strongly influenced by the soil permeability and the degree of saturation.

(4) The results also are significantly influenced by nonlinear soil behaviour for the full-scale simulations.

4.5   Air/Gas Content in Marine Soils

Air/gas bubbles

Air/gas content in marine soils is essential for momentary liquefaction, as discussed in the preceding paragraphs. There is direct and indirect evidence of the presence of air/gas bubbles in the coastal and offshore environment, e.g., in tidal areas in coastal regions where the seabed is exposed to air and water alternately.

First, we shall consider the question "in what form are air bubbles present in the soil with respect to the soil grains?" Anderson, Abegg, Hawkins, Duncan and Lyons (1998) suggested three possible scenarios: the air/gas bubbles are either (1) much smaller than soil grains and therefore they are contained in the interstitial spaces of the sediment framework of solid particles, or (2) they exceed the size of the individual pore spaces, but do not alter the soil fabric; or (3) they are much larger than the soil grains and are essentially free. The latter authors indicated that most bubbles observed through the clear walls of the coring tubes are of the third type. They also noted that, when a recovered seafloor core is sectioned and examined, it is not unusual to note evidence of Type 2 bubble occurrence as well. They pointed out that the existence of free gas bubbles *in situ* is difficult to extrapolate from samples that have been raised to the sea surface because of changes resulting from the release of hydrostatic pressure. Furthermore, they pointed out that evidence for Type 1 bubbles is rarely observed but part of the reason for this is their extremely small size. They emphasized that such very small bubbles would be ephemeral constituents of a core sample and in most cases would be lost in processing the sample for microscopic examination.

Field evidence of air/gas presence

Mory *et al.* (2007) measured pore water pressure distribution across the soil depth on a tidal beach (the measurement depth being across the top 1 m layer), located in Capbreton, France. They, in a follow-up publication (Michallet, Mory and Piedra-Cueva, 2009), reported further analysis of the data collected in Mory *et al.* (2007).

Mory *et al.* in their 2007 study found that momentary liquefaction occurred over significant portions of time, suggesting that the soil was unsaturated. This prompted Mory *et al.* (2007) (see also Michallet *et al.*, 2009) to

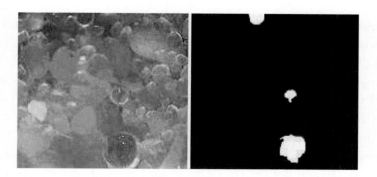

Figure 4.11:   Image of air bubbles from geoendoscopic video recording of the soil near the bed surface on a tidal beach with an accompanying picture showing the processed image where the bubble images were singled out (Mory *et al.*, 2007). Digital image: By courtesy of Professor Mathieu Mory of the University of Pau, France.

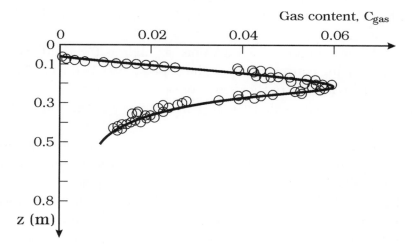

Figure 4.12:   Gas content versus depth, measured in Michallet *et al.* (2009). The gas content was calculated from the image analysis of the geoendescopic video camera recording.

conduct a geoendoscopic video recording of the soil near the bed surface to observe visually whether or not the soil contained air bubbles. To this end, images of a 25 mm^2 area with a magnification of 10 were acquired. This study showed the presence of significant quantities of air inside the soil, down to 0.50 m, and vanishing with the depth beyond this level. Fig. 4.11 illustrates a recorded image of air bubbles with an accompanying picture showing the processed image where the bubble images were singled out. As seen, the bubble size is in the same order of magnitude as the grain size, d_{50} being $d_{50} = 0.35$ mm. It was found that the size of air bubbles decreased progressively from $z = 25$ cm to 50 cm below the soil surface, and gas was hardly detected deeper than $z = 50$ cm (Michallet *et al.*, 2009). Fig. 4.12 is reproduced from Michallet *et al.* (2009), illustrating the gas content versus the depth. The gas content is defined by

$$C_{gas} = \frac{V_g}{V_v} \tag{4.22}$$

in which V_g is the volume of gas and V_v the total pore volume (Fig. B.1, Appendix B). C_{gas} in Michallet *et al.*'s (2009) study was calculated from the image analysis.

As seen, the gas content reaches values as high as 6% inside the soil, the maximum value being concentrated at the depth $z = 0.2 - 0.3$ m. This, $C_{gas} = 6\%$, corresponds to a value of the degree of saturation of $S_r = 1 - C_{gas} = 0.94$. The diagram also indicates that the top 0.10 m layer of the soil contains only a very little amount of air.

Mory *et al.* (2007) link the presence of air bubbles to the tidal variations: air is introduced inside the soil at low tide when the water level is below the soil level. When the water level rises with the high tide, the soil saturation will not be fully completed; some air bubbles will be trapped inside the soil, and remain there until the next tide cycle. Mory *et al.* (2007) explain, on the other hand, the very small air content within the top 0.10 m soil layer in terms of surface-sediment mobility; the fact that the soil is constantly being reworked by sediment transport will help air bubbles escape the bed.

In this context, it is also interesting to note the following finding of Michallet *et al.* (2009). Mory *et al.*'s (2007) measurements (also reported in Michallet *et al.*, 2009) of pore pressure, which took place on two consecutive days (namely on September 24 and September 25), indicated that the soil properties were modified during the tidal period of September 24. The wave conditions were severe during this tide, and the change was attributed to the fact that some air escaped the soil due to the enhanced mobility of the bed sediment under these severe waves.

An indirect piece of evidence of the potential presence of air comes from Tørum's (2007) analysis of a set of field data obtained from extensive measurements of pore water pressure carried out by de Rouck (1991), de Rouck and van Damme (1996), and de Rouck and Troch (2002). These measurements were carried out in connection with the planning of the extension of the Zeebrugge Harbour, Belgium. They were conducted in two kinds of soils, sand and clay. Tørum (2007) compared the measured pressure distributions with those calculated from Mei and Foda's (1981) solution (see also Sakai et al., 1992 a), discussed in Section 4.3.1 above.

(de Rouck also developed an analytical model, based on essentially a 2-D consolidation theory, on wave-induced pore pressures. Tørum (2007), notes, however, that the purpose of his analysis was to compare de Rouck's measurements with a more advanced theory, referring to Mei and Foda's, 1981 work.)

Fig. 4.13 displays this comparison for the sand case, and Fig. 4.14 for the clay case. The field data match with the theory (Mei and Foda, 1981) only when the degree of saturation, S_r, in the Mei and Foda solution is set equal to 0.97 (Fig. 4.13 b, and Fig. 4.14 b), indicating that it is most likely that the soil contained air/gas in de Rouck's field measurements.

Tørum (2007), relating the recent measurements of Mory et al. (2007) to his analysis, argues that de Rouck's measurements were done in deeper water than Mory et al.'s measurements, and were not influenced by beach draining, the most likely cause of the air bubbles in Mory et al.'s measurement site. Tørum (2007) argues that a possible reason for the presence of air/gas bubbles in de Rouck's site may be: "... deterioration of organic material, or air trapped in the pores during a lower water level thousands of years ago". In relation to the latter, he refers to Hovland et al.'s (2001) study where the authors discuss effects of slope stability of gas hydrates and seeps, and remark that many of the known historical large underwater slides have occurred in areas with gas hydrates.

Hattori, Sakai and Hatanaka (1992) and Sakai, Mase, Cox and Ueda (1992 b) also fitted Mei and Foda's (1981) solution to field data, "tuning" the quantity K' in Eq. 4.13 so that they could obtain a reasonable match between the data and the Mei and Foda (1981) solution. Their results suggest the presence of air in seabed soil, similar to Tørum (2007).

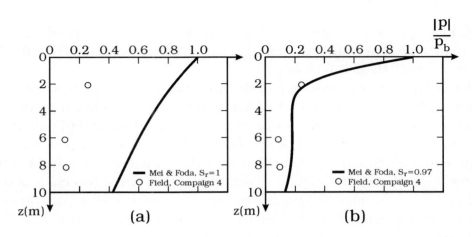

Figure 4.13:   Field measurements of de Rouck of pore pressure distribution, compared with Mei and Foda's (1981) analytical solution.   The degree of saturation in Mei and Foda's solution is taken as $S_r = 1$ in (a), and $S_r = 0.97$ in (b). Taken from Tørum (2007).

Sills, Wheeler, Thomas and Gardner (1991) stated that gas bubbles may form in the offshore environment where the methane is generated around nuclei of bacteria locally within a soft, consolidating soil. They report that the gas bubbles produced in this way are considerably larger than the fine-grained soil particles, and the resulting soil structure consists of large bubble "cavities" within a matrix of saturated soil. (Incidentally, the latter authors developed a laboratory technique to mimic as closely as possible the process of bubble formation in the offshore environment.) Fig. B.1a (Appendix B), adapted from Sills *et al.* (1991), illustrates a schematic representation of a gassy soil.

Measurement of air/gas content

Various attempts have been made in the past to develop samplers for measurement of gas content in soils. These efforts have indicated that this is not a straightforward task. A review of the existing work has been given in Sandven, Foray and Long (2004). Recently, a new sampler that enables in-situ measurement of gas content in the seabed has been developed under the EU research program "Liquefaction Around Marine Structures (LIMAS)"

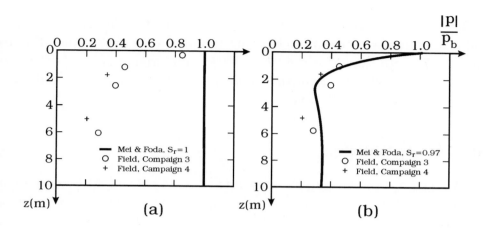

Figure 4.14: Field measurements of de Rouck of pore pressure distribution, compared with Mei and Foda's (1981) analytical solution. The degree of saturation in Mei and Foda's solution is taken as $S_r = 1$ in (a), and $S_r = 0.97$ in (b). Taken from Tørum (2007).

by Sandven *et al.* (2007). The latter authors pointed out that previously developed samplers for gassy soils utilize sealing methods such as inflatable membranes, ball valves for sealing pressurized core barrels, and plate valves or core catchers for the same purpose, while the aim of their study, they stressed, was to develop a sampler enabling measurement of the gas content, and at the same time retrieving a representative soil sample, making it possible to determine both the relative gas content and the degree of saturation in the soil.

The developed sampler was used successfully at the LIMAS research site in Capbreton, France, the site Mory *et al.* (2007) (also reported in Michallet *et al.*'s, 2009) did their measurements (see the previous subsection). The obtained results showed that the gas content could vary significantly in the upper 0.3–0.6 m top layer of the soil, consistent with Mory *et al.*'s (2007) (see also Michallet *et al.*'s, 2009) geoendoscopy measurements. Sandven *et al.* (2007) report that the smallest gas contents, representing 0.2–0.65% of the sample volume (corresponding to the range of $S_r = 0.988 - 1$, S_r being the degree of saturation) were measured under *falling tide conditions* after repeated cycles of strong waves.

The previously mentioned air/gas content measurements by Mory *et al.* (2007) and Sandven *et al.* (2007), were not conducted at the same time, and therefore they are not directly comparable, as the degree of saturation is influenced by the constantly changing tide and wave conditions. Sandven *et al.* (2007) note that Mory *et al.*'s (2007) measurements were carried out during calm sea whereas Sandven *et al.*'s (2007) sampling was undertaken under rougher wave conditions.

4.6   References

1. Anderson, A., Abegg, F., Hawkins, J.A., Duncan, M.E. and Lyons, A.P. (1998): Bubble populations and acoustic interaction with the gassy floor of Eckernfjorde Bay. Continental Shelf Research, vol. 18, No. 14–15, 1807–1838.

2. Dean, R.G. and Dalrymple, R.A. (1984): Water Wave Mechanics for Engineers and Scientists. Prentice Hall, xii+353 p. Englewood Cliffs, New Jersey 07632.

3. de Rouck, J. (1991): De stabiliteit van storsteengolfbreakers. (Stability of rubble mound breakwaters. Slope stability analysis; a new armour unit). PhD. thesis, Dep. Burgerlijke Bouwkunde, Laboratorium voor hydraulica, Katholieke Universiteit Leuven, Faculteit Toegepaste Wetenschapen, Leuven, Belgium.

4. de Rouck, J. and Troch, P. (2002). Pore water pressure response due to tides and waves based on prototype measurements. PIANC Bulletin No. 110, 9–31.

5. de Rouck, J. and van Damme, L. (1996): Overall stability analysis of rubble mound breakwaters. Proceedings of the 25th International Conference on Coastal Engineering, Orlando, FL, ASCE, Reston, VA, 1603–1616.

6. Gratiot, N. and Mory, M. (2000): Wave induced sea bed liquefaction with application to mine burial. Proceedings of the Tenth International Offshore and Polar Conference, ISOPE-2000, May 28–June 2, 2000, Seattle, WA, vol. 2.

7. Hattori, A., Sakai, T. and Hatanaka, K. (1992): Wave-induced porewater pressure and seabed stability. Proceedings of the 23rd Conference on Coastal Engineering (ICCE), vol. 2, 2095–2107.

8. Hovland, M., Orange, D. and Gudmestad, O.T. (2001): Gas hydrates and seeps–Effect on slope stability: The hydraulic model. Proceedings of the 11th International Conference on Offshore and Polar Engineering, Stavanger, Norway.

9. Hsu, J.R.S. and Jeng, D.S. (1994): Wave-induced soil response in an unsaturated anisotropic seabed of finite thickness. International Journal for Numerical and Analytical Methods in Geomechanics, vol. 18, No. 11, 785–807.

10. Isaacson, M. (1979): Wave-induced forces in the diffraction regime. In: Mechanics of Wave-induced Forces on Cylinders, (ed. T.L. Shaw). Pitman Advanced Publishing Program, San Francisco 68–89.

11. Jeng, D.S. and Hsu, J.R.S. (1996): Wave-induced soil response in a nearly seabed of infinite thickness. Géotechnique, vol. 46, No. 3, 427–440.

12. Law, A.W.K. (1993): Wave-induced effective stress in seabed and its momentary liquefaction. Discussion to the paper by Sakai *et al.* (1992 a), Journal of Waterway, Port, Coastal and Ocean Engineering, ASCE, 119, No. 6, 694–695.

13. Liu, X. and Garcia, M.H. (2007): Numerical investigation of seabed response under waves with free-surface water flow. International Journal of Offshore and Polar Engineering, vol. 17, No. 2, 97–104.

14. Mei, C.C. and Foda, M.A. (1981): Wave-induced responses in a fluid filled poroelastic solid with a free surface A boundary layer theory. Geophysics, Journal of the Royal Astronomical Society, vol. 66, No. 3, 597–631.

15. Merxhani, A. and Liang, D. (2012): Investigation of the poro-elastic response of seabed to solitary. Proceedings of the 22nd (2012) International Offshore and Polar Engineering Conference, Rhodes, Greece, June 17–22, 2012, 101–108.

16. Michallet, H., Mory, M., and Piedra-Cueva, I. (2009): Wave-induced pore pressure measurements near a coastal structures, Journal of Geophysical Research, vol. 114, No. C06019, 1–18.

17. Mory, M., Michallet, H., Bonjean, D., Piedra-Cueva, I., Barnoud, J.M., Foray, P., Abadie, S. and Breul, P. (2007): A field study of momentary liquefaction caused by waves around a coastal structure. Journal of Waterway, Port, Coastal and Ocean Engineering, ASCE, vol. 133, No. 1, 28–38.

18. Sakai, T., Hatanaka, K. and Mase, H. (1992 a): Wave-induced effective stress in seabed and its momentary liquefaction. Journal of Waterway, Port, Coastal and Ocean Engineering, ASCE, vol. 118, No. 2, 202–206. See also Discussions and Closure in vol. 119, No. 6, 692–697.

19. Sakai, T., Mase, H., Cox, D.T. and Ueda, Y. (1992 b): Field observation of wave-induced porewater pressures. Proceedings of the 23rd Conference on Coastal Engineering (ICCE), vol. 2, 2397–2410.

20. Sandven, R., Foray, P. and Long, M. (2004): Review of test methods for evaluation of liquefaction potential, with special emphasis on wave induced liquefaction and determination of air/gas content. Project Report, EU LIMAS Project EVK3-CT-2000_00038. Final review report, revision date: 20045.04.30. NTNU Geotechnical Division, Trondheim, Norway. April 2004. Electronic copy obtainable from B. Mutlu Sumer.

21. Sandven, R., Husby, E., Husby, J.E., Jønland, J., Roksvåg, K.O., Staehli, F. and Tellugen, R. (2007): Development of a sampler for measurement of gas content in soils. Journal of Waterway, Port, Coastal and Ocean Engineering, ASCE, vol. 133, No. 1, 3–13.

22. Sills, G., Wheeler, S.J., Thomas, S.D. and Gardner, T.N. (1991): Behaviour of offshore soils containing gas bubbles. Géotechnique, vol. 41, No. 2, 227–241.

23. Sumer, B.M., Sen, M.B., Karagali, I., Ceren, B., Fredsøe, J., Sottile, M., Zilioli, L. and Fuhrman, D.L. (2011): Flow and Sediment transport induced by a plunging solitary wave. Journal of Geophysical Research, vol. 116, No. C01008, doi:10.1029/2010JC006435, 1–15.

24. Svendsen, I.A. (2006): Introduction to Nearshore Hydrodynamics. World Scientific, Singapore, xxii+722 p.

25. Tørum, A. (2007): Wave-induced pore pressures–Air/gas content. Journal of Waterway, Port, Coastal and Ocean Engineering, ASCE, vol. 133, No. 1, 83–86.

26. Yamamoto, T., Koning, H.L., Sellmeijer, H. and van Hijum, E. (1978): On the response of a poro-elastic bed to water waves. Journal of Fluid Mechanics, vol. 87, No. 1, 193–206.

27. Young, Y.L., White, J.A., Xiao, H. and Borja, R.I. (2009): Liquefaction potential of coastal slopes induced by solitary waves. Acta Geotechnica, vol. 4, 17–34.

28. Zen, K. and Yamazaki, H. (1990 a): Mechanism of wave-induced liquefaction and densification in seabed. Soils and Foundations, vol. 30, No. 4, 90–104.

29. Zen, K. and Yamazaki, H. (1990 b): Oscillatory pore pressure and liquefaction in seabed induced by ocean waves. Soils and Foundations, vol. 30, No. 4, 147–161.

30. Zen, K. and Yamazaki, H. (1991): Field observation and analysis of wave-induced liquefaction in seabed. Soils and Foundations, vol. 31, No. 4, 161–179.

Chapter 5

Floatation of Buried Pipelines

Field observations show that "undrained" granular soils such as silt or fine sand can be present on the ocean floor in the loose state. Indeed, cone penetration tests reveal that the relative density can be as low as $D_r = O(0.2)$. (This may be because the top layer of the seabed, $O(1\,\mathrm{m})$ or more, is being reworked by sediment transport processes, as discussed in Example 2 under Section 3.1.4.) As is known, these types of soils are prone to liquefaction under waves due to buildup of pore-water pressure.

"Undrained", fresh, granular backfill soils in pipeline trenches also are prone to liquefaction.

Similarly, in the case when a pipeline is buried by plough or by jetting, the soil surrounding the pipe will also be in the loose state, and therefore will be vulnerable to liquefaction if the material is silt/fine sand.

The stability of pipelines buried in loose granular soils is of major concern in practice. Of particular interest is the potential for floatation of gas pipelines. The specific gravity of a gas pipeline can be as low as 1.6. When buried in a soil which is vulnerable to liquefaction, the pipeline can float to the surface of the soil simply because its density is less than that of the liquefied soil. Similar floatation problems may also arise for sea outfalls. Therefore, it is important to determine the "critical" pipeline density below which the pipeline floatation occurs. It is also equally important to determine the density of liquefied soil so that assessments could be made as to whether or not there is potential for pipeline floatation for a given set of soil, wave' and pipeline parameters.

181

There are reported incidents in the literature:

Christian *et al.* (1974) report that a 3.05 m-diameter steel pipeline in Lake Ontario (with a backfill of 2.14 m deep over the top of the pipe) has failed several times, apparently because of liquefaction, where sections of the pipeline floated to the surface of the soil during storms.

Likewise, Herbich *et al.* (1984) report that a 3.05 m-diameter pipeline, under construction, was found on the surface after a rather severe storm.

In another incident (which has led to a major dispute and where the author acted as a technical expert), a sea outfall to transport treated effluent/wastewater about 5 km offshore was raised to the surface of the bed over two stretches. The pipe diameter was 1.4 m and the pipe density was 1.34 (including concrete collars designed for the pipeline stability). The specific gravity was small because the pipe was plastic. A cutter suction dredger excavated a trench; the material was pumped to a temporary stockpile area; the pipe was laid in the trench; and subsequently the dredger re-dredged stockpile, pumped backfill to the trench, and discharged the material around the pipe. The area was exposed to a severe storm about one month after the backfill was completed, resulting in the observed failure. Here, too, all the "ingredients" for liquefaction failure were evidently present, namely a fine (mostly silty sand), fresh (loose, not compacted yet) backfill, and severe waves.

Fig. 1.4 illustrates a floatation accident, an aerial view of two pipelines which floated to the surface on a mud flat, from Damgaard *et al.* (2006). This accident happened near Fao Island in Shatt Al Arab Delta, in 1960. As described in Chapter 1, the pipeline was installed in a predredged trench by a bottom pull operation. The seabed material was described as very soft sandy silt with d_{30} of 0.02 mm. Due to waves and tidal motion, the seabed soil around the trench turned into a dense liquid that flooded the trench, and as a result the pipe floated (Damgaard *et al.*, 2006).

Apart from the reported incidents, there are also pipeline failures for which information never entered into the public domain.

Fig. 5.1 displays an image illustrating three model pipes which floated to the surface of the bed in a laboratory test where the soil was subject to wave-induced residual liquefaction (Sumer *et al.*, 2006 b). The pipes, 2 cm in diameter, were initially buried at 4.5 cm. The pipe densities were 1.6, 1.7, and 1.8. Upon the liquefaction of the soil, the initially buried pipes floated to the surface of the bed. Similar laboratory results were also reported in Sumer *et al.* (1999) and Teh *et al.* (2003 and 2006). Video 5 on the CD-ROM accompanying the present book illustrates an example where an initially

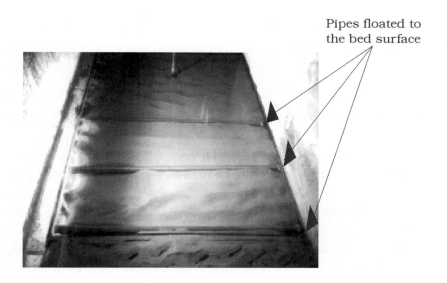

Pipes floated to
the bed surface

Figure 5.1: Pipes which floated to the surface of soil in Test 2 of Sumer *et al.*'s (2006 b) wave experiments. Viewed from onshore end of silt box.

buried pipe floats to the surface of the liquefied sediment bed.

The previously described field observations and laboratory experiments suggest that liquefaction of soil for trenched/buried pipelines must be a design condition to check for pipeline stability for floatation. This is particularly important for light pipelines.

Main pipeline design codes underline the importance of liquefaction for the stability of buried pipelines. BSI (1993, Section 4.54 and Annex B) advises that seabed liquefaction should be considered when assessing vertical stability. Although no such requirement is included in the code for wave action, it is stated that the possibility of soil liquefaction should be investigated with regard to seismic action. However, wave action is included as one of the causes of seabed liquefaction among other causes such as seismic action, tidal action, and river discharge. The DNV (1988, Section 3.4) code also states that buried pipelines should be checked for sinking/floatation in soils which may be liquefied. The code specifically states that the density of gas- or air-filled pipes should be larger than, or equal to, that of the liquefied soil.

The focus of the present chapter is the instability (floatation) of pipelines in a failed soil where the failure is due to liquefaction, with special emphasize on the critical pipe density for floatation (Section 5.2), and the density of liquefied soil (Sections 5.3 and 5.4), along with a numerical example (Section 5.5), which will be followed by Sections 5.6 and 5.7, on a new stability-design concept developed by Teh *et al.* (2006), also summarized in Damgaard *et al.* (2006), and on floatation due to momentary liquefaction, respectively.

5.1   Existing Work and Problem Statement

The pipeline floatation has been the subject of extensive research. Early research which dates back to 1949 (see ASCE Pipeline Floatation Research Council, 1966) involved the agitation of the soil in laboratory tests by different means (different from waves). Silvis (1990) reports the results of a study where liquefaction potential has been assessed for Zeepipe, a 810 km long and 1.25 m diameter gas pipeline laid from Sleipner in the North Sea to Zeebrugge at the Belgian coast. de Groot and Meijers (1992) present a study undertaken in connection with a 0.91 m gas pipeline off the Dutch coast; the main concern was to study the risk of the floatation of the pipeline placed in a trench which was filled with sand from a nearby source.

Sumer *et al.* (1999) made a laboratory study of liquefaction due to the buildup of pore-water pressure (the residual liquefaction), and the sinking/floatation of marine objects such as pipelines, spheres, and cubes in the liquefied soil. The model marine objects were free to move in the vertical direction. Displacements of the models were monitored simultaneously with the pore-water pressure measured across the soil depth. The focus of the tests was the sinking of the model objects. In one experiment, however, a model pipe with a specific gravity of $s_p = 1.0$ was tested; this pipe, upon the liquefaction of the soil, floated to surface of the bed.

Note that, for convenience, throughout this chapter (unless otherwise stated), the specific gravity, s, rather than the density, ρ, will be used to characterize the unit weight of the material (pipe or liquefied soil): for instance, s_p, the specific gravity of the pipe,

$$s_p = \frac{\gamma_p}{\gamma}. \tag{5.1}$$

in which γ is the specific weight of water, and γ_p the specific weight of the pipe,

$$\gamma_p = \rho_p g \tag{5.2}$$

in which ρ_p is the density of the pipe, and g the acceleration due to gravity; likewise, s_{liq}, the specific gravity of the liquefied soil

$$s_{liq} = \frac{\gamma_{liq}}{\gamma} \tag{5.3}$$

in which γ_{liq} is the specific weight of the liquefied soil,

$$\gamma_{liq} = \rho_{liq} g \tag{5.4}$$

in which ρ_{liq} is the density of the liquefied soil.

In a recent experimental program, Teh *et al.* (2003, 2006) did the same exercise as in Sumer *et al.* (1999), but also with model pipelines free to move in two directions, vertical and horizontal. For the cases where the pipe acted as a hydrometer (with two degrees of freedom, Fig. 5, Teh *et al.*, 2006), the test results indicated that the specific gravity of the liquefied soil was in the range $s_{liq} = 1.86 - 2.08$, s_{liq} increasing with the depth. A model pipe with $s_p = 1.5$ in Teh *et al.*'s experiments floated to the surface.

Teh *et al.* (2006) demonstrated that the conventional pipeline stability concept may be challenged if the seabed (when undrained, e.g. for silt and fine sand, the material vulnerable to liquefaction) is liquefied before the pipeline becomes unstable, and clearly, in this case, the conventional stability concept will be violated because the pipeline will sink in the seabed and therefore will become unstable before the "conventional-instability" point is reached. On this premise, Teh *et al.* (2006) proposed a method to determine a minimum specific gravity for the pipeline to become self-buried to a preferred pipe embedment when the bed becomes unstable/liquefied. They developed a model to predict the required specific gravity. The model includes the density of liquefied soil. Teh *et al.* (2006) expressed the latter as $s_{liq} = (s + e_{cr})/(1 + e_{cr})$ in which s is the specific gravity of soil grain, and e_{cr} the critical void ratio which, the authors argued, may correspond to the maximum void ratio, $e_{\max}$, at very low effective stress.

de Groot and Meijers (1992) stated that completely liquefied sand behaves like a fluid with a unit weight of $s_{liq} = 1.8$. However, they did not elaborate on the latter value. Gravesen and Fredsøe (1983) stated: "... Only a pipe with a specific gravity of approximately $s_p = 1.9$ is totally safe against liquefaction in sand/silt...". They also did not elaborate on this "critical" value of s_p, 1.9.

The specific gravity values of liquefied soils as small as $s_{liq} = 1.3 - 1.4$ have been reported in the literature (ASCE Pipeline Floatation Research

Council, 1966, Table 3, Figs. 7 and 11). The latter values are obtained for soft silty clay soil samples where, from the given values of liquid limit, it is gathered that very large void ratio values were experienced, leading to rather small values of s_{liq}.

From the preceding review, which covered the work until the mid-2000s, there appeared to be a great deal of uncertainty on the values of the critical pipe density for floatation, and on the density of liquefied soil. The questions appeared to be: (1) what is the critical floatation density of pipelines buried in a soil where the soil is undergoing wave-induced liquefaction? (2) Is the latter a constant set of values, or does it vary across the soil depth? (3) What is the density of liquefied soil? (4) What are the parameters which govern the density of liquefied soil? These questions have been addressed in a recent study by Sumer *et al.* (2006 b). The following paragraphs will mainly summarize the results obtained in Sumer *et al.* (2006 b), along with a numerical example.

5.2   Critical Density of Pipeline for Floatation

Fig. 5.2 displays two kinds of time series (taken from Sumer *et al.*, 1999): (1) the time series of the pore-water pressure (in excess of static pressure) at four depths (Fig. 5.2 a), and (2) that of the displacement of a buried pipeline (Fig. 5.2 b), recorded simultaneously with the pressure time series in Fig. 5.2 a. The specific gravity of the pipe in this test was $s_p = 1.0$. The pipe had a one-degree-of-freedom of movement, namely in the vertical direction. The pipe diameter was $D = 4$ cm. The pore pressures were measured away from the pipe, representing the far-field values. (It may be noted that the value of the relative density of the soil reported in Sumer *et al.*, 1999, is the after-the-test value; the before-the-test value was not measured in Sumer *et al.*, 1999.)

It is seen from the figure that the pipe begins to rise as soon as the soil is (nearly) liquefied, which is characterized by the fact that the pressure reaches a substantial value comparable to σ_0', the initial mean normal effective stress (see 3.1.2). See also the detailed discussion in Chapter 6 in conjunction with the process of sinking of marine objects.

Fig. 5.3 illustrates the results of the pipe stability experiments of Sumer *et al.* (2006 b). In the experiments, the soil was silt with $d_{50} = 0.078$ mm.

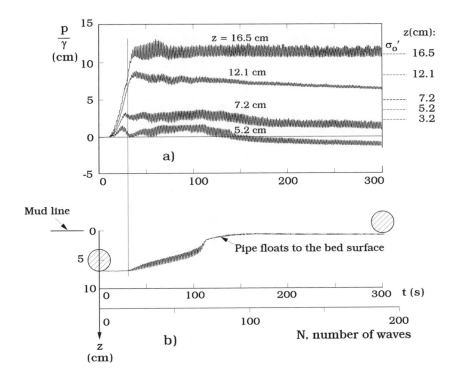

Figure 5.2:   Time series of pipe displacement (floatation) in the vertical (bottom diagram) and excess pore pressure (top diagram) recorded simultaneously in Sumer *et al.*'s (1999) experiment. H=16.6 cm, T=1.6 s, h=42 cm. Pipe diameter, D=4 cm, and specific gravity of pipe, $s_p = 1$.

Pipeline models of 2 cm diameter were used. They were buried in the soil at different depths in the range 3–15.5 cm, the total soil depth being 17.5 cm. The pipes were completely free (two degrees of freedom of movement, in contrast to the pipeline model used in Sumer *et al.*, 1999, which had one degree of freedom of movement, namely in the vertical direction). Waves (with 17 cm wave height and 1.6 s wave period, the water depth being 42 cm) were used to liquefy the soil. The quantity z on the vertical axis in Fig. 5.3 is the initial burial depth of the model pipe, measured from the mudline. Line A represents the "critical" specific gravity of the pipeline for floatation, $s_{p,cr}$. As seen from the figure, this quantity is not a constant set of values, but rather a function of the burial depth, z, with $s_{p,cr} = 1.85$ at the mudline,

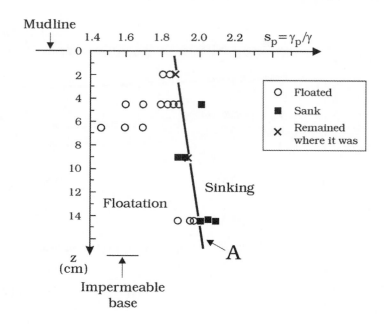

Figure 5.3:   Results of pipe floatation tests. Line A: critical curve for pipeline floatation. Sumer *et al.* (2006 b).

increasing to a value of $s_{p,cr} = 2.03$ at the impermeable base. The pipe floats when $s_p < s_{p,cr}$, and it sinks when $s_p > s_{p,cr}$.

Fig. 5.4 displays another set of data from the same study (Sumer *et al.*, 2006 b), namely the burial depth at the termination of the pipe travel (not only for floating pipes but also for sinking pipes as well). Also plotted in the figure is the critical specific gravity for floatation, $s_{p,cr}$, taken from the previous figure (Line A). From the figure, the two sets of data collapse on a single line. This implies that the pipe stops (in its travel upwards or downwards) at the depth where its specific gravity is equal to the critical specific gravity for floatation, $s_p = s_{p,cr}$.

In connection with the above-mentioned experiments, it is important to note the following. In Sumer *et al.*'s (2006 b) experiments, the pipeline motion was not recorded during floatation/sinking. Therefore, whether or not the soil was still in the liquefied regime when the pipe stopped during its travel is unknown. However, the time during which the pipe was in motion was estimated in Sumer *et al.* (2006 b), calculating the floatation/sinking velocity of the pipeline using the drag coefficient diagram given in Sumer *et al.* (1999), and subsequently calculating the travel time from the pipe's

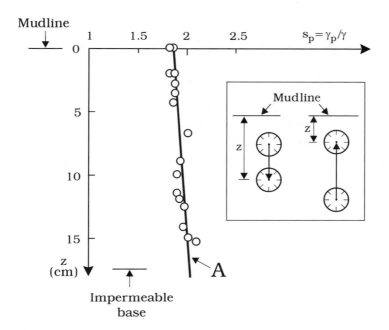

Figure 5.4:   Data corresponding to depth at which the floating/sinking pipe stopped. Line A: from the previous figure. Sumer *et al.* (2006 b).

displacement. This indicated that the soil was still in the liquefaction state (E in Fig. 3.11) when the pipe motion stopped.

5.3   Density of Liquefied Soil

In Sumer *et al.*'s (2006 b) tests presented above, the pipe can be considered to have acted as a hydrometer, the instrument to measure density of liquids. The pipe remained where it was when its density was equal to the density of the surrounding liquefied soil; or it stopped in its upward or downward motion at the point where its density was equal to that of the surrounding liquefied soil. Therefore, the specific gravity of the liquefied soil, $s_{liq}(=\gamma_{liq}/\gamma)$, could be obtained from

$$s_{liq} = s_{p,cr} \qquad\qquad (5.5)$$

where $s_{p,cr}$ is given in Fig. 5.3 or Fig. 5.4, represented by Line A. The latter data imply that the density of the liquefied soil is not constant, but increases with the depth.

The solid concentration of the liquefied soil in the liquefaction stage (E in Fig. 3.11) is, for the most part, determined by the gravitational fall out, and therefore the concentration will increase with increasing depth. This explains why the specific gravity of liquefied soil increases with increasing depth.

The density of the liquefied soil in Sumer $et\ al.$'s study (2006 b) was also determined from the force balance for the pipe (in the vertical direction) corresponding to the instant just prior to the floatation. This force balance involves two forces, the submerged weight of the pipe, and the lift force due to the upward-directed pressure gradient generated by the accumulated pore-water pressure. If the submerged weight is larger than the pressure-gradient force, the pipe will sink, if it is the opposite, it will float. Therefore the critical condition can be determined by setting the submerged weight equal to the pressure-gradient force:

$$\frac{\pi D^2}{4} L_p (\gamma_p - \gamma) = \frac{\pi D^2}{4} L_p \frac{\partial \overline{p}}{\partial z} \tag{5.6}$$

in which D is the pipe diameter, L_p the pipe length, γ_p the specific weight of the pipe, and $\overline{p}$ the accumulated pore pressure in excess of the static pressure, the overbar indicating the period-averaged value. The details of the calculations can be found in Sumer $et\ al.$ (2006 b). Although the vertical distribution of the density was not resolved, the result obtained in this way, $s_{liq} = 1.93$, is in good agreement with Fig. 5.4.

In Sumer $et\ al.$ (2006 b), the specific gravity of the liquefied soil was also calculated from

$$\gamma_{liq} = \frac{s + e_{\max}}{1 + e_{\max}} \gamma \tag{5.7}$$

(Eq. B.9, Appendix B), corresponding to the maximum void ratio $e_{\max}$ (see the discussion in Section 5.1 in conjunction with Teh $et\ al.$'s 2006 work). With this, it is essentially assumed that the void ratio in the liquefaction stage of the liquefied soil is approximately equal to the maximum void ratio. The value found from this approach was $s_{liq} = 1.89$ for $s = 2.721$ and $e_{\max} = 0.941$ (Table 1 in Sumer $et\ al.$, 2006 b). This value is slightly on the smaller side, reflecting more of the density near the surface of the bed rather than the average value.

Sumer $et\ al.$'s data, obtained from $s_{liq} = s_{p,cr}$ (where $s_{p,cr}$ is given in Fig. 5.4), is replotted in Fig. 5.5. The solid line in the figure represents this data. The depth z in the figure is normalized by d, the soil depth, because, as already mentioned, the solid concentration of the liquefied soil

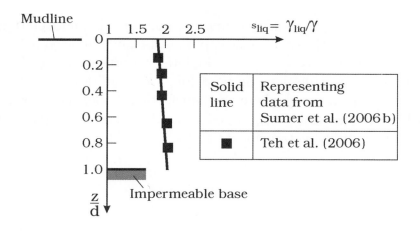

Figure 5.5: Specific gravity of liquefied soil as function of depth. Solid line (Sumer *et al.*, 2006 b, data), reproduced from the previous figure. The variation of s_{liq} with depth can be represented by the empirical expression $s_{liq} = 0.18(z/d) + 1.85$.

is largely determined by the gravitational fall out, and therefore the greater the liquefied soil depth, the smaller the variation in the solid concentration over the vertical. Hence, the vertical distance should be scaled with the soil depth d.

In Fig. 5.5, Teh *et al.*'s (2006) data is also plotted (square symbols). Of the latter data, only the data corresponding to the model pipeline with two-degrees-of-freedom-of-movement are plotted in Fig. 5.5 as this model pipeline acts as a hydrometer more realistically, although there is little difference between the results corresponding to the one-degree-of-freedom-of-movement and the two-degrees-of-freedom-of-movement models in terms of the resulting density of the liquefied soil. The test conditions corresponding to Teh *et al.* (2006) data and those of Sumer *et al.* (2006 b) are given in Table 5.1. (Note that the void ratio, e, of Teh *et al.*'s tests is actually not reported in the original publication, and therefore it is determined from the information given for the initial total soil density, and the specific gravity of soil grains. Likewise, the value of the relative density, D_r, is obtained from the determined value of e.)

Returning to Fig. 5.5, it is noteworthy to observe the striking agreement between the two sets of data, although the soil properties as well as the soil depths are substantially different (Table 5.1). This supports the argument regarding the soil depth as the scaling parameter. The data plotted in the latter figure are from the small-scale experiments, and still need to be checked against large-scale experiments and field data. Nevertheless, the plotted data can, to a first approximation, be used for field conditions to estimate the density of the liquefied soil where the liquefaction is induced by waves. The relationship between $s_{liq}(= \gamma_{liq}/\gamma)$ and z/d in Fig. 5.5 can be represented by the following empirical expression:

$$s_{liq} = 0.18\frac{z}{d} + 1.85. \tag{5.8}$$

Table 5.1.   Test conditions in Sumer *et al.* (2006 b) and Teh *et al.* (2006) data, plotted in Fig. 5.5.

Quantity:	Sumer *et al.* (2006 b)	Teh *et al.* (2006)
Soil depth, d (cm)	17.5	30
Grain size, d_{50} (mm)	0.078	0.033
Void ratio, e	0.77	0.80
Max. void ratio, $e_{\max}$	0.941	1.18
Min. void ratio, $e_{\min}$	0.499	0.39
Relative density, D_r	0.387	0.48
Initial total specific gravity of soil, γ_t/γ	1.97	1.95
Specific gravity of soil grain, γ_s/γ	2.72	2.71
Coefficient of lateral earth pressure, $k_0(= 1 - \sin\phi)$	0.40	0.41

Caution must be observed that s_{liq} is also governed by other nondimensional parameters such as the initial total specific gravity of soil, γ_t/γ, the specific gravity of soil grain, γ_s/γ, and the coefficient of lateral earth pressure, k_0, as will be demonstrated in the following mathematical modelling exercise. The diagrams that are produced from this mathematical model may be used to assess the influence of these parameters if such a "refinement" in the assessment is required.

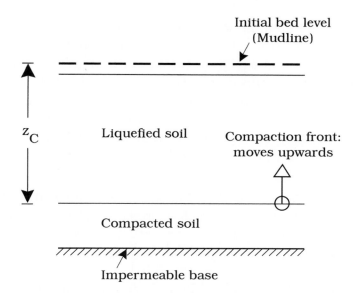

Figure 5.6: Definition sketch. Snapshot of soil with the compacted soil and the liquefied soil layers.

5.4 Mathematical Model for Density of Liquefied Soil

5.4.1 Model equations

As seen in Chapter 3, with the buildup of excess pore pressure (followed by liquefaction), an upward-directed pressure gradient will be generated. This pressure gradient will gradually drive the pore water upwards while the soil grains settle due to gravity, leading to the compaction of the soil (see Section 3.1.4). This compaction process will progressively develop in the upward direction, as illustrated in Fig. 5.6.

Now, the force balance equation for a soil grain settling in the liquefied soil (Fig. 3.38) reads

$$\frac{\pi d_{50}^3}{6}\gamma_s - \frac{\pi d_{50}^3}{6}\gamma - \frac{1}{2}\rho C_D \frac{\pi d_{50}^2}{4}w^2 - \frac{\pi d_{50}^3}{6}\frac{\partial \overline{p}}{\partial z} = 0. \tag{5.9}$$

(Note that the distance between the grains in the liquefied-soil layer in Fig. 3.38 is grossly exaggerated for illustration purposes). The first term in the

above equation represents the weight of the grain, the second term the buoyancy force, the third term the drag on the grain and the fourth term the
pressure-gradient force on the grain (an upward force in excess of the static
situation). Here, it is assumed that the grains have spherical shapes (with
the diameter equal to d_{50}). In Eq. 5.9, γ_s is the specific weight of sediment
grains, C_D the drag coefficient and w the fall velocity of grains. The drag
coefficient and the fall velocity will be influenced heavily by the very high
solid concentration, as will be detailed later.

It may be noted that the drag force is actually $\frac{1}{2}\rho C_D \frac{\pi d_{50}^2}{4}(w-V)^2$ in which
V is the vertical component of the velocity corresponding to the seepage-flow
driven upwards by the upward-directed pressure gradient. As discussed in
Sumer *et al.* (2006 b), the seepage velocity V is one order of magnitude
smaller than the fall velocity w, and therefore can be neglected for the test
conditions reported in Sumer *et al.* (2006 b).

Soil which can undergo residual liquefaction, i.e., the liquefaction caused
by the buildup of pore-water pressure as in the present application, is normally an "undrained" soil, such as very fine sand and silt. Therefore, the drag
on the grains can be assumed to be in the Stokes regime. Hence, replacing
the drag force with the Stokes-regime drag force (see e.g. Sumer and Fredsøe,
1997), the preceding equation reads

$$\frac{\pi d_{50}^3}{6}\gamma_s - \frac{\pi d_{50}^3}{6}\gamma - 6\pi\mu\left(\frac{d_{50}}{2}\right)w - \frac{\pi d_{50}^3}{6}\frac{\partial \overline{p}}{\partial z} = 0 \qquad (5.10)$$

in which μ is the viscosity. From Eq. 5.10

$$\left(\frac{\gamma_s}{\gamma} - 1\right) - \frac{\partial(\overline{p}/\gamma)}{\partial z} - \frac{18\nu w}{g d_{50}^2} = 0 \qquad (5.11)$$

in which g is the acceleration due to gravity and ν is the kinematic viscosity
coefficient.

Now, the kinematic viscosity ν and the fall velocity w are functions of
solid concentration, c, in which c is the volume concentration. As pointed
out earlier, the drag coefficient in Eq. 5.9 is influenced heavily through the
viscosity. For the kinematic viscosity, this function can be approximated by

$$\nu = \frac{2}{2 - 3c}\nu_0 \qquad (5.12)$$

in which ν_0 is the kinematic viscosity coefficient with dilute concentrations,
where the concentration does not influence the viscosity (Cheng, 1997).

As for the fall velocity, this quantity is a function of concentration and the grain Reynolds number (or alternatively the nondimensional grain size $d^* = ((s-1)gd^3/\nu^2)^{1/3}$), and can be approximated by (see e.g., Fredsøe and Deigaard, 1992, p. 200, Raudkivi, 1998, p. 19 and Cheng, 1997, Fig.1)

$$w = (1-c)^n w_0 \tag{5.13}$$

in which the exponent n is a function of the grain-size parameter, $n = n(d^*)$. Here, w_0 is the fall velocity for individual grains in water with dilute concentrations where the concentration does not influence the fall velocity. Experimental data indicates that, for small values of d^* ($d^* < 40$), the exponent $n(d^*)$ goes to a constant value. Baldock et al. (2004, Fig. 5) give $n = 4.5$ for sieve grain size $d < 0.1$ mm (for sands). In the present case, for an "undrained" soil (fine sand or silt), it can be shown that d^* is always $d^* < 40$, and therefore n in Eq. 5.14 can be taken as a constant. (Incidentally, from the preceding equation, it is seen that the fall velocity in a heavy concentration will be reduced drastically, the so-called hindered settlement.)

Now, at the onset of liquefaction, the accumulated period-averaged excess pore pressure reaches the initial mean normal effective stress (Section 3.1.2, and also see Fig. 3.11, Point B):

$$\bar{p} = \gamma' z \frac{1 + 2k_0}{3} \tag{5.14}$$

in which γ' is the submerged specific weight of soil $\gamma' = \gamma_t - \gamma$ and k_0 is the coefficient of lateral earth pressure.

For a single grain settling in water with zero solid concentration, Eq. 5.11 reduces to

$$\left(\frac{\gamma_s}{\gamma} - 1\right) - \frac{18\nu_0 w_0}{g d_{50}^2} = 0. \tag{5.15}$$

From Eqs 5.11–5.15, the following equation is obtained

$$\left(\frac{\gamma_s}{\gamma} - 1\right) - \left(\frac{\gamma_t}{\gamma} - 1\right)\frac{1 + 2k_0}{3} - \frac{2}{2 - 3c}(1-c)^n\left(\frac{\gamma_s}{\gamma} - 1\right) = 0. \tag{5.16}$$

This is the first equation of the model. The specific gravity of the liquefied soil, on the other hand, can be written in terms of the solid concentration as

$$s_{liq} = (1-c) + c\frac{\gamma_s}{\gamma}. \tag{5.17}$$

Given the quantities γ_s/γ, γ_t/γ and k_0, the solid concentration c can be calculated from Eq. 5.16, and subsequently the specific gravity of the liquefied soil can be obtained from Eq. 5.17.

The above analysis also shows that, from Eqs 5.16 and 5.17, the density of liquefied soil is governed by three nondimensional parameters, namely γ_s/γ, γ_t/γ and k_0.

5.4.2   Calibration of the model

In the literature, the reported values of the exponent n in Eq. 5.13 have been determined on the basis of sedimentation and fluidization experiments. As mentioned in the preceding subsection, Baldock et al.'s (2004) prediction (verified by the existing data) indicates that $n = 4.5$ for sieve grain size for sand $d < 0.1$ mm. However, in the present situation, the grains settle in a field with an upward-directed pressure gradient, and therefore the value $n = 4.5$ cannot be used.

Cheng (1997) demonstrated that n is also a function of the grain specific gravity, $s(= \gamma_s/\gamma)$, and found that n decreases with decreasing s. For example, for $c = 0.1$, n is reduced from 5.5 to 3.9 when s is changed from 2.65 to 1.5. It may be noted that, with decreasing values of n, the reduction in the fall velocity with respect to w_0, the fall velocity for individual grains in water with dilute concentrations, will become moderate, and, in the limiting case of neutrally buoyant grains ($s = 1$), the reduction will completely disappear, i.e., $w \to w_0$ (or alternatively $n \to 0$) as $s \to 1$. Now, the present upward-directed pressure gradient plays essentially a similar role; namely, it is as if the grains have a reduced specific gravity due to the upward-directed pressure gradient. Therefore, it is expected that n should assume values smaller than 4.5.

In the present application, n is taken as the tuning coefficient of the model, and calibrated against the present experimental data regarding the specific gravity of liquefied soil. n was tuned so that the model result for the specific gravity of liquefied soil, found from Eqs 5.16 and 5.17, match with the measured $s_{liq} = 1.94$, the depth-average value corresponding to the range $s_{liq} = 1.85 - 2.03$ measured in the experiments of Sumer et al. (2006 b) (see Section 5.3, Fig. 5.5).

Taking $\gamma_s/\gamma = 2.721$, $\nu_0 = 0.01 \, \text{cm}^2/\text{s}$, $d_{50} = 0.078$ mm, $\gamma_t/\gamma = 1.97$ and $k_0 = 0.41$, the values reported for the tests of Sumer et al. (2006 b), this calibration exercise indicated that the calculated and measured values of s_{liq} would match if the exponent n was taken as $n = 2.697$. The solid concentration found from this exercise is $c = 0.54$ and the fall velocity $w = 0.067\,\mathrm{cm/s}$.

The calibration has been checked for a number of items, and against an independent set of data, namely the data from Teh et al. (2003) (see Sumer et al., 2006 b). Teh et al. (2006) report that the average silt density was $1946\,\mathrm{kg/m^3}$ before a test. The specific gravity of soil grains was $\gamma_s/\gamma = 2.71$. The average silt specific weight (before test) was $\gamma_t = \rho_t g = 19.09\,\mathrm{kN/m^3}$, and therefore $\gamma_t/\gamma = 1.95$. For this value of the specific gravity and for the value of $k_0(= 1 - \sin\phi) = 0.41$, the model predicts $s_{liq} = 1.94$. The measured range of s_{liq} in Teh et al. (2003) is 1.86–2.08 with an average value of 1.97 (see Fig. 5.5). Hence, the value predicted by the model is 1.5% smaller than the measured average value 1.97. It may be noted that the measured range of s_{liq} in Teh et al. (2003) was erroneously reported to be 1.73–1.86 in the original publication of Sumer et al., 2006 b, which has been corrected in the above discussion.

Finally, it is to be noted that, with the value of the tuning parameter $n = 2.697$, (1) the ratio of the pressure-gradient force to the submerged weight of the grain (Eq. 5.9) is found to be 0.34, and (2) the ratio of the drag force to the submerged weight of the grain (Eq. 5.9) is found to be 0.66, indicating that both the pressure-gradient force and the drag force are significant in the process of settlement of soil grains in liquefied soil.

5.4.3   Implementation of the model

The mathematical model validated and verified by Sumer et al. (2006 b) has been further implemented to study the influence of the three governing parameters, γ_s/γ, γ_t/γ and k_0, on the specific gravity of liquefied soil.

Fig. 5.7 shows the results of the model calculations for the solid concentration as a function of the initial specific gravity of soil for two different values of k_0, namely $k_0 = 0.4$ and 0.5. Fig. 5.8, on the other hand, displays the model results for the specific gravity of liquefied soil, s_{liq}, as a function of the initial specific gravity of soil γ_t/γ for the same values of k_0. The various categories of soil with different relative densities indicated at the bottom of the diagrams correspond to a silty sand. For this type of granular soil, the maximum and minimum void ratios can be taken as $e_{\max} = 0.90$ and $e_{\min} = 0.30$ (Table B.2 in Appendix B, reproduced from Lambe and Whitman, 1969, p. 31). Hence, the specific gravity of soil γ_t/γ can be determined

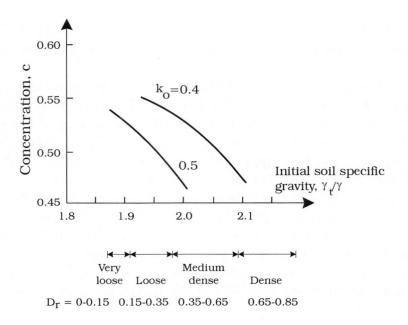

Figure 5.7:   Concentration (in volume) of solid in liquefied soil. Results of mathematical model. $\gamma_s/\gamma = 2.65$. Various soil categories added to bottom of the diagram correspond to silty sand with $e_{\max} = 0.9$ and $e_{\min} = 0.3$. Sumer $et\ al.$ (2006 b).

from the following set of equations for different values of relative density D_r:

$$\gamma_t = \frac{s+e}{1+e}\gamma, \quad \text{and} \quad D_r = \frac{e_{\max} - e}{e_{\max} - e_{\min}} \tag{5.18}$$

(Eqs B.9 and B.15, respectively).   The γ_t/γ values calculated from the preceding equations are as follows:

Soil	D_r	γ_t/γ
Very loose	0.00–0.15	1.87–1.91
Loose	0.15–0.35	1.91–1.98
Medium dense	0.35–0.65	1.98–2.09
Dense	0.65–0.85	2.09–2.19

In Figs 5.7 and 5.8, these soil categories (very loose, loose, etc.)   are indicated along the axis of γ_t/γ.

From Figs 5.7 and 5.8, the following conclusions can be drawn.

First of all, the concentration of solid grains in liquefied soil decreases with increasing γ_t/γ, the initial soil specific gravity (Fig. 5.7). This is explained as follows. The force that keeps the particles in "suspension" in the liquefied soil is the upward-directed pressure-gradient force; this force is proportional to γ'/γ, or alternatively γ_t/γ (see Eq. 5.14). The larger the value of γ_t/γ, the larger the pressure-gradient force, the more "dilute" the mixture of soil and water, and therefore the concentration should decrease with increasing γ_t/γ.

Secondly, the concentration decreases with increasing k_0 (Fig. 5.7). This also can be explained as in the preceding.

Thirdly, s_{liq} decreases with increasing γ_t/γ (Fig. 5.8). This is because the concentration decreases with increasing γ_t/γ, and therefore the specific gravity of liquefied soil should decrease with increasing γ_t/γ.

Fourthly, s_{liq} decreases with increasing k_0, Fig. 5.8, and this can be explained in the same way.

Finally, if the specific gravity of pipeline, $s_p(=\gamma_p/\gamma)$, is smaller than the specific gravity of the liquefied soil, s_{liq}, the pipeline will float; otherwise, it will sink. This is marked in the diagrams in Figs. 5.8 a and b as floatation and no floatation areas. In the case when any combination of the pipe specific gravity, s_p, and the initial soil specific gravity, γ_t/γ (or alternatively the soil category, very loose, loose, and so on), falls in the areas marked Floatation, the pipeline floatation will occur.

Fig. 5.9 displays the influence of the specific gravity of soil grains on the specific gravity of liquefied soil. Fig. 5.9 shows that the specific gravity of liquefied soil increases with an increase in the specific gravity of soil grains, as expected.

As implied by Figs 5.8 and 5.9, the specific gravity of liquefied soil may vary, depending on:

1. the soil category (i.e., whether or not the soil is a very loose soil, or a loose soil, or a medium dense soil, etc.; this feature of the soil is characterized by the initial value of the relative density D_r);

2. the soil "class" (whether or not the soil is silty sand, or micaceous sand, or silty sand and gravel, etc.; this feature of the soil may be characterized by the maximum and minimum void ratios, $e_{\max}$ and $e_{\min}$, see, e.g., Lambe and Whitman, 1969, p. 31);

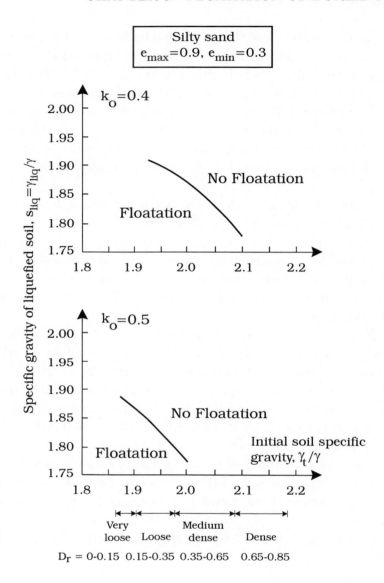

Figure 5.8: Specific gravity of liquefied soil. Results of mathematical model. $s = \gamma_s/\gamma =$ 2.65. Influence of γ_t/γ and k_0. Various soil categories added to bottom of the diagram correspond to silty sand with $e_{max} = 0.9$ and $e_{min} = 0.3$. Sumer *et al.* (2006 b).

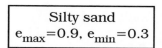

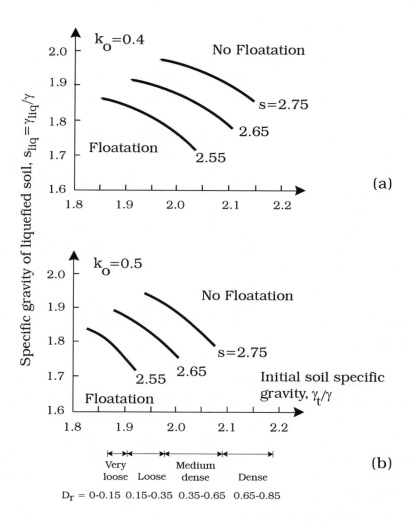

Figure 5.9:   Specific gravity of liquefied soil. Results of mathematical model. Influence of $s = \gamma_s/\gamma$. Various soil categories added to bottom of the diagram correspond to silty sand with $e_{\max} = 0.9$ and $e_{\min} = 0.3$. Sumer *et al.* (2006 b).

3. the initial soil specific gravity, γ_t/γ;

4. the specific gravity of soil grains, $s = \gamma_s/\gamma$; and

5. the coefficient of lateral earth pressure, k_0.

All these factors may, to some extent, explain why the reported values of the specific gravity of liquefied soil (or alternatively, the critical value of the pipe specific gravity for floatation) differ so widely; see the discussion in Section 5.1.

As a final remark, the present mathematical model gives a constant (an average) value for the density of liquefied soil. It does not resolve the depth variation (albeit weak), revealed by the floatation tests illustrated in Figs 5.4 and 5.5). The reason for this is that the fall velocity and the pressure gradient in Eq. 5.10 are not resolved as a function of depth, z, because no theoretical information is available to reveal these dependencies. Nevertheless, the average value predicted by the model for the specific gravity of liquefied soil would be adequate for most engineering problems.

5.5   Assessment of Time of Travel for Floating Pipe

The time of travel of a floating pipe may also be a concern in practice. No study is yet available investigating this aspect of the problem. However, the following approach may be adopted to make an estimate of this quantity.

Fig. 5.2 shows the time series of the displacement of the floating pipe in Sumer et al.'s (1999) laboratory tests discussed in Section 5.2. The figure indicates that the pipe motion reaches a steady state (in which the pipe moves with a practically constant velocity) shortly after the onset of the floatation. This continues to be the case until the pipe comes close to the surface of the bed. Clearly, the motion is not steady near the initiation and termination of the motion. Nevertheless, we may, for the most part, assume that the pipe moves in a steady state.

In this steady-state pipe motion, the forces acting on the pipe in the vertical direction should be in balance:

$$\frac{\pi D^2}{4} L_p(\gamma_{liq} - \gamma_p) = \frac{1}{2}\rho_{liq}C_D D L_p w^2 \qquad (5.19)$$

in which L_p is the length of the pipe, D the pipe diameter, γ_p the specific weight of the pipe, γ_{liq} the specific weight of the liquefied soil (i.e., the specific weight of the mixture of soil and water in the liquefied state), ρ_{liq} the density of the liquefied soil, $\rho_{liq} = \gamma_{liq}/g$, and w the pipe's velocity upwards. The term on the left-hand side of the equation is the buoyancy force on the pipe in the liquefied soil minus the pipe's weight, and the term on the right-hand side of the equation is the downward-directed resistance (drag) force. The preceding equation should not be confused with Eq. 5.6. The ambient fluid implied by Eq. 5.6 is water whereas, that implied by Eq. 5.19 is liquefied soil, and thus the pressure-gradient force, already accounted for by $\dot{\gamma}_{liq}$, will not be included in Eq. 5.19.

Eq. 5.19 may be used to determine the floatation velocity, w, using the drag coefficient data given in Section 6.3 (Fig. 6.10). Once the floatation velocity is determined, the time of travel will then be determined from $T_{float} = L_{float}/w$ in which L_{float} is the vertical distance travelled by the floating pipe. The next example illustrates this with reference to a real-life problem.

Example 1 *Floatation of a marine pipeline. A numerical example*

Given: a sea outfall (a 1.4 m outer diameter pipeline, Fig. 5.10) transporting treated effluent into the deep ocean is trenched, with the native soil pumped back into the trench after the placement of the pipeline, as sketched in Fig. 5.11. The specific gravity of the pipeline (including the concrete collars designed for the stability of the pipeline in Fig. 5.10) is $s_p = 1.34$. (The reason why the specific gravity of the pipeline is so small is that the pipeline material is plastic.)

The backfill soil is liquefied when it is exposed to storm waves. As the specific gravity of the pipe, $s_p = 1.34$, is smaller than that of the soil, taken as $s_{liq} = 1.94$, an average value from Eq. 5.8, clearly the pipe will float.

Question: find the time of travel of the pipe until the top of the pipe reaches the surface of the backfill.

From Eq. 5.19, the pipe velocity

$$w = \left(\frac{\pi D g}{2 C_D} \frac{s_{liq} - s_p}{s_{liq}} \right)^{1/2} \tag{5.20}$$

and taking $C_D = 10^7$, an asymptotic value corresponding to large Reynolds numbers (see Section 6.3 and Fig. 6.10), the pipe velocity is found to be $w =$

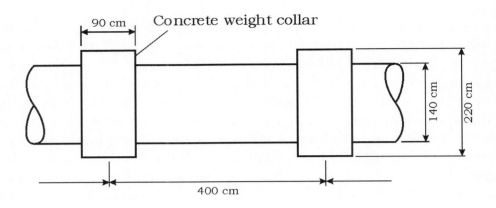

Figure 5.10:   Pipeline considered in the numerical example.

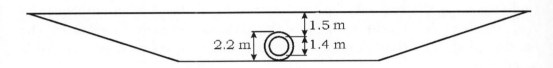

Figure 5.11:   Pipeline placed in the trench in the numerical example.

0.82 mm/s, and the time of travel for the top of the pipe to reach the surface of the backfill (Fig. 5.11) is found to be $T_f = L/w = (1.5 \text{ m})/(0.82 \times 10^{-3}$ m/s) $= 1,836 \text{ s}$, or approximately half an hour.

It is important to note that the above calculation assumes that the soil remains in the liquefaction state over the entire period of the floatation time, 1,836 s. Assuming that the pipe begins to move upwards when the liquefaction front reaches the bottom of the trench, then the question is whether or not the floatation velocity is larger than the velocity of the compaction front (see Chapter 3, Section 3.4, and the numerical example, Example 9, given under Section 3.4). If this is so, i.e., $w > U_c$, then the soil will remain in the liquefaction state during the pipe's upward motion, and therefore the above calculation will be valid. If the compaction front in its upward motion "overtakes" the pipe, then the pipe's upward travel will obviously come to an end at that point.

Finally, we note that the specific gravity of the pipe in the above example is $s_p = 1.34$, a very small value. Clearly, the pipe specific gravity should be large enough to provide stability. If there is a potential for liquefaction of the backfill, then the specific gravity of the pipe should certainly be larger than that of the liquefied soil to avoid floatation.

5.6 Stability Design

Teh *et al.* (2006), also summarized in Damgaard *et al.* (2006), demonstrated that the conventional pipeline stability concept may be challenged if the seabed is liquefied before the pipeline becomes unstable. Clearly, in this case, the conventional stability concept (which is based on the force balance between the agitating- and resisting-friction forces) is no longer valid, as the seabed has become already unstable. On this premise, Teh *et al.* (2006) proposed a method to determine a minimum specific gravity for the pipeline to become self-buried to a preferred pipe embedment when the bed becomes unstable/liquefied. They developed a model to predict the required specific gravity. This model is based on the equilibrium of vertical forces on the pipe, these forces being the weight of the pipe, and the buoyancy force associated with both the liquefied soil and the water. The required specific weight of the pipe is given by

$$s_p = 1 + \left[\frac{\psi}{2\pi} + \left(\frac{2e}{\pi D} - \frac{1}{\pi} \right) \sin \frac{\psi}{2} \right] (s_{liq} - 1) \tag{5.21}$$

in which ψ is

$$\psi = 2 \arccos \left(1 - \frac{2e}{D} \right) \tag{5.22}$$

in which e is the embedment of the pipe. This relationship is shown in Fig. 5.12. Note that, for a design embedment of $e \geqslant D$, we get the simple relationship $s_p = s_{liq}$ (e.g., Eq. 5.5).

Eq. 5.21 and the design chart in Fig. 5.12 are based on the assumption of full liquefaction and, as a result, zero shear strength. If this is not the case, then it is necessary to add a term characterizing the bearing capacity of the soil:

$$s_p = 1 + \left[\frac{\psi}{2\pi} + \left(\frac{2e}{\pi D} - \frac{1}{\pi} \right) \sin \frac{\psi}{2} \right] (s_{liq} - 1) + 4N \frac{S_u}{\pi \gamma D} \tag{5.23}$$

where S_u is the undrained shear strength and N the Brich Hansen bearing capacity factor (Damgaard *et al.*, 2006). For cohesionless soil, S_u is given as $\sigma' \tan \varphi$, where σ' is the effective stress (which will be influenced by the excess

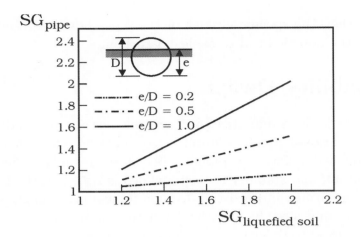

Figure 5.12:   Specific gravity of a pipeline as function of specific gravity of liquefied soil and pipe embedment. Teh *et al.* (2006), also summarized in Damgaard *et al.* (2006).

pore pressure) and φ the angle of shearing resistance, which can be obtained from triaxial or direct shear tests. N, on the other hand, is tabulated in most standard geotechnical codes and textbooks.

Regarding the density of liquefied soil, Teh *et al.* (2006) expressed it as

$$s_{liq} = \frac{s + e_{cr}}{1 + e_{cr}}$$

in which s is the specific gravity of soil grain, and e_{cr} the critical void ratio. They stated that e_{cr} may correspond to the maximum void ratio at very low effective stress, as already pointed out in Section 5.1. Assuming that $e_{cr} = e_{max}$, they predicted the specific gravity of the soil for a set of experimental data, and obtained a good agreement between the observed embedment ratio e/D vs. s_p, and the predicted one, predicted from Eq. 5.21.

5.7   Floatation due to Momentary Liquefaction

In the previously mentioned instabilities, the liquefaction failure of the soil is due to the buildup of pore pressure, the residual liquefaction. The liquefaction failure of the soil may also occur due to the presence of an upward-directed pressure gradient during the passage of the wave trough, i.e., the momentary liquefaction (see Chapter 4).

Only a few studies are available, investigating the floatation/sinking of pipelines/marine objects under the momentary liquefaction. This may be due to the fact that it is difficult to produce the momentary liquefaction in a laboratory-sized wave flume, using a normal sand bed, as pointed out by Sakai *et al.* (1992).

Maeno, Magda and Nago (1999) studied the floatation of a pipeline in the laboratory where the effect of a progressive wave was simulated by an oscillating water table. A model pipeline buried in the soil floated to the surface of the soil, presumably due to the lift force on the pipeline induced by the oscillating water table. The lift force on the pipeline (and therefore the upward displacement of the pipe) was found to be associated with the "trough" half period of the motion of the water table (Maeno *et al.*, 1999, Fig. 4). With the repeated passage of the wave trough, the pipe is lifted each time, and as a consequence, some silt/sand fall under it. When the crest passes, the pipe rises in the silt/sand, and is slightly higher in position than before. With this, the pipe eventually floats to the surface, what might be called the jacking effect.

It may be noted that, in the study of Maeno *et al.* (1999) (also see Sakai *et al.*, 1992), the degree of saturation plays a significant role; even a slight reduction in the degree of saturation (a reduction from the value of unity) may lead to the momentary liquefaction (Sakai *et al.*, 1992, also discussed in Chapter 4, Section 4.3), or to a considerable uplift on the pipe (Maeno *et al.*, 1999).

The following practice may be recommended to make assessments about floatation of pipelines:

1. Use the simple formula given in Eq. 4.10 with Eqs 4.11–4.13 as a first screening to assess whether there is a potential for momentary liquefaction. See Example 10 in Chapter 4.

2. If it turns out that there is a potential, then conduct a detailed analysis, using the finite-depth solution of Hsu and Jeng (1994), following the procedure outlined in Section 4.3.2. See Example 11 in Chapter 4.

3. The above exercise will give the depth of the soil layer in which the momentary liquefaction occurs, z_ℓ.

4. If this depth is below the pipeline, then the pipe will float if its density is smaller than the density of the liquefied soil.

5.8 References

1. ASCE Pipeline Floatation Research Council (1966): ASCE preliminary research on pipeline floatation. J. Pipeline Division, ASCE, vol. 92, No. PL1, 27–71.

2. Baldock, T.E., Tomkins, M.R., Nielsen, P. and Hughes, M.G. (2004): Settling velocity of sediments at high concentrations. Coastal Engineering, vol. 51, No. 1, 91–100.

3. BSI (1993): British Standards Institute, Code of Practice for pipelines-Part 3. Pipelines subsea: Design, construction and installation. BS8010, London, UK.

4. Cheng, N.-S. (1997): Effect of concentration on settling velocity of sediment particles. Journal Hydraulic Engineering, ASCE, vol. 123, No. 8, 728–731.

5. Christian, J.T., Taylor, P.K., Yen, J.K.C. and Erali D.R. (1974): Large diameter underwater pipeline for nuclear plant designed against soil liquefaction. Offshore Technology Conference, May 6–8, 1974, Houston, TX, OTC 2094, 597–606.

6. Damgaard, J., Sumer, B.M., Teh, T.C., Palmer, A.C., Foray, P. and Osorio, D. (2006): Guidelines for pipeline on-bottom stability on liquefied noncohesive seabeds. Journal of Waterway, Port, Coastal and Ocean Engineering, ASCE, vol. 132, No. 4, 300–309.

7. de Groot, M.B. and Meijers, P. (1992): Liquefaction of trench fill around a pipeline in the seabed. BOSS 92: Behaviour of Offshore Structures, London, 1333–1344.

8. DNV (1988): Det Norske Veritas. On-bottom stability design of submarine pipelines. DNV RP E305.

9. Fredsøe, J. and Deigaard, R. (1992): Mechanics of Coastal Sediment Transport, World Scientific, Singapore, 369 p.

10. Gravesen, H. and Fredsøe, J. (1983): Modelling of liquefaction, scour and natural backfilling processes in relation to marine pipelines. Offshore Oil and Gas Pipeline Technology, 1983 European Seminar, 2–3. February, 1983, Copenhagen, Denmark.

11. Herbich, J.B., Schiller, R.E., Dunlap, W.A. and Watanabe, R.K. (1984): Seafloor Scour, Design Guidelines for Ocean-Founded Structures, Marcel Dekker, Inc., New York and Basel.

12. Hsu, J.R.S. and Jeng, D.S. (1994): Wave-induced soil response in an unsaturated anisotropic seabed of finite thickness. International Journal for Numerical and Analytical Methods in Geomechanics, vol. 18, No. 11, 785–807.

13. Lambe T.W. and Whitman, R.V. (1969): Soil Mechanics. John Wiley and Sons, Inc., New York, 553 p.

14. Maeno, S., Magda, W. and Nago, H. (1999): Floatation of buried pipeline under cyclic loading of water pressure. Proceedings of the 9th International Offshore and Polar Engineering Conference and Exhibition, Brest, France, May 30–June 4, 1999, vol. II, 217–225.

15. Raudkivi, A.J. (1998): Loose Boundary Hydraulics. A.A. Balkema, Rotterdam/Brookfield, 496 p.

16. Sakai, T., Hatanaka, K. and Mase, H. (1992): Wave-induced effective stress in seabed and its momentary liquefaction. Journal of Waterway, Port, Coastal and Ocean Engineering, ASCE, vol. 118, No. 2, 202–206. See also Discussions and Closure in vol. 119, No. 6, 692–697.

17. Silvis, F. (1990): Wave induced liquefaction of seabed below pipeline. The 4th Young Geotechnical Engineers' Conference, Delft, The Netherlands, 18–22 June 1990.

18. Sumer, B.M. and Fredsøe, J. (1997): Hydrodynamics Around Cylindrical Structures, World Scientific, Singapore, 530 p. Second edition 2006.

19. Sumer, B.M., Fredsøe, J., Christensen, S. and Lind, M.T. (1999): Sinking/floatation of pipelines and other objects in liquefied soil under waves. Coastal Engineering, vol. 38, No. 2, 53–90.

20. Sumer, B.M., Hatipoglu, F., Fredsøe, J. and Sumer, S.K. (2006 a): The sequence of soil behaviour during wave-induced liquefaction. Sedimentology, vol. 53, 611–629.

21. Sumer, B.M., Hatipoglu, F., Fredsøe, J. and Hansen, N.-E. O. (2006 b): Critical floatation density of pipelines in soils liquefied by waves and density of liquefied soils. Journal of Waterway, Port, Coastal and Ocean Engineering, ASCE, vol. 132, No. 4, 252–265.

22. Teh, T.C., Palmer, A. and Damgaard, J. (2003): Experimental study of marine pipelines on unstable and liquefied seabed. Coastal Engineering, vol. 50, No. 1–2, 1–17.

23. Teh, T.C., Palmer, A., Bolton, M.D. and Damgaard J. (2006): Stability of submarine pipelines on liquefied seabeds. Journal Waterway, Port, Coastal and Ocean Engineering, ASCE, vol. 132, No. 4, 244–251.

Chapter 6

Sinking of Pipelines and Marine Objects

As mentioned previously, with the soil liquefied, buried pipelines may float to the surface of the seabed, pipelines laid on the seabed may sink (self-burial of pipelines), large individual blocks (like those used for scour protection) may penetrate into the seabed, sea mines may enter into the seabed and eventually disappear, and caisson structures may burrow into the seabed. Floatation of pipelines in a liquefied soil has been studied in the previous chapter. This chapter focuses on sinking of pipelines, and marine objects such as individual blocks (stones, etc.) in a liquefied seabed.

Incidents of this nature have been quoted in the literature. For example, Dunlap, Bryant, Williams and Suheyda (1979) report storm-induced pore pressures in soft, clayey sediments in the Mississippi Delta where sinking of several of the measuring instruments up to 1.83–4.27 m was noted.

Sinking of pipelines is a concern in practice, as large longitudinal forces can be induced by the pipe deflection (Brown, 1975, Herbich, 1981). Herbich (1981) notes, however, that these stresses are unimportant unless generated near the pipe riser.

Miyamoto, Yoshinaga, Soga, Shimizu, Kawamata, and Sato (1989) report the subsidence of offshore breakwaters composed of concrete blocks at Niagata Coast, Japan (see also Goda, 1994).

Likewise, there have been many unreported cases where structures have suffered considerable damages as a consequence of liquefaction failure of the soil and the resulting sinking.

In this chapter, we shall first describe the process of sinking of marine objects in liquefied soils, including the physics governing the termination of the sinking motion. The latter is important to determine the depth at which the sinking motion comes to an end, the sinking depth. Subsequently we shall turn our attention to the drag on a marine object moving in a liquefied soil. This analysis will enable us to develop a simple viscous-fluid model of the liquefied soil, a model which can eventually be used to estimate the time of sinking. The chapter will end with an analysis of the case of momentary liquefaction.

6.1 Description of the Process

Sumer *et al.* (1999) investigated the sinking/floatation of pipelines and marine objects (cube- and spherical-shaped) in a liquefied soil. In a later study, Sumer *et al.* (2010) studied sinking of cover stones. In both studies, the liquefaction was due to the buildup of pore pressure. Similar experiments were carried out by Teh *et al.* (2003 and 2006). The following description is mainly based on Sumer *et al.* (1999 and 2010).

Fig. 6.1 displays two kinds of time series (Sumer *et al.*, 1999); the top ones are for the pore-water pressure (in excess of static pressure) at several depths in the soil (Fig. 6.1a), while the bottom one is the displacement of a model pipeline initially buried in the bed, recorded simultaneously with the pressure time series (Fig. 6.1b). The specific gravity of the pipe $s_p(= \gamma_p/\gamma)$ is $s_p = 3.1$. The pipe diameter is $D = 4$ cm. The pore-water pressure was measured in the test away from the pipe, thus representing the far-field values. Fig. 6.2 illustrates the close-up picture of the initial stage of the process in Fig. 6.1. The following two observations can be made from Fig. 6.1 and Fig. 6.2: (1) the pipe begins to sink as soon as the soil is (nearly) liquefied (Figs. 6.1 and 6.2); and (2) sinking of the pipe stops not at the impermeable base, but somewhat earlier (Fig. 6.1).

A close examination of Fig. 6.2 indicates that the pipe begins to sink in the soil before the pore pressure reaches the value of the initial mean normal effective stress, σ_0'. However, for the onset of sinking, the accumulated pore pressure apparently needs to reach a substantial value. For example, $\bar{p}$ reaches a value of about 70% of σ_0' for $z = 16.5$ cm, about 60% of σ_0' for $z = 12.8$ cm, and about 40% of σ_0' for $z = 7.2$ cm in the test presented in Fig. 6.2.

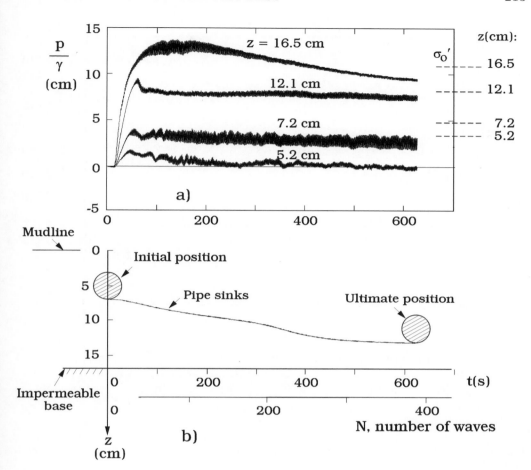

Figure 6.1: Time series of pipe displacement in the vertical and excess pore pressure recorded simultaneously. Pipe specific gravity, $s_p = 3.1$. Sumer *et al.* (1999). (It may be noted that the value of the relative density of the soil reported in Sumer *et al.*, 1999, is the after-the-test value; the before-the-test value was not measured in Sumer *et al.*, 1999.)

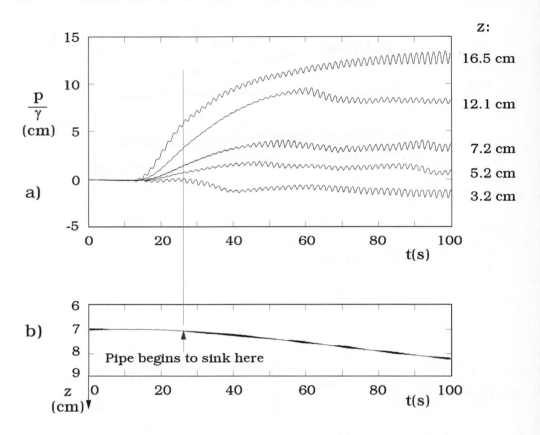

Figure 6.2:   Close-up of the time series depicted in the previous figure. Sumer *et al.* (1999).

The reason why the pipe begins to sink before the pore pressure reaches the value of σ_0', i.e., before the soil is fully liquefied, may be due partly to the considerable reduction of the soil strength caused by a high degree of pore-pressure buildup in the soil, and partly to the additional accumulation of the pore pressure in the neighbourhood of the pipe (Sumer *et al.*, 2006 b), which may presumably lead to the soil liquefaction earlier around the pipe than in the undisturbed-flow situation. Sumer *et al.* (1999) note, however, that, in six tests out of the total 14 in their study, $\overline{p}$ at $z = 16.5\,\text{cm}$ and $12.8\,\text{cm}$ reached the σ_0' value at the time when the pipe began to sink, and that the latter behaviour did not reveal any correlation with the parameters such as the wave height, the initial pipe position, the specific gravity of the pipe, and the soil depth.

Similar experiments were carried out by Sumer *et al.* (1999) with various wave heights, both in the liquefaction regime and in the no-liquefaction regime. No displacement of the pipe was observed in the case of the no-liquefaction regime. In one no-liquefaction test ($H = 9.1$ cm, $D = 4$ cm, and e_0, the burial depth from the mudline to the pipe bottom $= 7$ cm), the pipe was replaced with a much heavier pipe ($s_p = 8.9$) with the same outcome, namely no sinking.

The experiments with sphere- and cube-shaped bodies in Sumer *et al.* (1999), and with cover stones in Sumer *et al.* (2010), showed precisely the same behaviour as that described in the preceding paragraphs.

6.2 Termination of Sinking

6.2.1 Related physics

As seen from the above description, the sinking of the object terminates at a certain depth, but not necessarily at the impermeable base (Fig. 6.1b). The question is: what is the mechanism which controls the termination of the downward motion of the sinking object? The experimental constraints in Sumer *et al.* (1999) concealed this feature of the process, and therefore no explanation was given in Sumer *et al.* (1999). This question has been addressed in Sumer *et al.* (2010), where sinking of cover stones (initially sitting on the bed) was studied (Fig. 6.3), as mentioned previously. The following description is based on Sumer *et al.* (2010).

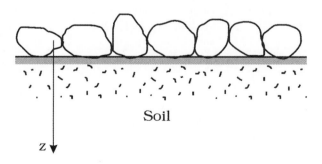

Figure 6.3: Cover stones on a liquefiable seabed, studied in Sumer *et al.* (2010).

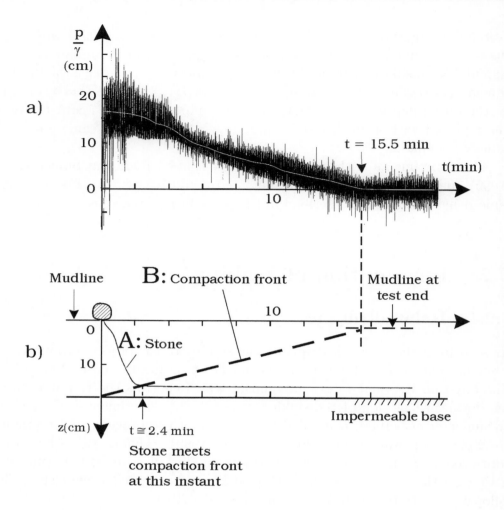

Figure 6.4:   (a): Time series of excess pore pressure at $z = 16$ cm. (b): Time series of the downward displacement of a single stone cover layer, marked A; and the time series of the compaction front, marked B, Test 3 in Sumer *et al.* (2010).

In one of the tests of Sumer *et al.* (2010), the downward displacement of an arbitrarily selected stone of the cover, one-layer deep and made up of 4 cm stones, was recorded. The selected stone was mounted to a frame which was free to move in the vertical direction. The displacement of the stone was measured. The set-up was the same as that used in Sumer *et al.* (1999). Fig. 6.4 gives three time series obtained in this test:

1. The time series of the pore-water pressure (in excess of hydrostatic pressure) at $z = 16$ cm (Fig. 6.4a);

2. The time series of the vertical displacement of the monitored stone (A in Fig. 6.4b), cf. this time series and that in Fig. 6.1; and

3. The time series of the compaction front (B in Fig. 6.4b). The time series B in Fig. 6.4b represents the upward progression of the compaction front.

A close inspection of Fig. 6.4b indicates that the monitored stone on its way downwards in the liquefied soil practically stops when it meets the compaction front which is on its way up. The figure indicates that this occurs at time $t = 2.4$ min.

Inferring from Sumer *et al.*'s (2010) cover stone experiments, the mechanism whereby the sinking of a marine object, for example, a pipeline, comes to a complete stop can be described as follows (Fig. 6.5). The object begins to sink in the soil as soon as the liquefaction sets in (Fig. 6.5, Panel 1). Upon the arrival of the liquefaction front at the impermeable base, the compaction process begins at the impermeable base, and it gradually progresses in the upward direction (Fig. 6.5, Panel 1), see Section 3.1 for the description of the entire liquefaction and compaction process. The downward motion of the object in the liquefied soil eventually terminates when the object meets the compaction front (Fig. 6.5, Panel 2). The object will eventually be stopped completely as the compaction front continues to progress upwards (Fig. 6.5, Panel 3).

We note that the description given in Teh *et al.* (2006) in connection with their pipeline experiments is in complete accord with the above description.

Hydrometer case. At this juncture, we consider the variation of the specific weight of the liquefied soil with the depth, discussed in Chapter 5, Section 5.3, and recall that the specific weight of the liquefied soil increases with the depth (Fig. 5.5). Now, the above discussion covers the case where the specific weight of the marine object is larger than that of the liquefied soil at the impermeable base, i.e., where $s_p > s_{liq}(z = d) = 2.03$ in which

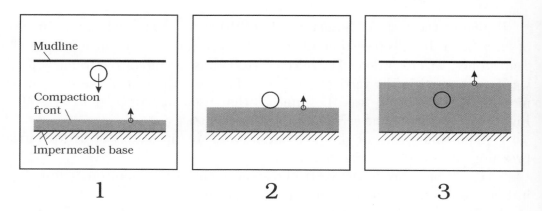

Figure 6.5:   Sequence of sketches illustrating the process of sinking of an object (a pipe in the present illustration). Panel 1: the pipe begins to sink in the liquefied soil. Panel 2: the pipe's downward motion terminates when the pipe meets the compaction front, which is travelling upwards. Panel 3: the pipe is fully stopped in the compacted soil, as the compaction front continues to move upwards.

$s_{liq}(z = d)$ is the specific gravity of the liquefied soil at the impermeable base, and s_p is the specific gravity of the marine object.

If the specific weight of the marine object, s_p, is such that $s_{liq}(z = 0) < s_p < s_{liq}(z = d)$, with $s_{liq}(z = 0)$ being the specific gravity of the liquefied soil at the mudline (namely 1.85 for $z = 0$, Eq. 5.8), the marine object, when released from a depth below the mudline, will begin to sink to a depth where the specific gravity of the liquefied soil is the same as that of the object, and, as a result, its sinking motion will come to a stop, as described in Section 5.2, the hydrometer case. Clearly, this mechanism is different from that described in the previous subsection.

For this mechanism to be valid, the sinking time in this case, T_s, should be smaller than (or equal to) the time associated with the progression of the compaction front. The latter time is the sum of two times, $T_1 + T_2$, in which T_1 is the time for the liquefaction front to travel from the initial location of the object to the impermeable base, and T_2 is that for the compaction front to travel from the impermeable base to the depth where it meets the object. See Section 3.1 for a complete description of the liquefaction and compaction process.

Another scenario where the termination of the sinking motion of the marine object is controlled by the "hydrometer mechanism" involves structures sitting on the seabed. As the structure begins to sink upon the onset of liquefaction, the bottom portion of the structure will enter into the liquefied seabed. As the structure moves downwards, larger and larger portions of the structure will be buried in the liquefied soil. This will obviously affect the buoyancy force on the structure, as the liquefied soil has a density of a factor of two larger than that of water, resulting in progressive reduction of the specific weight of the structure with the sinking. The sinking motion of the structure will come to a stop when the specific weight of the structure becomes equal to the specific weight of the liquefied soil. To illustrate this, a numerical example is worked out for a caisson structure initially sitting on the seabed (Example 14, later in the chapter).

The following example discusses rock covers (over trenched pipelines), and therefore relates the preceding subsection to an interesting real-life project. The reader is also referred to Section 3.6 in this context.

Example 13. *Over-dumping of rock in the form of a rock berm. The Tangguh Project.*

The Tangguh LNG Project in the Papua region of Indonesia features 24-inch (60 cm) buried pipelines which operate at elevated temperatures. The design solution adopted to mitigate upheaval buckling was to trench the pipelines, then backfill and bury the pipelines with quarried rock, since the required download to prevent upheaval buckling substantially exceeded the self weight of the pipe. The required rock cover height varied along the pipelines depending on the surveyed curvature, with various berm configurations depending on the seabed soil properties. Due to strong tidal currents, the project was concerned that mobile sand could be washed back into the trench, forming an intervening layer between the pipe and the quarried rock. Much of the routes therefore featured multi-layer filter/armour berms. The sand had a $d_{50} = 0.5$ mm, with filter rock $d_{50} = 20$ mm and armour rock $d_{50} = 85$ mm; further details including particle size distributions are included in the references Griffiths *et al.* (2011), and Marcollo *et al.* (2011).

The project area is seismically active, and in the event of seabed liquefaction, there is the risk of the cover above the pipeline settling into the liquefied seabed (in the manner described in Sections 6.1 and 6.2). This would reduce the uplift resistance and potentially result in an upheaval buckle. Since the pipeline was prone to upheaval buckling, the typical approach of requiring the pipe specific gravity to be equal to the liquefied soil was ineffective. The project considered a range of potential approaches to mitigating this risk, considering the benefits and disadvantages of each. The project eventually chose to adopt an increase on the required rock cover height to reduce the likelihood of the pipe buckling even if a seabed liquefaction event occurred and some rock settlement resulted. The project also chose to reduce the allowable weld defect sizes to reduce the likelihood of pipeline rupture in the event of an upheaval buckle. Of the implemented measures, the over-dumping of rock in the form of a rock berm across the entire width of the trench is particularly interesting from the point of view of the behaviour of rock cover over a liquefiable soil, as discussed in the previous paragraphs. See also the discussion in Griffiths *et al.* (2007).

The reader is referred to Griffiths *et al.* (2007, 2011), and Marcollo *et al.* (2011) for further information regarding the mitigation measures adopted in the project.

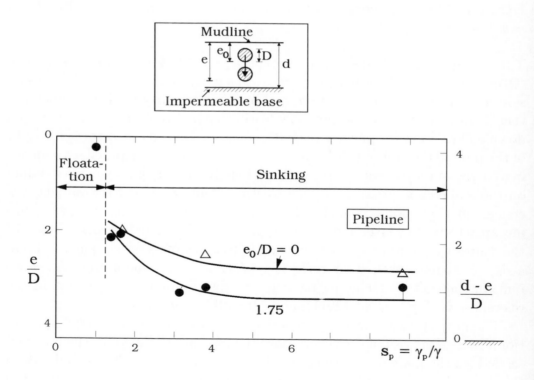

Figure 6.6:   Ultimate sinking/floatation depth of pipeline. Sumer *et al.* (1999).

6.2.2 The depth of sinking

The depth of sinking will be further elaborated in this sub-section *by reference to the pipeline case* studied in Sumer *et al.* (1999). (It may be noted that the value of the relative density of the soil reported in Sumer *et al.*, 1999, is the after-the-test value; the before-the-test value was not measured in Sumer *et al.*, 1999.)

From dimensional considerations, the depth of sinking of the pipe normalized by the pipe diameter, e/D (see Fig. 6.6 for definition sketch), depends on the following quantities:

1. The pipe properties

$$\frac{e_0}{D}, s_p, \frac{d}{D} \tag{6.1}$$

in which D is the pipe diameter, e_0 is the initial pipe position, s_p is the specific gravity of the pipe and d is the soil depth. (The third parameter in Eq. 6.1, d/D, appears through the effect of the presence of the pipe itself on the pore-pressure accumulation).

2. The quantities which govern the pore-pressure accumulation, given in Eq. 3.69.

Of these parameters, the first three parameters are of particular interest. Each parameter is now considered individually.

Effect of initial position e_0/D

Fig. 6.6 displays the results of the tests carried out with two different initial pipe positions; in one, the pipe was initially sitting on the bed $(e_0/D = 0)$, and in the other, it was completely buried in the soil with $e_0/D = 1.75$. The ultimate sinking position of the pipe is plotted against the specific gravity of the pipe.

Regarding the case of the bottom-seated pipe $(e_0/D = 0)$, scour will occur initially, as described in Sumer and Fredsøe (2002, Chapter 2). Sumer *et al.*'s (1999) analysis indicated that the pipe would sink to a depth of $O(1 \text{ cm})$ (or alternatively, $e/D = O(0.3)$) due to scour, before it sank due to liquefaction. Comparing the estimated sinking $e/D = O(0.3)$ with the data in Fig. 6.6, namely $e/D = 2 - 2.5$, indicates that the sinking in the tests of Sumer *et al.* (1999) is mainly governed by liquefaction.

Fig. 6.6 clearly shows that a pipe initially sitting on the bed sinks to a relatively shallower depth than a pipe initially buried in the soil. This is linked to the fact that the pipe initially sitting on the bed meets the compaction front somewhat later than the initially buried pipe (Fig. 6.5), resulting in a relatively shallower sinking depth.

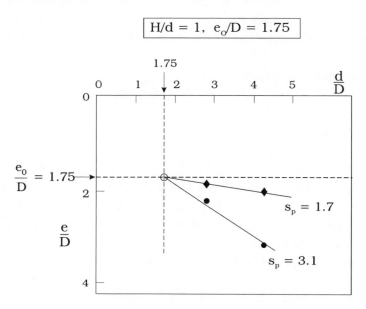

Figure 6.7:   Ultimate sinking depth of pipeline. Sumer *et al.* (1999).

Effect of pipe's specific gravity s_p

Fig. 6.6 shows that, for $e_0/D = 0$, the depth of sinking increases with increasing s_p. Fig. 6.7 also clearly shows the same trend, this time for $e_0/D = 1.75$. (It may be noted, however, that this trend is not captured for the $e_0/D = 1.75$ curve in Fig. 6.6 for very large values of s_p.) Nevertheless, from Figs 6.6 and 6.7, the general trend is that the depth of sinking increases with increasing s_p. This is obviously linked to the previously described mechanism, namely heavier pipes can travel larger distances before they meet the compaction front, and therefore their sinking depths will be larger.

Effect of soil depth d/D

Fig. 6.7 presents the results where the effect of the soil depth has been investigated. As seen, the soil depth is also an influencing parameter. The larger the soil depth, the larger the sinking depth. This behaviour is also not unexpected.

This is directly linked to the time required for the sinking pipe and the compaction front to meet. This time, T_s, is actually the sum of two times, $T_s = T_1 + T_2$, in which T_1 is the time for the liquefaction front to travel from the mudline to the impermeable base, and T_2 is that for the compaction front (that forms immediately after the liquefaction front reaches the impermeable base) to travel to the level where it meets the pipe (Section 3.1). Clearly, the larger the soil depth, the larger the time T_s, and therefore the sinking depth should be larger, as revealed by Fig. 6.7.

It is interesting to note that, in addition to the parameters in Eq 6.1, the effect of the wave height has also been investigated in Sumer *et al.* (1999), to observe if sinking would be different in a liquefied soil with different wave heights for a given water depth. The results showed that this effect is practically nil. This is because once the soil is liquefied, the pipe will not be able to differentiate whether the wave height is large or small. While the accumulated pore pressure increases with increasing wave height for the no-liquefaction regime, it remains constant, at $p_{\max} = (\gamma_{liq} - \gamma)z$, for the liquefaction regime (see Fig. 3.5 and the discussion in Section 3.1.2).

The above discussion is obviously for the case where the liquefaction occurs across the entire soil depth. If the liquefaction occurs down to a certain depth, then the wave height will have to affect the end results because the liquefaction depth will increase with increasing wave height.

Sphere- and cube-shaped bodies

Sumer *et al.* (1999) also studied the sinking of sphere- and cube-shaped bodies. (These objects may be considered to represent armour blocks for practical applications.) The test set-up was the same as that in the study of pipeline described in the previous sub-sections. Some supplementary tests were also made with cube-shaped bodies to see the shape effect. The study indicated that the process of sinking of these 3-D objects is basically the same as in the case of pipelines described in the previous section. The following paragraphs will summarize the highlights of these tests.

Sphere-shaped body. Fig. 6.8 shows the sinking depth as a function of the specific gravity of the sphere for an initially buried sphere where $e_0/D = 1.4$. The variation of e/D with the specific gravity s_p is quite similar to that given for the pipeline case (Fig. 6.6); e/D increases with s_p. However, as opposed

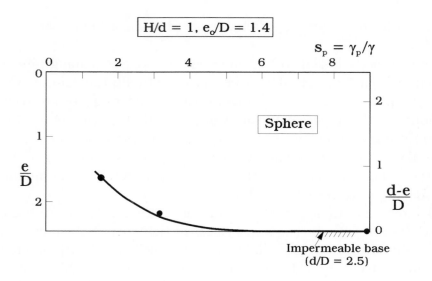

Figure 6.8:   Ultimate sinking depth of sphere in Sumer *et al.*'s (1999) experiments.

to the case of the pipeline, the sinking of the sphere for $s_p = 8.9$ stops at the impermeable base. This is linked to the very large sinking velocity of the $s_p = 8.9$ sphere, about 0.2 cm/s (in contrast to about $0.\ 03$ cm/s in the pipeline case).

Fig. 6.9 illustrates the effect of d/D on the depth of sinking (cf. Fig. 6.7, the pipeline case). The way in which the sinking depth varies with d/D can be interpreted in the same fashion as in Fig. 6.7.

Cube-shaped body. Two tests in Sumer *et al.*'s (1999) study were carried out with a cube the size $D = 5.5$ cm, its volume being equal to that of the sphere with $D = 6.9$ cm in the experiments, to observe the influence of the actual shape of the sinking object. The cube was initially buried at a depth of about $e_0/D = 1.4$. In one test $(s_p = 3.2)$, the sinking depth e was measured to be $e = 16$ cm (cf., $e = 15$ cm for the sphere with $s_p = 3.2$ for the same initial burial depth e_0/D), while in the other test $(s_p = 8.9)$, $e = 16.5$ cm (cf., $e = 17$ cm for the sphere with $s_p = 8.9$, again, for the same initial burial depth). As seen, given the volume of the sinking object, the actual shape of the object apparently does not have any dramatic influence on e.

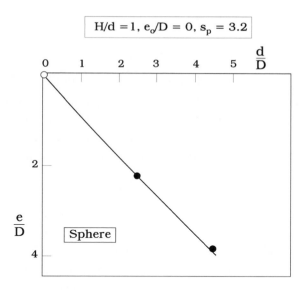

Figure 6.9:   Ultimate sinking depth of sphere in Sumer *et al.*'s (1999) experiments.

6.3   Drag on a Sinking Object

Pipelines

Fig. 6.1 shows that the pipe's downward motion reaches a steady state (in which the pipe moves with a practically constant velocity) shortly after the onset of motion. This continues until the pipe comes close to its ultimate position. In this steady-state pipe motion, obviously the forces acting on the pipe in the vertical direction should be in balance:

$$\frac{\pi D^2}{4} L_p (\gamma_p - \gamma_{liq}) = \frac{1}{2} \rho_{liq} C_D D L_p w^2 \qquad (6.2)$$

in which L_p is the length of the pipe, D the pipe diameter, γ_p the specific weight of the pipe, γ_{liq} the specific weight of the liquefied soil (i.e., the specific weight of the mixture of soil and water in the liquefied state), ρ_{liq} the density of the liquefied soil, $\rho_{liq} = \gamma_{liq}/g$, and w the fall (or the sinking) velocity of the pipe. The term on the left-hand side of the equation is the submerged weight of the pipe in the liquefied soil. The term on the right-hand side of the equation is the resistance (drag) force, which may be considered as a kind of viscous drag:

$$F_D = \frac{1}{2}\rho_{liq}C_D DL_p w^2. \tag{6.3}$$

Obviously the process of resisting force in a liquefied soil may be different from that in an ordinary fluid. Nevertheless, making an analogy to the steady motion of a body in a fluid (see, e.g., Sumer and Fredsøe, 1997), the resistance force can, to a first approximation, be written as that in Eq. 6.2 where C_D is the drag coefficient. It is expected that C_D is governed primarily by the Reynolds number, defined by

$$Re_p = \frac{wD}{\nu_{liq}} \tag{6.4}$$

in which ν_{liq} = the kinematic viscosity of the liquefied soil. Clearly, ν_{liq} is different from the water viscosity, ν. The viscosity of the liquefied soil ν_{liq} can be expressed as $\nu_{liq}/\nu = f$, in which f is a function of solid concentration (Happel and Brenner 1973, p. 453).

Now, from Eq. 6.2, the drag coefficient is

$$C_D = \frac{\pi}{2}\frac{g(s_p - s_{liq})D}{w^2 s_{liq}} \tag{6.5}$$

in which $s_p(= \gamma_p/\gamma)$ and $s_{liq}(= \gamma_{liq}/\gamma)$ are the specific gravities of the pipe and the liquefied soil, respectively.

Sumer et al. (1999) calculated the drag coefficient, using the fall velocity data they obtained in their experiments. However, in their calculation they assumed $s_{liq} = 1.3$, an incorrect value. This was due to the lack of knowledge/data on s_{liq} at the time of their study. Sumer et al.'s (1999) data have been recast with s_{liq} taken as 1.94 (a depth-averaged value; see Eq. 5.8, Section 5.3), and plotted in Fig. 6.10 against Re_p. The Reynolds number is taken as

$$Re_p = \frac{wD}{\nu} \tag{6.6}$$

to avoid uncertainties and also unnecessary complication related to the kinematic viscosity ν_{liq}. This is not a problem, however, since ν_{liq} in a real-life situation would not be radically different to that in the experiments in Fig. 6.10. Hence, ν_{liq} would not be a scaling parameter.

First of all, Fig. 6.10 shows that C_D decreases with increasing Re_p, a behaviour quite similar to that of fluid drag. However, the drag coefficient

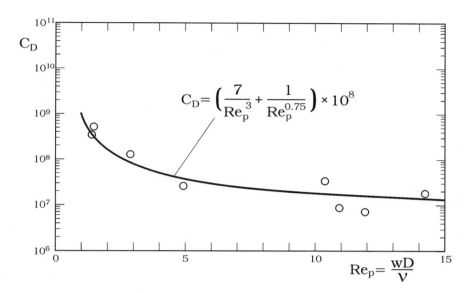

Figure 6.10: Drag coefficient for a pipeline in liquefied soil. ν is the kinematic viscosity coefficient for water. ν is taken as the kinematic viscosity coefficient of water rather than that of the liquefied soil for convenience.

in the present case is a multiple orders of magnitude larger than the ordinary fluid drag (Schlichting, 1979, p. 17). The variations of C_D with Re_p in Fig. 6.10 can be represented by the following empirical expression

$$C_D = \left(\frac{7}{Re_p^3} + \frac{1}{Re_p^{0.75}} \right) \times 10^8 \quad \text{for } 1 < Re_p < 15. \tag{6.7}$$

This equation can be used to estimate the sinking (or floatation) velocity w of a pipe in a liquefied soil. The latter information may prove useful in making assessments about the time scale of the sinking (or floatation) process. Caution must be exercised, however, when the preceding equation is extrapolated to larger Re_p numbers.

Viscosity of liquefied soil

A backward calculation where C_D is assumed to be given by the Lamb–Oseen relation (see e.g., Sumer and Fredsøe, 1997, p.219) (the creeping flow)

$$C_D = \frac{8\pi}{Re_p \ln(7.4/Re_p)} \tag{6.8}$$

with $Re_p = wD/\nu_{liq}$, gives the kinematic viscosity of the liquefied soil as $\nu_{liq} = 150$ m^2/s, a value which is eight orders of magnitude larger than the water value, i.e., 10^{-6} m^2/s at 20^0. From the above value of the kinematic viscosity, the dynamic viscosity of the liquefied soil can be obtained as follows

$$\mu_{liq} = \nu_{liq}\rho_{liq} = \nu_{liq}\frac{\gamma_{liq}}{g} = \nu_{liq}\frac{s_{liq}\gamma}{g} = 150 \times \frac{1.94 \times 9.81}{9.81} \simeq 300 \text{ kN.s/m}^2.$$

(6.9)

At this juncture, we note the following. Observing that the liquefied soil displays properties of a heavy viscous fluid, and indicating that such a model of liquefied soil was already suggested in Sumer and Fredsøe (2002, pp. 496–498) — an earlier version of the above analysis, Sawicki and Mierczynski (2009) carried out experiments with the purpose of determining the dynamic viscosity of liquefied soils. They did two kinds of experiments. In the first experiment, they conducted cyclic undrained triaxial tests on a sand the size 0.42 mm with a uniformity coefficient of 2.5. The cyclic shearing caused the pore-water pressure to build up, leading to liquefaction. By measuring the velocity of the top of the sample during what they called deformation, and from the known shearing, and assuming that the constitutive equation of the liquefied soil satisfies that of a Newtonian fluid, they obtained a value for the dynamic viscosity of $\mu_{liq} \simeq 760$ kN.s/m^2. In the second experiment, Sawicki and Mierczynski (2009) monitored the sinking of a vertical circular cylinder in the same sand placed in a container which was in turn placed on a shaking table. With the shaking of the table, the soil was liquefied and the cylinder began to sink in the liquefied soil. By measuring the sinking velocity, and from a relation similar to that used in Sumer $et\ al.$ (1999), the dynamic viscosity of the liquefied soil was found to be $\mu_{liq} \simeq 530$ kN.s/m^2. As seen, these values are not radically different from that found in Eq. 6.9. This provides confidence in the use of the previously obtained values of the viscosity of the liquefied soil, as well as the drag-coefficient data in Fig. 6.10.

It is to be noted, however, that viscosity of liquefied soil in connection with earthquake-induced liquefaction has been determined in many experiments. Hwang, Kim, Chung and Kim (2006) compiled data, including their own, and presented it as a function of the relative density. The data showed a very large scatter in which the viscosity of the liquefied soil varied from $\mu_{liq} \simeq 0.001$ kN.s/m^2 to $\mu_{liq} \simeq 10$ kN.s/m^2 although most of the data lies in a range of $\mu_{liq} \simeq 0.1 - 10$ kN.s/m^2, a range much below the range indicated in the preceding paragraphs. The data plotted in Hwang $et\ al.$ (2006) seem to indicate a trend in which the viscosity increases with increasing relative density, which is not unexpected.

The understanding of the viscous behaviour of liquefied soil is, at the present time, very poor. The process of flow of liquefied soil (undisturbed, or in the presence of a structure, a pipe, or a pile) is largely unknown. The stress–strain rate relationship, the constitutive equation, of liquefied soils is essentially unknown although there are studies which indicate that lique-fied soils behave like a pseudoplastic fluid, a non-Newtonian fluid, quoted in Hwang *et al.* (2006). The subject will not be pursued further as it is beyond the scope of the present treatment.

Sphere- and cube-shaped bodies

The data regarding the drag on a sphere and a cube in Sumer *et al.* (1999), recast and plotted in the same way as in Fig. 6.10, show that the drag coefficient defined by

$$F_D = \frac{1}{2}\rho_{liq}C_D A w^2 \tag{6.10}$$

is two orders of magnitude smaller than the case of the pipeline. Here, F_D is the drag force, and A is the cross-sectional area of the object projected on a horizontal plane, namely $A = \frac{\pi D^2}{4}$ for a sphere and $A = D^2$ for a cube, D being the diameter of the sphere, or the size of the cube. The data for the sphere case (very limited, though, only seven data points) can be represented by the following empirical equation

$$C_D = \frac{1.5 \times 10^6}{Re_b^{0.3}} \quad \text{for } 30 < Re_b < 150 \tag{6.11}$$

in which Re_b is defined as in Eq. 6.6 with D being the diameter of the sphere.

As regards the cube case, although it is, at best, suggestive (only two data points with $Re_b = 50$ and 210 in which Re_b is defined as in Eq. 6.6, with D being the size of the cube), practically no significant difference has been observed between C_D data for the sphere and that for the cube. Again, as in the case of the pipeline, caution must be exercised when the above equation is extrapolated to large Reynolds numbers.

Irregular-shape blocks

Kirca (2013) studied sinking of irregular-shape blocks, using stones with nominal spherical diameter of $D_n = 5.39$ cm and 8.64 cm in which

$$D_n = \left(\frac{6}{\pi}V_b\right)^{1/3} \tag{6.12}$$

in which V_b is the volume of the stone. Kirca (2013) also carried out reference experiments with cube- and sphere-shaped bodies with the cube size in the range $D = 6.5-8.7$ cm and the sphere diameter in the range $D = 4.0-7.1$ cm. The soil used was silt the size $d_{50} = 0.070$ mm. The soil was liquefied by waves with the wave height $H = 7.7$ cm and the wave period $T = 1.09$ s at a water depth of $h = 30$ cm.

Kirca (2013) measured the sinking of the test bodies, similar to Sumer et al. (1999). Based on the latter data, Kirca (2013) obtained the drag coefficient, and plotted it against the Reynolds number, Re_b, defined as in Eq. 6.6 with D being the nominal spherical diameter of the stone, D_n. In the case of spheres and cubes, he adopted the same definition of the Reynolds number as in the previous subsection.

Kirca's (2013) drag coefficient data for all three test bodies (including the stones) collapse on a single curve, indicating that there is no significant shape effect. This supports the conclusion in the previous sub-section that no significant difference has been observed between the sphere data and the cube data.

Kirca's (2013) results agree well with the expression given in Eq. 6.11. Kirca's sphere and cube data extend over a much broader range of the Reynolds number than in Eq. 6.11, namely with $1 < Re_b < 200$. Kirca (2013) gives the following empirical expression, representing his overall data (including the stone data):

$$C_D = \frac{4 \times 10^7}{Re_b} \qquad \text{for } 1 < Re_b < 200. \tag{6.13}$$

Example 14. *Sinking of a caisson. A numerical example.*

Given: we consider the concrete caisson depicted in Fig. 6.11 sitting on the seabed. The caisson dimensions are 41.7 m long (in the direction to the

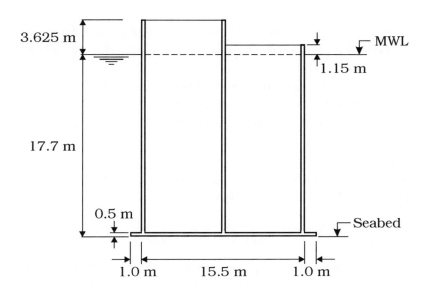

Figure 6.11:   Caisson studied in the numerical example.

picture plane) and 15.5 m wide with an extension of 1 m outside at the base
on each side, giving a total width of 17.5 m. It contains about 6,120 t of
concrete and about 244 t rebar. The height of the caisson at the front side
is 21.325 m, and 18.85 m at the back. The chambers inside the caisson are
filled with ballast water to the top of the walls for stability.

The seabed is liquefied by large waves during a storm. Find (1) the
displacement (sinkage) of the caisson in the seabed; and (2) the sinking time.

We will achieve this in three steps. (We note, however, the following. We
assume that the caisson moves in the liquefied soil with only one degree of
freedom of movement, namely in the vertical direction. It is important to
note that this obviously does not represent a possible scenario of what may
happen, but rather it represents an academic problem through which we can
make an order of magnitude estimate of the sinking of the caisson.)

1) Submerged weight of the caisson

As the caisson moves downwards, the buoyancy force on the caisson will
change because of the following two independent effects:

1. The caisson initially acts as an emerged (surface piercing) structure.
However, with its sinking, the rear part of the caisson will be submerged

first in the water, and subsequently the front part will also be submerged. Clearly, whether it is emerged, or submerged (partially or fully) will affect the buoyancy force on the caisson.

2. As the caisson begins to sink, the bottom portion of the caisson will enter the liquefied soil. As the caisson moves downwards, a larger and larger portion of the caisson will be buried in the liquefied soil. This will also affect the buoyancy force on the caisson as the liquefied soil has a density of a factor of two larger than the water density.

Both effects are included in the calculation of the buoyancy force on the caisson. The total force on the caisson, i.e., the submerged weight, is then calculated by

$$F(z) = W - B(z) \tag{6.14}$$

in which z is the depth of sinking (or sinkage), W is the weight of the caisson (including the ballast water), taken as $W = 1.65 \times 10^5$ kN, and B is the buoyancy force. In the calculations, the specific gravity of the liquefied soil is taken as 1.85, or alternatively the specific weight $\gamma_{liq} = 18.1$ kN/m^3. The values of the submerged weight from Eq. 6.14 are calculated, and tabulated as a function of z in Table 6.1.

Table 6.1.   Submerged weight of the caisson as function of the sinkage, z.

Sinkage, z (m)	Submerged weight of the caisson, $F(z)$ (kN)
0.0	5.2×10^4
2.0	3.1×10^4
4.0	1.5×10^4
6.0	0.4×10^4
6.8	0.0

2) Depth of sinking and sinking velocity of the caisson

Table 6.1 shows that the submerged weight of the caisson, W, is balanced by the buoyancy force, B, when the caisson sinks to a depth of $z = 6.8$ m. When the caisson reaches that level, it will remain there, acting like a "hydrometer".

During the quasi-steady downward movement of the caisson, the submerged weight of the caisson, which is $F(z) = W - B(z)$, is balanced by the drag force:

$$F(z) = \frac{1}{2}\rho_{liq}C_D D L w^2 \tag{6.15}$$

in which C_D is the drag coefficient, L is the length of the caisson, 41.7 m, D is the width of the caisson, taken as 15.5 m, and w is the downward (or sinking) velocity of the caisson. The present case resembles the pipe case (rather than the cube case) studied above, as the caisson has a much larger size in the third dimension, similar to the pipe. Therefore, the pipe drag data is, to a first approximation, used in the present estimate, with a value of C_D as $C_D = 10^7$, an asymptotic value corresponding to large Reynolds numbers (Fig. 6.10).

With the above values, the sinking velocity can be obtained from Eq. 6.15, as a function of the sinkage z (Table 6.2).

Table 6.2.   Sinking velocity of the caisson as function of the sinkage.

Sinkage, z (m)	Sinking velocity, w (mm/s)
0.0	3.0
2.0	2.3
4.0	1.6
6.0	0.9
6.8	0.0

3) Sinking time of the caisson

As seen, the sinking velocity is not constant, and it decreases as the caisson moves downwards, and therefore the time of sinking will also change with the caisson displacement/sinkage. This time is calculated from

$$T_s = \int_0^z \frac{dz}{w(z)}. \tag{6.16}$$

The results are summarized in Table 6.3.

Table 6.3.   The time of caisson sinking as function of the sinkage.

Sinkage, z (m)	Time of sinking, T_s (min)
0.0	0.0
2.0	13.8
4.0	32.4
6.0	61.5
6.8	89.9

It is important to note that the above calculations assume that the soil remains in the liquefaction state over the entire period of 89.9 minutes, and therefore the compaction front has not yet reached to the depth $z = 6.8$ m (cf., the mechanism related to the termination of the sinking of blocks, pipelines, etc., discussed in Section 6.2.1 above.)

Finally, we note that Sawicki and Mierczynski (2009) have adopted a similar approach, stressing that this approach can be used in engineering calculations for sinking of various structures such as buildings, breakwaters, piers, pipelines, and other heavy objects in liquefied soils with the liquefaction induced by earthquakes or severe storms.

6.4   The Case of Momentary Liquefaction

Pipelines and short pipes

No study is yet available investigating the sinking/floatation of pipelines under momentary-liquefaction regime. This may be due to the fact that it is difficult to produce the momentary liquefaction in a small-sized wave flume using a normal sand bed, as pointed out by Sakai *et al.* (1992) (see Section 4.3.1). However, Maeno *et al.* (1999) have studied the floatation of a pipeline in the laboratory where the effect of a progressive wave was simulated by an oscillating water table. A detailed review of this work is given in Chapter 5, Section 5.7.

Regarding the short pipes, Chowdury, Dasari and Nogami (2006) studied the sinking of these types of objects. They used a cylindrical column (with a diameter of 0.8 m), to get around the problem associated with small-scale wave flume facilities, mentioned in the preceding paragraph.

The tests were designed to mimic the unsaturated state of the seabed, leading to momentary liquefaction, with two short cylinders, one with a diameter of 5 cm and a length of 20 cm, and with a specific gravity of 1.05, and the second one with a diameter of 10 cm and a length of 20 cm, and with a specific gravity of 2.40. The authors note that these objects represent typical sea mines in the conducted model tests. In all the tests, the sand ($d_{50} = 0.25$ mm) bed depth was 1.4 m while the water depth was 1.1 m. The authors note that the soil would be unsaturated with normal water, which normally contains air bubbles. They determined the degree of saturation by fitting their pressure distribution to that obtained by theoretical solution, in much the same way as in the study of Tørum (2007), discussed in Section 4.5.

With a load (similar to that created by a wave with a 10 s wave period and a 4 m wave height at a water depth of 8 m), the soil underwent momentary liquefaction, and the heavy pipe sank while the light pipe remained on the seabed. The sinking depth of the heavy pipe was as much as 0.92 m.

Chowdury *et al.* (2006) also studied the effect of changing wave loading, and also the displacement pattern during "loading" and "unloading" phases during one wave period.

Armour blocks

Sakai, Gotoh and Yamamoto (1994) report an extensive series of laboratory experiments where the block subsidence due partly to the momentary liquefaction and partly to the oscillatory flow action has been investigated.

Maeno and Nago (1988) present the results of an experimental study where the effect of a progressive wave is simulated by an oscillating water table. A concrete, rectangular-prism-shaped block sitting initially on the surface of the soil gradually sank, as the oscillating movement of the water table continued. A similar approach was also adopted by Zen and Yamazaki (1990) to observe sinking of a rectangular-prism-shaped heavy object under momentary-liquefaction conditions.

Gratiot and Mory (2000), from their study on momentary liquefaction referred to earlier in Section 4.3.1, draw conclusions in relation to sea mine burials.

Although the problem of sinking/subsidence of armour blocks, sea mines etc. in soils subject to momentary liquefaction has been recognized widely, no study is yet available investigating this problem in a systematic manner under laboratory/field conditions.

Similar to pipeline floatation (Section 5.7), the following practice may be recommended to make assessments about the burial/sinking depth of marine objects:

1. Use the simple formula given in Eq. 4.10 with Eqs 4.11–4.13 as a first screening to assess whether there is a potential for momentary liquefaction. See Example 10 in Chapter 4.

2. If it turns out that there is a potential, then conduct a detailed analysis, using the finite-depth solution of Hsu and Jeng (1994), following the procedure outlined in Section 4.3.2. See Example 11 in Chapter 4.

3. The above exercise will give the depth of the soil layer in which the momentary liquefaction occurs, z_ℓ.

4. This depth can, to a first approximation, be taken as the sinking depth of the marine object (pipelines, short pipes, armour blocks, stones, sea mines, etc.).

6.5   References

1. Brown, R.J. (1975): How to protect offshore pipelines. Pipeline Industry, vol. 42, No. 3, 43–47.

2. Chowdury, B., Dasari, G.R. and Nogami, T. (2006): Laboratory study of liquefaction due to wave-seabed interaction. Journal of Geotechnical and Geoenvironmental Engineering, ASCE, vol. 132, No. 7, 842–851.

3. Dunlap, W., Bryant, W.R., Williams, G.N. and Suheyda, J.N. (1979): Storm wave effects on deltaic sediments — Results of SEASWAB I and II. Port and Ocean Engineering Under Arctic Conditions (POAC 79), Norwegian Institute of Technology, 899–920.

4. Goda, Y. (1994): A plea for engineering-minded research efforts in harbor and coastal engineering. International Conference on Hydro-Technical Engineering for Port and Harbor Construction, Hydro-Port '94, October 19–21, 1994, Yokosuka, Japan, 1–21.

5. Gratiot, N. and Mory, M. (2000): Wave induced sea bed liquefaction with application to mine burial. Proceedings of the Tenth International

Offshore and Polar Conference, ISOPE-2000, May 28–June 2, 2000, Seattle, WA, vol. 2.

6. Griffiths, T., O'Brien, D. and Johnson, R. (2007): Risk Mitigation Options for UHB Design of Pipelines in Seismic Risk Areas. Proceedings of the 30th annual Offshore Pipeline Technology Conference & Exhibition, Amsterdam, 2007.

7. Griffiths, T., Marcollo, H., Johnson, R. and Mariatmo, D. (2011): Tangguh LNG Project Analysis of Full Scale Statistical Uplift Testing with Multiple Soil Layers. OMAE 2011-49153, Proceedings of the ASME 2011 30th International Conference on Ocean, Offshore and Arctic Engineering, Rotterdam, 2011.

8. Happel, J. and Brenner, H. (1973): Low Reynolds Number Hydrodynamics. Second Edition, Noordhoff, Leyden, The Netherlands.

9. Herbich, J.B. (1981): Offshore Pipeline Design Elements. Marcel Dekker, Inc., New York and Basel.

10. Hsu, J.R.S. and Jeng, D.S. (1994): Wave-induced soil response in an unsaturated anisotropic seabed of finite thickness. International Journal for Numerical and Analytical Methods in Geomechanics, vol. 18, No. 11, 785–807.

11. Hwang, J.I., Kim, C.Y., Chung, C.K. and Kim, M.M. (2006): Viscous fluid characteristics of liquefied soils and behavior of piles subjected to flow of liquefied soils. Soil Dynamics and Earthquake Engineering, vol. 26, No. 2–4, 313–323.

12. Kirca, V.S.O. (2013): Sinking of irregular shape blocks into marine seabed under wave-induced liquefaction. Coastal Engineering, vol. 75, 40–51.

13. Maeno, S. and Nago, H. (1988): Settlement of a concrete block into a sand bed under water pressure variation. Modelling Soil-Water-Structure Interactions. Kolkman *et al.* (eds.), 67–76, Balkema, Rotterdam, The Netherlands.

14. Maeno, S., Magda, W. and Nago, H. (1999): Floatation of buried pipeline under cyclic loading of water pressure. Proceedings of the 9th

International Offshore and Polar Engineering Conference and Exhibition, Brest, France, May 30–June 4, 1999, vol. II, 217–225.

15. Marcollo, H., Griffiths, T., Johnson, R. and Abbs, T. (2011): Development of a Full Scale Two Dimensional Pipeline Uplift Test Apparatus for Upheaval Buckling (UHB) Model Testing on the Tangguh LNG Project. OMAE2011-49152, Proceedings of the ASME 2011 30th International Conference on Ocean, Offshore and Arctic Engineering, Rotterdam, 2011.

16. Miyamoto, T., Yoshinaga, S., Soga, F., Shimizu, K., Kawamata, K. and Sato, M. (1989): Seismic prospecting method applied to the detection of offshore breakwater units settling in the seabed. Coastal Engineering in Japan, vol. 32, No. 1, 103–112.

17. Sakai, T., Hatanaka, K. and Mase, H. (1992): Wave-induced effective stress in seabed and its momentary liquefaction. Journal of Waterway, Port, Coastal and Ocean Engineering, ASCE, vol. 118, No. 2, 202–206. See also Discussions and Closure in vol. 119, No. 6, 692–697.

18. Sakai, T., Gotoh, H. and Yamamoto, T. (1994): Block subsidence under pressure and flow. Proceedings of the 24th Conference on Coastal Engineering (ICCE 94), 23–28 October, 1994, Kobe, Japan, 1541–1552.

19. Sawicki, A. and Mierczynski, J. (2009): On the behaviour of liquefied soil. Computers and Geotechnics, vol. 36, No. 4, 531–536.

20. Schlichting H. (1979): Boundary-Layer Theory. McGraw-Hill, New York.

21. Sumer, B.M. and Fredsøe, J. (1997): Hydrodynamics Around Cylindrical Structures, World Scientific, Singapore, 530 p. Second edition 2006.

22. Sumer, B.M. and Fredsøe, J. (2002): The Mechanics of Scour in the Marine Environment. World Scientific, Singapore, 552 p.

23. Sumer, B.M., Dixen, F.H. and Fredsøe, J. (2010): Cover stones on liquefiable soil bed under waves. Coastal Engineering, vol. 57, No. 9, 864–873.

24. Sumer, B.M., Fredsøe, J., Christensen, S. and Lind, M.T. (1999): Sinking/Floatation of pipelines and other objects in liquefied soil under waves. Coastal Engineering, vol. 38, No. 2, 53–90.

25. Sumer, B.M., Hatipoglu, F., Fredsøe, J. and Sumer, S.K. (2006 a): The sequence of soil behaviour during wave-induced liquefaction. Sedimentology, vol. 53, 611–629.

26. Sumer, B.M., Truelsen, C. and Fredsøe, J. (2006 b): Liquefaction around pipelines under waves. Journal of Waterway, Port, Coastal and Ocean Engineering, ASCE, vol. 132, No. 4, pp. 266–275.

27. Teh, T.C., Palmer, A. and Damgaard, J. (2003): Experimental study of marine pipelines on unstable and liquefied seabed. Coastal Engineering, vol. 50, No. 1–2, 1–17.

28. Teh, T.C., Palmer, A., Bolton, M.D. and Damgaard J. (2006): Stability of submarine pipelines on liquefied seabeds. Journal of Waterway, Port, Coastal and Ocean Engineering, ASCE, vol. 132, No. 4, 244–251.

29. Tørum, A. (2007): Wave-induced pore pressures–Air/gas content. Journal of Waterway, Port, Coastal and Ocean Engineering, ASCE, vol. 133, No. 1, 83–86.

30. Zen, K. and Yamazaki, H. (1990): Mechanism of wave-induced liquefaction and densification in seabed. Soils and Foundations, vol. 30, No. 4, 90–104.

Chapter 7

Liquefaction Under Standing Waves

A detailed account of seabed liquefaction under progressive waves has been given in the previous chapters. The seabed liquefaction is also a concern in standing-wave situations, e.g., areas in front of breakwaters, seawalls, and large caisson structures are all subject to standing waves.

The present chapter will study the seabed liquefaction in standing waves. Although a great many works have been devoted to the investigation of seabed liquefaction under progressive waves (as discussed in greater detail in Chapters 3–6), there have been very few investigations of seabed liquefaction under standing waves, Sekiguchi, Kita and Okamoto (1995), Sassa and Sekiguchi, 1999 and 2001 and Kirca, Sumer and Fredsøe (2012 and 2013).

The description of the liquefaction process given in the next section is mainly based on Kirca *et al.* (2013). (Earlier results of the latter work appeared in Kirca *et al.*, 2012.) As will be seen, the process of residual liquefaction of the seabed under standing waves has many similarities to that occurring in progressive waves (Kirca *et al.*, 2013). This will lead us to develop a mathematical model, which is essentially based on that given for progressive waves (Chapter 3, Section 3.2). Although the focus in the present chapter is the residual liquefaction under standing waves, a brief account of momentary liquefaction will also be included.

7.1 Residual Liquefaction Under Standing Waves

7.1.1 General description

Consider a vertical, rigid wall (Fig. 2.5), subjected to a progressive wave (the incident wave). As the incident wave impinges on the wall, a reflected wave moves in the offshore direction. As discussed in Section 2.2.2, the superposition of these two waves results in a standing wave. The surface elevation of this standing wave, measured from the mean water level, is given by

$$\eta(x,t) = \frac{(2H_i)}{2} \cos(\lambda x) \cos(\omega t) \tag{7.1}$$

$$\omega^2 = gk \tanh(kh) \tag{7.2}$$

in which ω is the angular frequency ($\omega = 2\pi/T$), λ is the wave number ($\lambda = 2\pi/L$), T is the wave period, L is the wave length, and h is the water depth. The quantity $2H_i$ is the height of the standing wave, and is twice the wave height (H_i) of each of the two progressive waves (i.e., the incident wave and the reflected wave) forming the standing wave (Dean and Dalrymple, 1984, Chapter 4). For convenience, the height of the standing wave, $2H_i$, in the analysis throughout this chapter will be denoted by H, the wave height measured from crest to trough of the standing wave.

This wave field is certainly different from that of a progressive wave, and therefore the process of residual liquefaction will be different from that caused by a progressive wave. This section will describe the process of liquefaction in standing waves, based on Kirca et al.'s (2013) study.

Kirca et al. (2013) did their experiments in a wave flume. The soil was placed in a 40 cm deep, 60 cm wide and 78 cm long pit (Fig. 7.1), made from the same material as the glass side walls of the flume. In the tests, the pit was filled with the soil so that the surface of the soil was flush with the bottom of the flume. Although the focus was liquefaction under standing waves, some experiments were also conducted with progressive waves to facilitate comparison.

The standing waves were obtained by a fully reflecting, vertical, plywood plate, Plate P, in the flume placed at the onshore end of the soil pit (Fig. 7.1). The plate was sealed at the edges to ensure the full reflection. The experiments were designed so that one half of the wave length, $L/2$, was equal to the length of the soil pit, 78 cm, as indicated in Fig. 7.1, in which L is the

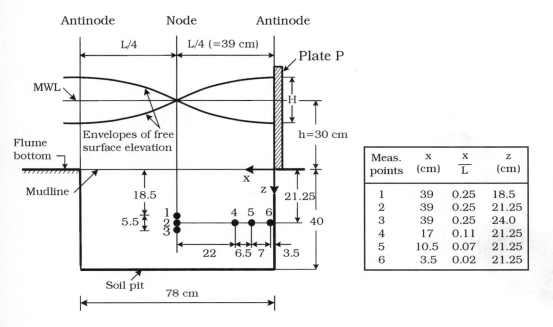

Meas. points	x (cm)	$\frac{x}{L}$	z (cm)
1	39	0.25	18.5
2	39	0.25	21.25
3	39	0.25	24.0
4	17	0.11	21.25
5	10.5	0.07	21.25
6	3.5	0.02	21.25

Figure 7.1:   Test set-up used in Kirca *et al.*'s (2013) experiment.

wave length, 156 cm. Hence, the nodal section coincided with the center of the soil pit, whereas the antinodal sections coincided with the two (offshore and onshore) ends of the soil pit. In this way, the soil in the pit was exposed to the standing wave fully. This wave length, $L = 156$ cm, corresponded to a wave period of $T = 1.09$ s.

Pore-water-pressure measurements were made at Points 1–6, indicated in Fig. 7.1. Similar measurements were also made in the middle of the soil pit across the soil depth (not shown in Fig. 7.1) to monitor the response of the soil in the vertical direction. All the tests were videotaped with a camera shooting through the glass side wall of the pit, synchronized with the pore-water-pressure measurements.

The soil used in the experiments was silt with $d_{50} = 0.070$ mm, with the relative density (or the density index) being $D_r = 0.28$. Other soil parameters can be found in Kirca *et al.* (2013).

Figs 7.2a and b show the time development of the pore pressure at Point 5 (Fig. 7.1) after a standing wave (with $H = 6.9$ cm and $T = 1.09$ s) was introduced (Fig. 7.2a) illustrating the response for the first 29 s, and Fig. 7.2b for the entire test period (a little more than 18 minutes).

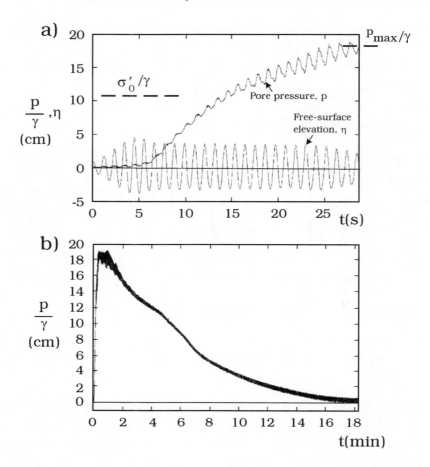

Figure 7.2: Time series of excess pore pressure in Kirca *et al.*'s (2013) experiment, at the measurement Point 5 indicated in the previous figure. The wave height of the standing wave (measured from crest to trough) $H = 6.9$ cm, and wave period $T = 1.09$ s. (a): Close-up of the first 29 s of the record, including the surface-elevation time series. (b): The entire record.

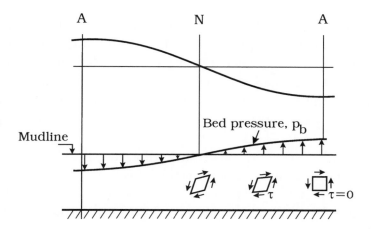

Figure 7.3:   Shear stress in the soil under a standing wave. Notice that the shear stress gradually decreases towards the antinode, and eventually ceases to exist at the antinode section.

As seen from Fig. 7.2a, although Point 5 is located at a section not exposed to any significant amount of shear deformation in the soil (caused by the wave-induced bed pressure), the pore pressure builds up, and eventually exceeds the value of the initial mean normal effective stress (Eq. 3.6),

$$\sigma_0' = \gamma' z \frac{1 + 2k_0}{3}, \tag{7.3}$$

in which γ' is the submerged specific weight of the soil, z the depth measured from the mudline (Figs 2.5 and 7.1), and k_0 the coefficient of lateral earth pressure (or the lateral stress ratio) at rest (Section 3.1.2). As seen from Fig. 7.2a, the pore pressure exceeds the value of σ_0' at around $t = 14\,\text{s}$, indicating that liquefaction occurs at Point 5. Visual observations in the test revealed this; indeed, these observations revealed that liquefaction did occur even at the antinodal section, Plate P (Fig. 7.1). See Video 4 on the CD-ROM accompanying the present book.

Regarding the long-time behaviour (Fig. 7.2b), the figure shows that the accumulated pressure, after experiencing a narrow plateau, begins to fall off, and eventually is dissipated over a period of about 18 minutes. The way in which the pore pressure builds up and subsequently is dissipated appears to be quite similar to that in the case of progressive waves, discussed in Chapter 3, Section 3.1.

However, the two processes are expected to differ from each other because of the differences in the bed pressure, and therefore in the shear strain in the soil caused by the waves: shear strains, the key element in the process of pore pressure accumulation, are largest under the node, and they gradually decrease towards the antinodes, and eventually cease to exist at the antinode section (Fig. 7.3), whereas they remain unchanged in the case of progressive waves.

Hsu and Jeng's (1994, Eq. 53) solution for the shear stress/strain in the soil under small-amplitude standing waves (Section 2.2.2, Eq. 2.81) reveals precisely the same kind of variation as sketched in Fig. 7.3. The implications of this variation for the buildup of pore pressure (and liquefaction) will be discussed next.

7.1.2   Buildup of pore pressure

Variation with x

Fig. 7.4 displays four time series of the pore pressure recorded at Points 2, 4, 5, and 6 (Fig. 7.1) in the same test as in Fig. 7.2. The time at which the accumulated pore pressure reaches the initial mean normal effective stress σ_0', i.e., the time corresponding to the onset of liquefaction (Section 3.1.2), is indicated in Fig. 7.4 with an arrow in each panel. Fig. 7.4 clearly shows that the onset of liquefaction is reached first at Point 2 (the nodal section), then at Point 4, and subsequently at Points 5 and 6. It takes 1.75 waves for the onset-of-liquefaction condition to reach Point 4; 3 waves to reach Point 5; and 3.5 waves to reach Point 6.

Fig. 7.5a displays the number of waves to cause liquefaction, N_ℓ, plotted against the horizontal distance, x, at the same depth as in Fig. 7.4 whereas Fig. 7.5b gives the number of waves for the accumulated pressure to reach the maximum pressure, plotted against x (see the insets in Fig. 7.5 for the definition sketches). Figs 7.4 and 7.5 imply that the pore pressure builds up at the nodal section more rapidly than any other location. More importantly, the figures also show that the pore pressure builds up, eventually leading to liquefaction, even at the antinodal section where there is zero shear deformation. We shall return to this point shortly.

First we address the question of the mechanism of buildup of pore pressure and liquefaction in the area around the node. By making an analogy to progressive waves (Chapter 3, Section 3.1), this mechanism is described as

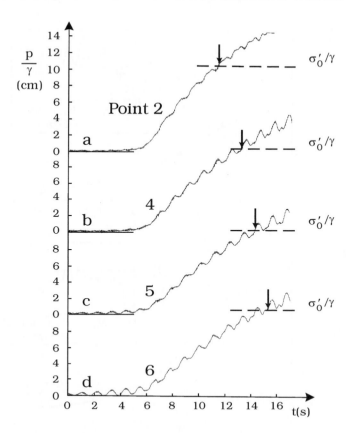

Figure 7.4: Time series of excess pore pressure at different sections. Point 2: just at the nodal section. Points 4–6: At sections between nodal and antinodal sections, Point 6 being very close to the antinodal section. See Fig. 7.1 for the precise locations of the measurement points. The wave height of the standing wave (measured from crest to trough) $H = 6.9$ cm, and wave period $T = 1.09$ s. Kirca *et al.* (2013).

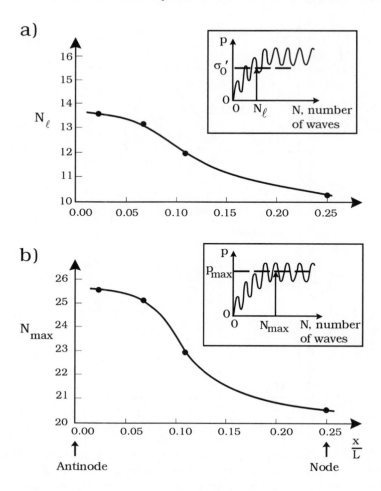

Figure 7.5: (a) Number of waves to cause liquefaction as function of x (the distance measured from the antinodal section offshore). (b) Number of waves for p to reach the maximum pressure as function of x. At the same depth z as in the previous figure. The wave height of the standing wave (measured from crest to trough) $H = 6.9\,\text{cm}$, and wave period $T = 1.09\,\text{s}$. Kirca *et al.* (2013).

follows. The soil in this area is subject to cyclic shear stresses (and therefore cyclic shear strains, or shear deformations) due to the cyclic pressure distribution exerted on the bed caused by the standing wave (Fig. 7.3). These shear strains will gradually rearrange the soil grains at the expense of the pore volume of the soil. The latter will pressurize the pore water, and, as a result, lead to the buildup of pore pressure. During this continuous buildup, the pore pressure may reach such levels that it may exceed the initial mean normal effective stress in which case the soil grains will become unbound and free, and the soil will begin to act like a liquid. This is what occurs in the area around the node where the shear stresses (and therefore the shear strains or shear deformations) are largest (Fig. 7.3), as discussed above.

Now, in the areas around the antinodes, no significant shear stresses, and therefore no significant shear strains, will develop as the bed pressure does not vary significantly with respect to the horizontal distance. Indeed, the x-gradient of the bed pressure $\partial p_1/\partial x$ is nil just under the antinodes, Fig. 7.3, implying that the shear stress in the soil should also be nil at this section. Thus, no significant buildup of pore pressure, and therefore no liquefaction, is expected to occur under or around the antinodes. However, the measurements described in the preceding paragraphs show the opposite.

Kirca *et al.* (2013) linked this behaviour to the "diffusion" of the accumulated pore pressure in the soil. Recall the analysis carried out in Section 3.2.2 in which it was shown that the accumulated pore pressure generated by shear deformations is transported in the soil by a diffusion mechanism, governed by a diffusion equation with a diffusion coefficient equal to the coefficient of consolidation (Eq. 3.28). In the present case, the pore pressures generated in areas other than antinodal sections spread out from these areas to the antinodal sections through the aforementioned diffusion mechanism. This explains the observed accumulation of the pore pressure, eventually leading to liquefaction, at the antinodal sections. See Video 4 on the CD-ROM accompanying the present book.

Sassa and Sekiguchi (1999, 2001) study

Sassa and Sekiguchi (1999), in their centrifuge wave testing described in Section 3.3.2, also conducted standing-wave experiments. In contrast to the test set-up used in Kirca *et al.*'s (2013) study, their test setup was designed such that the antinode of the standing wave formed in the middle of their sediment pit. Sassa and Sekiguchi (1999) measured the pore-water pressure

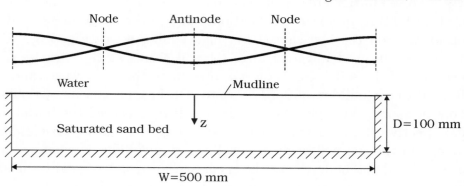

Figure 7.6:   Idealized standing-wave loading in Sassa and Sekiguchi's (2001) numerical experiment, simulating a centrifuge-experiment scenario, an ideal case covering nearly one full wave length.

across the soil depth at this section. Their pressure measurements indicated that liquefaction occurred at the antinodal section, in agreement with Kirca *et al.* (2013). However, they report that the liquefaction occurred not for the entire depth; whereas the pore pressure reached the "initial effective overburden stress" at $z = 66$ mm (liquefaction), it did not reach the "initial effective overburden stress" at $z = 93$ mm (no liquefaction), indicating that the antinodal section experienced a partial liquefaction, arguably due to the not very large wave height in the test. No pressure measurements were made in Kirca *et al.*'s (2013) study near the impermeable base, and therefore whether or not the soil column experienced liquefaction near the impermeable base is unknown. However, from the videos, the soil at a distance of 10 cm from the antinodal sections seemed to have exhibited oscillatory motion (albeit very small) (Fig. 7.9) down to a depth of about 10 cm from the impermeable base during the liquefaction stage.

    Sassa and Sekiguchi (2001), in a follow-up study, used a cyclic-plasticity constitutive model to account for the effect of stress axis rotation of sand. The model was incorporated into a finite element analysis procedure, which was then applied to soil responses to progressive wave and standing wave loading. Using the model, they simulated their centrifuge wave tests (both progressive- and standing-wave tests) from Sassa and Sekiguchi (1999).

    In addition to those tests, Sassa and Sekiguchi (2001) carried out an idealized test where they extended the sediment pit such that the entire

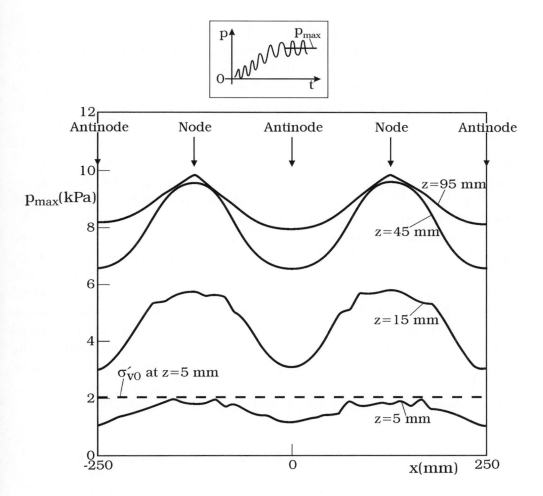

Figure 7.7:   Predicted horizontal distributions of p_{max} under ideal standing-wave loading illustrated in the previous figure. Sassa and Sekiguchi (2001).

length of the sediment pit covered nearly one full wave length (Fig. 7.6), with three antinodes (one in the middle of the sediment pit and the other two at the onshore and offshore ends of the pit), and two nodes between the three antinodes. This enabled them to observe what happens at the nodal and anti-nodal sections in their numerical simulation, with (nearly) free of end effects.

Fig. 7.7 displays the maximum accumulated pressure, p_{max}, at different depths, z, as function of the horizontal distance x. We note that the soil column, even at the nodal section is not liquefied as the maximum accumulated pressure near the mudline ($z = 5\,\text{mm}$) is smaller than the initial vertical effective stress σ'_{v0} (Fig. 7.7). The figure clearly shows, however, that the accumulated pore pressure is markedly larger at the nodal sections than at the antinodal sections, consistent with the description given in the previous subsection. Sassa and Sekiguchi (2001) note that, at each soil depth given in Fig. 7.7, the value of p_{max} is lowest under the antinodes of standing wave loading, where no rotations of the major principal stress direction are involved. They add that this soil behaviour is contrasted with that exhibited under the nodes of standing wave loading; in essence, the rotations of the principal stress axes become most significant at the nodes, allowing the value of p_{max} to peak there, as revealed in Fig. 7.7.

In a concluding remark, Sassa and Sekiguchi (2001) note that

> "...the liquefied zone develops first at the node and then extends laterally and vertically to neighbouring points. Thus soil behaviour at the antinode might be influenced by the liquefaction that (takes) place at points near the node, eventually undergoing liquefaction as well".

Variation with z

The above analysis is for the buildup of pore pressure and its variation with respect to the onshore–offshore direction. As for the variation of pore pressure buildup and liquefaction in the vertical direction, Kirca *et al.*'s (2013) experiments showed that liquefaction first occurs at the mudline and gradually spreads downward, similar to the case of progressive waves (Chapter 3, Section 3.1).

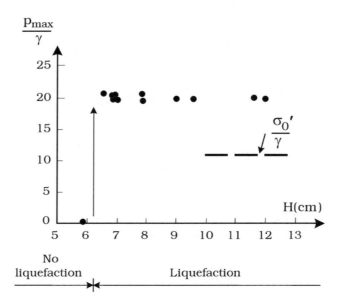

Figure 7.8: Maximum excess pore pressure versus the wave height in standing wave at nodal section, Point 2 (see Fig. 7.1), in Kirca *et al.*'s (2013) experiments. H is the wave height of the standing wave measured from crest to trough.

7.1.3   Liquefaction stage

Fig. 7.8 displays maximum pressure values attained by the period averaged pore pressure plotted against the wave height at Point 2 (Fig. 7.1), at the depth $z = 21.25$ cm, for all the standing waves tested in Kirca *et al.*'s (2013) experiments. The figure also includes the p_{max} value obtained in Kirca *et al.*'s (2013) study where the liquefaction phenomenon did not occur.

The behaviour exhibited in Fig. 7.8 is similar to that experienced in the case of progressive waves (cf. Fig. 3.8). There is apparently a critical wave height beyond which the soil is liquefied, as in the case of progressive waves. Of particular interest is the sudden transition at this critical wave height from the no-liquefaction regime to the liquefaction regime. This sudden transition is attributed to the fact that the soil is actually in two different states, the solid state and the liquid state, before and after the critical wave height, and presumably a smooth transition should, in fact, not be expected.

With the soil liquefied, the water column and the liquefied soil will form a two-layered system of liquids of different density, similar to that described

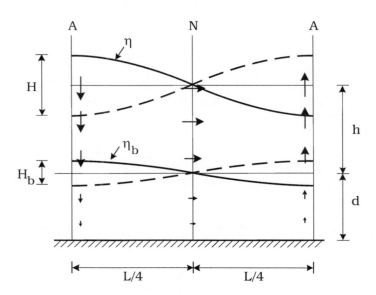

Figure 7.9:   The liquefied soil at the interface of the two-layered system of liquids (water and liquefied soil) experiences internal waves. Standing waves. cf. Fig. 3.9 for progressive waves. Oscillations in the present case in the two layers are in phase whereas oscillations in progressive waves are out of phase.

in progressive waves (Section 3.1.3). The interface between the layers of this system will experience an internal wave, again, similar to progressive waves. As seen in Section 3.1.3, the internal wave and the actual wave are out of phase in the case of progressive waves (Fig. 3.9). Kirca *et al.*'s (2013) visual observations showed, however, that, contrary to the progressive-wave case, the surface elevation of the internal wave, η_b, is apparently in phase with the free-surface elevation, η, as sketched in Fig. 7.9. Kirca *et al.* (2013) report that the oscillations in the water column and in the liquefied-soil column come in phase with each other a short while ($O(5 - 10)$ waves) after the onset of liquefaction whereby the oscillations of the internal wave begin to occur in "sympathy" with the oscillations of the water column. See Video 4 on the CD-ROM accompanying the present book.

To give a sense of the wave height of the internal waves, H_b (measured from crest to trough, defined in the same way as the wave height of the actual standing waves, Fig. 7.9), Kirca *et al.* (2013) report that $H_b = 1\,\mathrm{cm}$ and $H = 6.6\,\mathrm{cm}$ in one test, and $H_b = 1.9\,\mathrm{cm}$ and $H = 12\,\mathrm{cm}$ in another test.

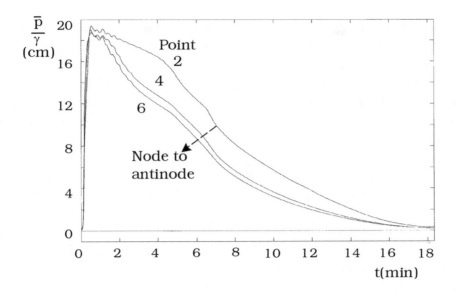

Figure 7.10:   Time series of period-averaged excess pore pressure in the standing wave experiments of Kirca *et al.*'s (2013) experiment. Point 2: just at the nodal section. Point 6: very close to the antinodal section. See Fig. 7.1 for the precise locations of the measurement points. The wave height $H = 6.9$ cm and wave period $T = 1.09$ s.

7.1.4   Dissipation of accumulated pressure and compaction

The process of buildup of pore pressure and resulting liquefaction is followed by dissipation of the accumulated pore pressure and compaction of the soil, similar to progressive waves (Section 3.1.4).

The way in which the pore pressure is dissipated appears to be similar to that in the case of progressive waves as already mentioned in conjunction with Fig. 7.2b.

Of particular interest is the variation of the pore-pressure dissipation in the onshore–offshore direction, x. Fig. 7.10 displays three time series of the *period-averaged* pore pressure, one monitored at the node (Point 2, $x = 39$ cm), and two others at Points 4 and 6 (Fig. 7.1), corresponding to $x = 17$ cm and 3.5 cm, recalling that the latter is very close to the antinode.

The wave height is $H = 6.9$ cm. As seen, the pore pressure at the node (Point 2) is dissipated at a markedly slower rate than that near the antinode (Point 6). The slower dissipation of the pore pressure at the node can be explained by the fact that the wave-induced shear stress/strain in the soil is largest (Fig. 7.3) at this location, and therefore will generate additional pore pressure, and this will delay the dissipation of the pore pressure at the nodal section.

Regarding the compaction process, the final stage of the sequence of the liquefaction/compaction process, Fig. 7.11 presents the time development of the compaction front in one of the tests in Kirca *et al.*'s (2013) study. The figure shows that the compaction front (and therefore the compaction process) is, for a given time, a function of x, in contrast to the case of progressive waves where the compaction-front contours for different times are horizontal lines, independent of x (Section 3.1.4 and Fig. 3.12). This result is expected, however, as the sequence of buildup of pore pressure (Fig. 7.4), the onset of liquefaction (Fig. 7.5a), and the dissipation of pore pressure (Fig. 7.10), are all dependent on x in standing waves.

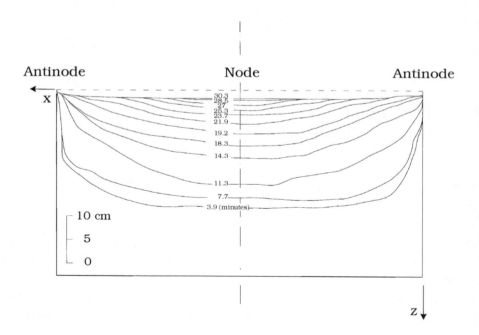

Figure 7.11:   Time development of compaction front in Kirca *et al.*'s (2013) standing-wave experiment, with wave height $H = 7.9$ cm and wave period $T = 1.09$ s.

Fig. 7.11 shows that compaction is delayed quite considerably at the node: while the soil is fully compacted at around $t = 11 - 14$ minutes at the antinodes, the compaction is reached at the node only after $t = 30$ minutes. The compaction is delayed at the node because there is quite a substantial amount of continuous pore pressure generation at the nodes where the shear stresses/strains are largest, while no pore pressure is generated at the antinodes because the shear stresses/strains are nil at these locations, as discussed previously.

Finally, we note that Kirca *et al.* (2013) comment that the mudline at the end of their tests moved about 1.7 cm downwards by the time when the compaction process was completed at around $t = 30$ minutes. They argue that this is due entirely to the compaction process itself. In the case of progressive waves, the downward displacement of the mudline is contributed (about 20–25% in the study of Sumer *et al.*, 2006) by the constant onshore transport of sediment due to the steady streaming present very close to the bed; see the discussion in Section 3.1.4. In the present standing-wave situation, the steady streaming near the bed "degenerates" into recirculation cells, directed from the antinode to the node near the bed, and from the node to the antinode away from the bed (Sumer and Fredsøe, 2002, p. 334). Kirca *et al.* (2013) argue that, in any event, the sediment in the present case (standing waves) will not be carried away from the soil pit, and remain within the soil-pit area, and therefore the contribution of the steady-streaming-induced sediment transport to the downward displacement of the mudline will be practically nil.

7.1.5 Comparison with the progressive wave

Figs 7.12 and 7.13 compare the standing-wave results at the *nodal section,* at Point 2 ($z = 21.25$ cm, Fig. 7.1), with the progressive-wave results, in terms of (1) the number of waves to cause liquefaction, N_ℓ (Fig. 7.12); and (2) the number of waves for the pore pressure to reach the maximum pressure, $N_{\max}$ (Fig. 7.13).

Fig. 7.14, on the other hand, compares the standing-wave (at the nodal section) and progressive-wave results for the maximum pressures experienced at the same measurement point, Point 2.

Figs 7.12–7.14 show that the results collapse. This is not unexpected, however, because the shear stress/strain at the *nodal section* in the case of the standing wave is precisely the same as that in the case of the progressive wave

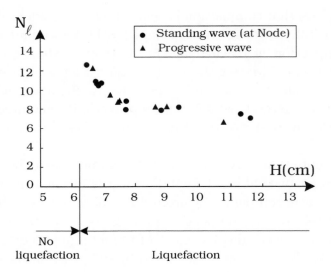

Figure 7.12:   Comparison between the standing-wave (at the nodal section) and the progressive-wave results in Kirca *et al.*'s (2013) experiment. The number of waves to cause liquefaction versus the wave height (measured from crest to trough). Point 2 in Fig. 7.1.

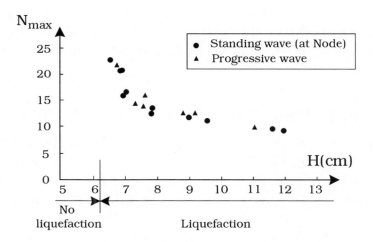

Figure 7.13:   Comparison between the standing-wave (at the nodal section) and the progressive-wave results in Kirca *et al.*'s (2013) experiment. The number of waves for the period-averaged excess pore pressure to reach $p_{\max}$ versus the wave height (measured from crest to trough). Point 2 in Fig. 7.1.

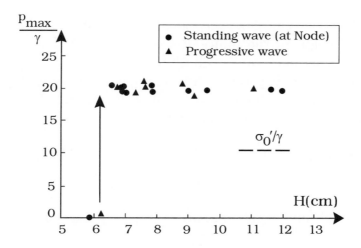

Figure 7.14:   Comparison between the standing-wave (at the nodal section) and the
progressive-wave results in Kirca *et al.*'s (2013) experiment.   The maximum period-
averaged excess pore pressure versus the wave height (measured from crest to trough).
Point 2 in Fig. 7.1.

provided that the wave heights in the two waves are the same. As mentioned
previously, analytical solutions developed by Hsu and Jeng (1994) reveal
this (Section 2.2.2). Hence, the characteristic quantities associated with the
buildup of pore pressure and liquefaction under the node of the standing
wave should match with the corresponding quantities of the progressive wave
as long as the wave height of the two waves are the same. This feature
would enable us to implement mathematical models for seabed liquefaction
developed for progressive waves for the present standing wave case. We shall
do this exercise in the next section, implementing the mathematical model
developed in Section 3.2.4.

7.2   Mathematical Modelling
of Residual Liquefaction

We have, in Section 3.2.2, developed a mathematical model describing the
residual liquefaction in a soil exposed to progressive waves. An analytical
solution to this model corresponding to the case of a soil with a finite depth
has been presented in Section 3.2.4. As the model is developed for progressive waves, it is obviously valid for every x-section in the onshore–offshore

direction. The model can, however, be adapted to standing waves, provided that the wave height in the solution is replaced with that of the standing wave. The following paragraphs will discuss this.

In the case of standing waves, Hsu and Jeng (1994) developed solutions to the Biot equations for finite-depth soils (Section 2.2.2). From Eqs 2.81 and 2.88, the amplitude of the shear stress, τ, is

$$\tau = |\tau_y| = p_b\{(C_1 - C_2\lambda z)e^{-\lambda z} - (C_3 - C_4\lambda z)e^{\lambda z} + \lambda\delta(C_5 e^{-\delta z} - C_6 e^{\delta z})\}$$
$$\times \sin(\lambda x). \tag{7.4}$$

Here p_b, the amplitude of the pressure on the bed, is given by

$$p_b = \gamma\frac{H}{2}\frac{1}{\cosh(\lambda h)} \tag{7.5}$$

in which H is the height of the standing wave (measured from crest to trough), which is twice the wave height (H_i) of each of the two progressive waves (i.e., the incident wave and the reflected wave) forming the standing wave (Fig. 2.5). It can be readily seen that the standing-wave solution at the nodes, e.g., $x = L/4$, (Eqs 7.4 and 7.5), not surprisingly, coincides with the progressive-wave solution (Eqs 3.39 and 2.46).

Now, τ has a sinusoidal variation with respect to x, Eq. 7.4. However, we can approximate the variation in Eq. 7.4 by

$$\tau = |\tau_y| \simeq p_b\{(C_1 - C_2\lambda z)e^{-\lambda z} - (C_3 - C_4\lambda z)e^{\lambda z} + \lambda\delta(C_5 e^{-\delta z} - C_6 e^{\delta z})\} \tag{7.6}$$

in the neighbourhood of the nodal section. This is precisely the same expression as that used for the solution to the equation of buildup of pore pressure for progressive waves (Eq. 3.39 see Section 3.2.4). Therefore, the progressive wave solution for the buildup of pore pressure can be applied for the present case, Eqs 3.41 and 3.42 with τ given by Eq. 7.6 in which p_b, the amplitude of the pressure on the bed, is to be calculated from Eq. 7.5 where H is the height of the standing wave (measured from crest to trough). The solution obtained in this way will, to a first approximation, be valid for every x-section around the nodal sections (Fig. 7.15); to give a sense of how broad this area around the nodal sections can be, we give the following figures: for example, the shear stress will change only by about 20% over

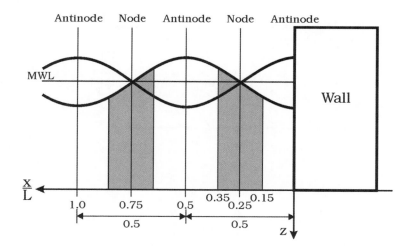

Figure 7.15: Nodal areas where the progressive-wave solution for buildup of pore presure (and liquefaction) can, to a first approximation, be applied.

an area of $O(0.15) < x/L < O(0.35)$; and by about 5% over an area of $O(0.2) < x/L < O(0.3)$ around the nodal section nearest the wall.

Example 15. *A numerical example for the assessment of liquefaction potential in front of a caisson.*

A berth at a port container terminal is constructed from precast concrete caissons. The caisson dimensions are 21 m (height) × 41.7 m (length) × 17.5 m (width).

(The way in which the berth is constructed is that the caissons towed to the construction site are positioned, seated down by ballasting onto their prepared bases, and permanently weighted down by underwater concrete and stone or sand fill. The reader is referred to Gerwick, 2007, p. 346, for constructional details.)

Now, in the present example, the caissons are towed, not to the construction site, but to a location nearby and placed temporarily on the seabed to serve, and act, as a temporary breakwater for the marine plant. The caissons are weighted down by ballast water.

The caissons are, at their temporary positions, exposed to normally-incident storm waves (Fig. 7.16). The seabed soil is loose and silty sand,

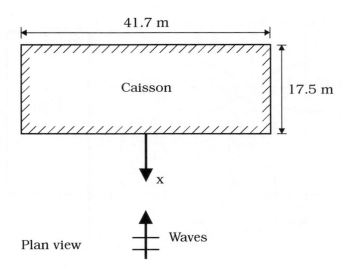

Figure 7.16: Definition sketch for the example.

and therefore susceptible to liquefaction. Assess whether or not there is liquefaction potential for the following set of wave and soil conditions.

The soil properties are given as follows: the soil depth, $d = 10$ m, the submerged specific weight, $\gamma' = 9.03$ kN/m³, the shear modulus, $G = 6600$ kN/m², the coefficient of permeability, $k = 9 \times 10^{-6}$ m/s, the porosity, $n = 0.44$, the degree of saturation, $S_r = 1$, the coefficient of lateral earth pressure, $k_0 = 0.5$, Poisson's ratio, $\nu = 0.33$, the relative density, $D_r = 0.18$, and the empirical constants in the Seed equation (Eq. 3.21), $\alpha = 0.145$ and $\beta = -0.393$. Using Eq. 3.26, the coefficient of consolidation is found as $c_v = 0.024$ m²/s.

The water properties, on the other hand, are: the specific weight of water, $\gamma = 9.81$ kN/m³, and the bulk modulus of elasticity of water, $K = 1.9 \times 10^6$ kN/m².

The properties of the *incident wave* are: the wave height, $H_i = 1.5$ m, the period $T = 5.3$ s, and the water depth, $h = 17.7$ m. The standing wave which will develop in front of the caisson will have a wave height of $H = 2H_i = 2 \times 1.5 = 3$ m.

From the linear wave theory (Appendix A), the wave length is $L = 43$ m, and the wave number $\lambda = 2\pi/L = 0.145$ m⁻¹. As discussed in the above paragraphs, the solution adapted to the present standing-wave case is, to a

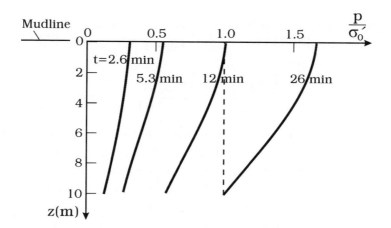

Figure 7.17: Time development of the accumulated excess pore pressure in the example. Liquefaction reaches 10 m depth within less than 26 minutes.

first approximation, valid for every x-section around the nodes, for example, around the node nearest the wall, over an area of $O(0.15) < x/L < O(0.35)$, or considering the wave length being $L = 43$ m, over $O(6$ m$) < x < O(15$ m$)$ with x being the distance from the caisson (Figs 7.15 and 7.16).

The accumulated pore pressure is calculated from Eqs 3.41 and 3.42, where the quantity H is taken as $H = 3$ m, the wave height of the standing wave. The results are given in Fig. 7.17. As seen from the figure, the pore pressure gradually builds up, and the pressure at the mudline reaches the initial-mean-normal-effective-stress value, σ'_0, within 12 minutes, meaning that the liquefaction condition is reached within 12 minutes at the mudline. The figure further shows that the liquefaction subsequently spreads across the depth, and reaches the 10 m depth within 26 minutes. Hence, the seabed in front of the caisson at the node nearest the wall over an area of $O(6$ m$) < x < O(15$ m$)$ will be liquefied within 26 minutes.

Clearly, the buildup of pore pressure and the resulting liquefaction will spread out towards the antinodal areas rather rapidly due to the diffusion mechanism described in the previous section, Section 7.1.2. The time scale of this diffusion process can be calculated from the relation

$$\ell^2 = 2c_v t \tag{7.7}$$

in which ℓ is the length in the x-direction over which the accumulated pore pressure spreads over the time t, and c_v is the coefficient of consolidation.

As mentioned in the previous section, c_v acts as the diffusion coefficient associated with this diffusion process. Taking the length ℓ as $\ell = L/4$, the time scale is found to be

$$t = \frac{\ell^2}{2c_v} = \frac{(L/4)^2}{2c_v} = \frac{(43/4)^2}{2 \times 0.024} = 2{,}400\,\text{s}. \tag{7.8}$$

As seen, the accumulated pore pressure at the nodal section will be diffused to the antinodal section at the wall within 2,400 s or 40 minutes. In reality, however, the accumulated pore pressure will be diffused to this section somewhat earlier as the pressures generated all over the area between the nodal section and the wall will be diffused to the antinodal section constantly from the time they begin to accumulate, and this will reduce the time scale quite substantially (Kirca *et al.*, 2013). The entire area in front of the wall will presumably be liquefied over a period of finite length, the liquefaction first starting at the nodal section, and then spreading gradually out towards the wall, with the liquefaction front arriving at the wall 40 minutes (at longest) after the onset of liquefaction at the nodal section.

At this junction, we note that, in a physical modelling study of liquefaction in front of a breakwater with a bedding layer, Tomi, Zen, Chen, Kasama and Yahiro (2009) observed seabed liquefaction in front of the breakwater, with the model breakwater and its bedding layer undergoing a substantial amount of settlement, and seaward tilting.

The present example deals with the liquefaction that occurs in front of a caisson. However, the standing wave in front of the caisson may cause the caisson structure to rock. This rocking motion of the structure itself may induce buildup of pore-water pressure in the soil, particularly underneath the structure, and, when the correct conditions exist, the soil at and underneath the structure may be liquefied. This process will be studied in the next chapter (Chapter 8).

7.3   Momentary Liquefaction Under Standing Waves

The momentary liquefaction occurs at the antinodal section during the time when the trough of the standing wave is present at the antinode (Fig. 7.18). At this moment, under the antinode (e.g., at $x = 0$, Fig. 7.18), the pore-water pressure (in excess of the hydrostatic pressure) has a negative sign.

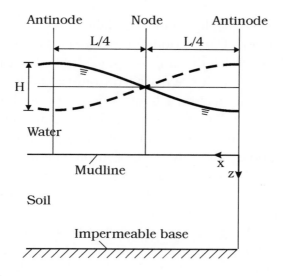

Figure 7.18:   Definition sketch.

Therefore, the pore-pressure distribution across the depth at this section will be similar to that in Fig. 1.7a for a completely saturated soil, and like in Fig. 1.7b for an unsaturated soil. As discussed earlier (Chapters 1 and 4), the soil in the latter case (Fig. 1.7b) contains some air/gas, and therefore the pore pressure is "dissipated" at a very fast rate with the depth.

The large pressure gradient in the unsaturated-soil case at shallow depths, z, creates a substantial amount of lift at the top layer of the soil. As in progressive waves, if this lift exceeds the submerged weight of the soil, the soil at the *antinodal section* will fail, and, as a result, it will be liquefied, the momentary liquefaction. Clearly, if the standing wave is taking place in front of a structure (for example, a caisson), the bed at the antinodal section $x = 0$ will coincide with the toe of the structure, and therefore occurrence of liquefaction at this location will have a strong implication for the stability of the structure.

The above analysis is for the antinodes of the standing wave. As for the nodes (Fig. 7.18), there will be no pore-pressure changes under the nodes because the surface elevation at the nodes remains unchanged as the standing wave continues, and therefore these sections will, theoretically, be unsusceptible to the momentary liquefaction. (We note that the pore pressure will spread in space due to a diffusion mechanism, and hence some variation of pore pressure will take place under the nodes.) With the antinodal section being most susceptible, and the nodal sections least susceptible to the momentary liquefaction, the sections between the nodes and the antinodes will be susceptible to the momentary liquefaction to a varying degree, the susceptibility increasing with the distance from the nodes to the antinodes.

Clearly, checks for momentary liquefaction for standing-wave situations should be carried out for the antinodal section as this section represents the location which is most susceptible to the momentary liquefaction.

As mentioned above, the physics of the momentary liquefaction at the antinodal section under standing waves is exactly the same as that in progressive waves. The mathematical analysis of the pore pressure at the antinodal sections in standing waves also, not surprisingly, reveals that the pressure obtained from the Biot equations is precisely the same as that obtained from the same equations for progressive waves, provided that the wave height in the standing-wave situation is taken as $2H_i$ (in which H_i is the wave height of each of the two progressive waves, i.e., the incident wave and the reflected wave) forming the standing wave. See the discussion in Section 2.2.2.

The above considerations allow us to implement the procedure summarized in Section 4.3.2 for the present case (obviously, at the antinodal section), provided that the wave height in Section 4.3.2 is to be replaced with the wave height (from crest to trough) of the standing wave (Fig. 7.18). In the analysis, pore pressure p is to be predicted from the analytical solution of Hsu and Jeng (1994) for the case of unsaturated soil where the Biot equations are solved with S_r different from unity, as described in greater details in Section 4.3.2.

Finally, we note that Tsai and Lee (1995) developed an analytical solution for the Biot equations for the case of a soil with an infinitely large depth exposed to a standing wave. They also carried out standard wave-flume experiments to measure pore-water pressure with sand the size $d_{50} = 0.187$ mm. They obtained good agreement between the experiments and the analytical solution, which provides further confidence in the use of the Biot equations, e.g. the analytical solution of Hsu and Jeng (1994) for the case of a soil with a finite depth.

7.4   References

1. Dean, R.G. and Dalrymple, R.A. (1984): Water Wave Mechanics for Engineers and Scientists. Prentice-Hall, Inc., New Jersey.

2. Hsu, J.R.S. and Jeng, D.S. (1994): Wave-induced soil response in an unsaturated anisotropic seabed of finite thickness. International Journal for Numerical and Analytical Methods in Geomechanics, vol. 18, No. 11, 785-807.

3. Gerwick, B.C. (2007): Construction of Marine and Offshore Structures. Third Edition. CRC Press, Taylor & Francis Group, Boca Raton, London, New York.

4. Kirca, V.S.O., Sumer, B.M. and Fredsøe, J. (2012): Residual liquefaction under standing waves. Proceedings of the 22nd International Offshore (Ocean) and Polar Engineering Conference, Rhodes, Greece, June 17–22, 2012, 1392–1398.

5. Kirca, V.S.O., Sumer, B.M. and Fredsøe, J. (2013): Seabed liquefaction under standing waves. Journal of Waterway, Port, Coastal and Ocean Engineering, ASCE. vol. 139, No. 6, 489–501.

6. Sassa, S. and Sekiguchi, H. (1999): Wave-induced liquefaction of beds of sand in a centrifuge. Géotechnique, vol. 49, No. 5, 621–638.

7. Sassa, S. and Sekiguchi, H. (2001): Analysis of wave-induced liquefaction of sand beds. Géotechnique, vol. 51, No. 2, 115–126.

8. Sekiguchi, H., Kita, K. and Okomoto, O. (1995): Response of poroelastoplastic beds to standing waves. Soils and Foundations. Soils and Foundations, vol. 35, No. 3, 31–42.

9. Sumer, B.M. and Fredsøe, J. (2002): The Mechanics of Scour in the Marine Environment. World Scientific, Singapore, 552 p.

10. Sumer, B.M., Hatipoglu, F., Fredsøe, J. and Sumer, S.K. (2006): The sequence of soil behaviour during wave-induced liquefaction. Sedimentology, vol. 53, 611–629.

11. Tomi, Y., Zen, K., Chen, G., Kasama, K. and Yahiro., Y. (2009): Effect of relative density on the wave-induced liquefaction in seabed around a

breakwater. Proceedings of the ASME 28th International Conference on Ocean, Offshore and Arctic Engineering, OMAE2009, May 31–June 5, 2009, Honolulu, HI, Paper number: OMAE2009-79601.

12. Tsai, C.-P. and Lee, T.-L. (1995): Standing wave induced pore pressures in a porous seabed. Ocean Engineering, vol. 22, No. 6, 505–517.

Chapter 8

Liquefaction at Gravity Structures

Gravity structures are used invariably in marine civil engineering, e.g., caisson breakwaters (or vertical-wall breakwaters), gravity-base offshore platforms, gravity-base foundations of offshore wind turbines, gravity-base caissons (or box caissons), used as "building blocks" of berths at port terminals, or bridge piers, etc.

Buildup of pore-water pressure around and under such structures may strongly affect the processes associated with the failure of foundations of these structures. Both the slip-surface failure and the excess settlement failure, two of the most important failure modes (Coastal Engineering Manual, 2006, Chapter 2, Part 6), may be affected by the accumulation of pore-water pressure. If the buildup of pore-water pressure eventually leads to liquefaction, this will obviously lead to direct failure of the structure.

This chapter, for the most part, discusses the processes related to buildup of pore-water pressure in front of and under a gravity structure under wave loading. The topic is developed in a sequential order, first with a general description of the liquefaction process which involves essentially two zones, one in front of the structure and the other under the structure. The liquefaction in front of the structure has been studied in Chapter 7 in the context of liquefaction in standing waves, and therefore attention in this chapter is concentrated on the liquefaction process under the structure. The physics related to the liquefaction under the gravity structure (involving the wave-induced overturning moment and rocking motion) is described next. This is followed by two liquefaction assessment exercises, one with a caisson structure and

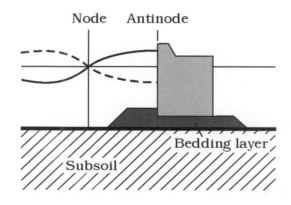

Figure 8.1:   Caisson breakwater.

the other with an offshore gravity-base oil tank. The chapter ends with a short analysis of momentary liquefaction in which the process is described by reference to a composite breakwater.

8.1   Description of Pressure buildup

8.1.1   General description

The process of buildup of pore pressure could possibly be best described by reference to a caisson breakwater (Fig. 8.1).

Consider a progressive wave (the incident wave) approaching the breakwater. As this wave impinges on the offshore face of the breakwater, a reflected wave will be created moving in the offshore direction, and, as a result, these two waves will form a standing wave in front of the breakwater, as described in Section 7.1.1.

The mechanism of the pressure buildup and eventual liquefaction under standing waves has been discussed in Chapter 7. This analysis has shown that, when the correct conditions exist, the pore pressure will build up under the nodes and spread out towards the areas under the antinodes.

The pore pressure may also build up under the structure. This is generated by two different mechanisms, namely (1) by wave motion, and (2) by caisson motion. In the former, wave-induced pressure will be transferred onto the seabed through the rubble-mound bedding layer (Fig. 8.1). In the latter, the waves will generate cyclic overturning moments, resulting in a rocking

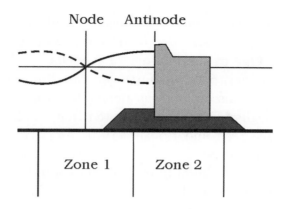

Figure 8.2: Potential liquefaction zones. Schematic.

motion of the structure, which will be transferred to the seabed in the form of cyclic bed pressure underneath the breakwater.

Both of these effects will generate cyclic shear stresses/strains in the subsoil, presumably causing pore-water pressures to build up, in the same manner as described previously in conjunction with the buildup of pore-water pressure in progressive waves (Section 1.2.1 and Section 3.1.1). In extreme conditions where this accumulated pore-water pressure exceeds the initial mean normal effective stress, liquefaction will occur.

To sum up at this point, the potential areas for buildup of pore-water pressure, and eventual liquefaction, will be (1) the area in front of the breakwater, centered at the nodal sections and extending towards the antinodal sections, as schematically illustrated in Fig. 8.2 (Zone 1); and (2) the area under the breakwater centered at the middle of the width and extending towards the onshore and offshore ends of the breakwater foundation (Fig. 8.2 Zone 2).

The mechanisms associated with the pressure buildup and liquefaction in Zone 1 have been studied in detail in Chapter 7. Therefore, we will, in the remainder of this section, turn our attention to the processes in Zone 2.

8.1.2 Pressure buildup due to rocking motion

Kudella, Oumeraci, de Groot and Meijers (2006), in a large-scale wave-flume experiment (the flume length being 307 m, the width 5 m, and the depth 7 m) carried out tests with a model caisson breakwater placed on sand simulating the subsoil with a thin clay layer on the surface (Kudella *et al.*, 2006, Fig. 9), the subsoil being sand the size $d_{50} = 0.21$ mm with the average relative density $D_r = 0.23$. One of the findings of Kudella *et al.* (2006) was that the accumulated (residual) pore pressures generated in the subsoil under the breakwater were due to caisson motions alone (the rocking motion), the wave contribution being negligible. It may be noted that similar results were also obtained by Rahman, Seed and Booker (1977, p. 1,428) for a gravity-base offshore oil tank. This result is important in the sense that the processes related to the pore pressure buildup can be studied in the laboratory with a test set-up where the model breakwater executes a rocking motion induced by an "overturning" moment, in the absence of waves, as illustrated in Fig. 8.3. (Incidentally, in Kudella *et al.*'s, 2006, experiments, the rocking motion, large enough to generate pressure buildup, could only be induced by severe breaking wave impacts. Non-breaking waves did not generate pressure buildup.)

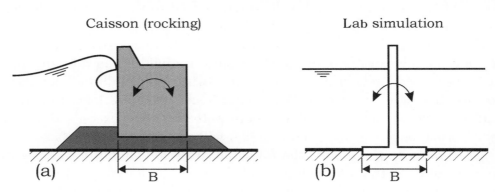

Figure 8.3:   Laboratory simulation of caisson breakwater to investigate the buildup of pore pressure. Sumer *et al.* (2008).

The above finding prompted Sumer, Sumer, Dixen and Fredsøe (2008) to employ a simple breakwater-foundation model, schematically illustrated in Fig. 8.3, to study the buildup of pore pressure in the subsoil. This enabled Sumer *et al.* (2008) to make a detailed study of the buildup of pore pressure in the subsoil in a simple, idealized environment. The idea in Sumer *et al.*'s (2008) study was not to simulate what occurs in the field under the complex, combined action of waves and the rocking motion of the breakwater, but rather to single out the process of buildup of pore pressure due to the cyclic overturning moment (and therefore rocking motion) alone, and to get a clear understanding of the process.

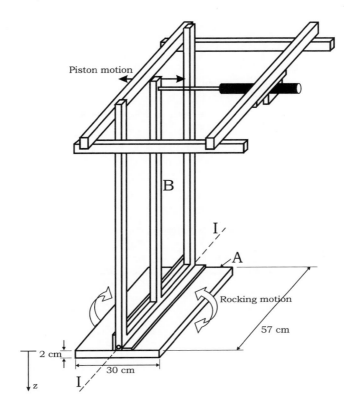

Figure 8.4: Model caisson breakwater used in Sumer *et al.*'s (2008) experiments.

Fig. 8.4 shows the model foundation in Sumer *et al.*'s (2008) experiment while Fig. 8.5 shows the experimental set-up with the measurement points (pore-pressure measurements), Points 1 to 10.

The soil used in Sumer *et al.*'s (2008) experiments was silt, the same as that used in Sumer *et al.* (1999). The grain size was $d_{50} = 0.045$ mm. Although not directly measured, an estimate of the soil relative density (before-the-test value), D_r, is given as $D_r = 0.38$. Sumer *et al.* (2008) point out that the D_r value reported in Sumer *et al.* (1999) was the after-the-test value of the relative density, and they add that the before-the-test value of the relative density was not measured in Sumer *et al.* (1999).

The size of the rocking motion of the model breakwater is characterized by the vertical displacement of the bottom corners of the model foundation (Fig. 8.6, with $y = A\sin(\omega t)$).

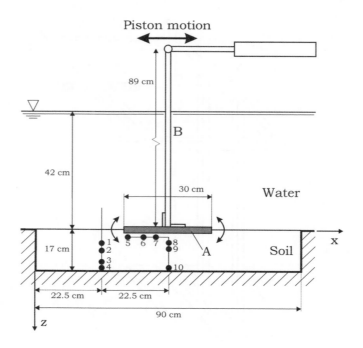

Figure 8.5:   Model caisson breakwater, soil box and pore-pressure measurement points in Sumer *et al.*'s (2008) experiments.

Fig. 8.7 displays the measured pore-pressure (in excess of static pore pressure) time series for the tested caisson-motion amplitudes at three characteristic locations, namely in the "far field" (at Points 2 and 4), and underneath the caisson, at Point 9 (Fig. 8.5). The values of the initial mean normal effective stress, σ'_0 (Eq. 3.6), are also plotted in the figure.

Fig. 8.7 shows that the way in which the pore pressure builds up, and subsequently dissipates, the sequence of liquefaction and compaction, is essentially the same as in the case of soils exposed to waves (Chapters 1, 3, and 7). Drawing an analogy to the wave-induced pressure buildup, the accumulated pressures generated in the soil in the present case drive the pore water upwards while the soil grains "settle", leading to a progressive compaction of the soil. Sumer *et al.* (2008) note that the soil compaction was revealed very clearly in their experiments when the soil was recovered after a typical test to prepare the bed for the next test.

Fig. 8.7 clearly shows that the buildup of pore pressure is a strong function of the amplitude of the caisson motion. Small-amplitude motions lead

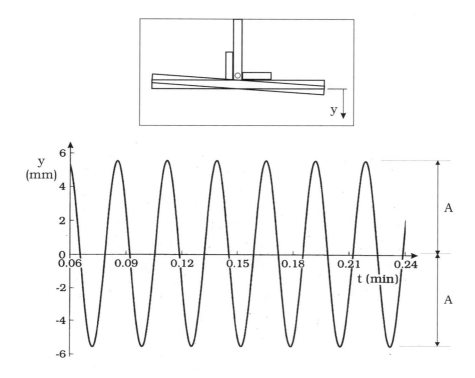

Figure 8.6: Time series of the vertical displacement of the bottom corner (offshore/onshore) of the model structure, an example. Sumer *et al.* (2008).

to buildup of pore pressure, but not necessarily to liquefaction. Indeed, no liquefaction occurred in the subsoil with amplitudes $A = 1.4\,\text{mm}$. From Fig. 8.7, it is seen that, even with the largest caisson motion, there are areas where liquefaction does not occur (for example, at Point 4 in the far field, p never reaches σ_0'). By contrast, at Point 9 underneath the caisson, except the case of $A = 1.4\,\text{mm}$, liquefaction occurs even with moderate amplitudes (Fig. 8.7c).

Fig. 8.8 displays the time development of liquefaction–no-liquefaction zones for the first 30 seconds of the process, determined from both the pressure measurements and the visual observations in a test in Sumer *et al.*'s (2008) study (the process was videotaped, synchronized with the pressure measurements). The figure shows that, not surprisingly, the buildup of pore pressure and the resulting liquefaction first start underneath the caisson and spreads sideways. The buildup of pore pressure starts underneath the caisson

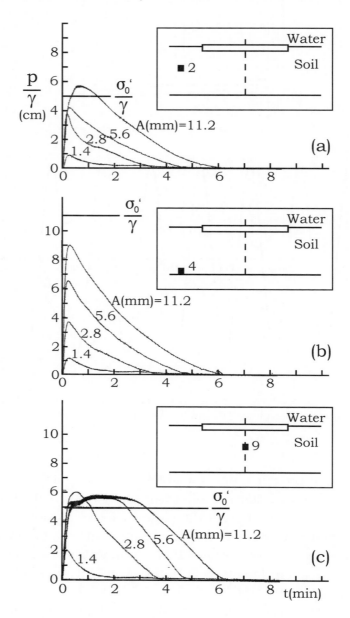

Figure 8.7:   Time series of excess pore pressure. Effect of amplitude of the caisson rocking motion, A. T = 1.6 s. Burial depth, e = 1.5 cm. Sumer *et al.* (2008).

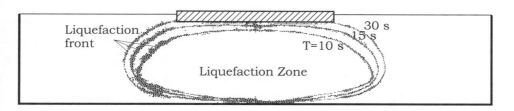

Figure 8.8: Time development of liquefaction front, determined from the pressure measurements and visual observations.

simply because this is where the largest cyclic shear deformations are generated. Despite the fact that the caisson motion is maintained throughout the test, the accumulated pore pressures are eventually dissipated, even in the liquefaction zone as the buildup/dissipation sequence continues whereby the pore water is driven upwards by the generated excess pressures, similar to the process experienced in the wave case.

Sumer *et al.* (2008) also investigated (1) the effect of period of the rocking motion, and (2) the effect of size and shape of the caisson structure, namely the effect of the width of the foundation, and the effect of the caisson shapes (rectangular, or circular shape, in plan view). The reader is referred to the original publication for detailed results.

8.1.3   Does pressure buildup reach liquefaction levels?

de Groot, Kudella, Meijers and Oumeraci (2006) give a detailed account of the studies on the buildup of pore-water pressure under gravity structures, considering various case histories. Of particular interest is the size of the accumulated pore pressure compared with, for example, the initial effective stress, to get a sense of whether liquefaction occurred in the studied cases. de Groot *et al.*'s (2006) results are summarized in the following.

(1) In Kudella *et al.*'s (2006) large-scale model experiments, referred to in the previous subsection, pore pressures in the subsoil under a model caisson breakwater reached as much as 50% of the initial vertical effective stress values, but not the value corresponding to the onset of liquefaction even though the buildup of pore pressure was "promoted" by placing a thin layer of clay on the surface of the subsoil.

(2) In the case of Ekofisk oil storage tank in the North Sea (field observation) reviewed in de Groot *et al.* (2006), pore pressures built up; however, the maximum value of the accumulated pore pressure experienced was limited to 7% of the initial vertical effective stress value. See also Example 17 under Section 8.2.2.

(3) de Groot *et al.* (2006) also included the results of a centrifuge test for a typical North Sea gravity platform; the results of two different storms indicated that the accumulated pore pressure reached only 40% and 20%, respectively, of the initial vertical effective stress.

(4) For the well-documented Niigata-breakwater case, the widely reported case history in connection with the damage that occurred on 29 October 1976, where the breakwater was hit by an extreme storm, the pore pressure buildup was, again, limited; the pore pressure reached only 30% of the initial vertical effective stress value. (Note that this figure was obtained from calculations.)

Soil liquefaction has been suggested as one of the causes of failure of vertical-wall breakwaters and other marine structures (Zen *et al.*, 1986, and Chaney and Fang, 1991), recognized also by Kudella *et al.* (2006).

Kudella *et al.* (2006) note, however, that for complete or even partial liquefaction to occur under storm waves, very unfavourable loading and drainage conditions are required, which are rarely encountered for a sandy seabed (de Groot and Meijers, 2004).

Kudella *et al.* (2006) further note that "... this is confirmed by the results of the analysis of more than 20 vertical breakwater failures, Oumeraci (1994)..." which concluded that (1) the contribution of geotechnical failure modes constitutes an important component, but (2) a failure due to complete residual liquefaction is unlikely.

They pointed out, however, that, particularly under caisson motions, large soil deformations are expected to occur, which may lead to a considerable buildup of pore pressure beneath the caisson structure (Oumeraci *et al.*, 2001). Likewise, of the case histories analysed by de Groot *et al.* (2006), only in one case (a case involving a North Sea platform), did the excess pore pressure reach the initial effective stress (i.e., liquefaction occurred), a result obtained from calculations.

Although limited, the above evidence suggests that the failure of caisson breakwaters due to complete liquefaction is arguably unlikely. However, there have been some recent incidents with very strong evidence that the failure in these cases has been caused by liquefaction. One such case history is given in Example 15 (Chapter 7) and another one in Example 16

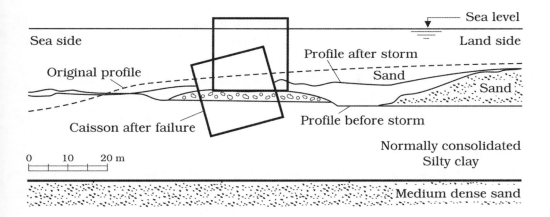

Figure 8.9: Cross-section through caisson 3 before and after failure. Four caissons at the North end of the Barcelona Harbour breakwater failed due to liquefaction. Adapted from Puzrin *et al.* (2010).

(this chapter) (the name and the locality of the structure is not disclosed due to confidentiality). A third example is the incident reported by Puzrin, Alonso and Pinyol (2010, pp. 85–148) in which four caissons failed, evidently due to liquefaction. The following paragraphs describe this latter incident.

The above-mentioned caissons were essentially four members of a shore-parallel vertical-wall breakwater at the North end, protecting the Barcelona Harbour (Puzrin *et al.*, 2010). These caissons (each 19.6 m wide, 19.5 m high and 33.75 m long) were towed to their intended positions where a bedding layer (made from coarse granular fill) was prepared. Once *in situ*, the total weight was increased by sand fill (cf., Example 15 in Chapter 7).

The time history of the failure is as follows: the caissons (filled with water) sank for the first two weeks after their placement, which was followed by sand filling. One week after this, the caissons were hit by a severe storm with maximum significant wave height of 4 m and period of 8–9 s.

All four caissons failed and were deeply buried in the soil, and tilted seaward (Fig. 8.9). Of the four caissons, the two center ones were severely damaged, as the wall reinforcement was not intended to withstand large amounts of tilt. Puzrin *et al.* (2010) note that "the failed caissons were demolished *in situ* by repeated hammering by a falling dead weight operated from a floating barge." Subsequently, a rock cover layer was installed, and a rubble-mound breakwater was finally built along the length of these four caissons. (The buried part of failed caissons remained in place.) The rest of the breakwater comprising seventeen similar caissons was built, as planned, and the caissons after their installation were constantly monitored for an extended period of time, for more than 400 days (Puzrin *et al.*, 2010).

The details of the failure and a comprehensive analysis related to the failure can be found in Puzrin *et al.* (2010). The analysis indicates that the failure was due to liquefaction of the subsoil.

Finally, we note that a detailed discussion on the subject of pressure buildup and liquefaction in the subsoil due to rocking motion of gravity-base structures can be found in the review paper by de Groot *et al.* (2006), which also covers the recent Japanese and European research.

8.2   Assessment of Residual-Liquefaction Potential

8.2.1   First screening

de Groot *et al.* (2006) define the following four types of failure, which are essentially an extension of two alternative types of failure mentioned by Oumeraci (1994), all four types of failure being related to liquefaction phenomena:

1. *Liquefaction flow failure*, i.e., a large continuous deformation of the foundation reached during one or a few extreme wave loads due to reduction of the soil strength caused by accumulated pore pressures;

2. *Stepwise liquefaction failure*, i.e., failure through stepwise residual caisson displacements in the presence of significant residual accumulated pore pressures;

3. *Stepwise failure*, i.e., failure through stepwise residual caisson displacements without significant residual accumulated pore pressures. Significant phase-resolved pore pressures may play an important role. Recall the phase-resolved pore pressure (cf., Fig. 3.23), $p - \bar{p}$ being the phase-resolved component, $\bar{p}$ the accumulated, or residual, component. This type of failure may be attributed to the failure mechanism associated with momentary liquefaction; and

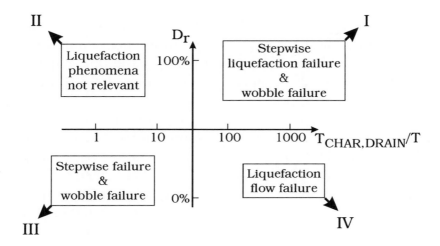

Figure 8.10:   Potential failure types as a function of relative density and relative characteristic drainage period. de Groot *et al.* (2006).

4. *Wobble failure*, i.e., failure by very large phase-resolved (not residual, or accumulated) deformations due to large cyclic shear strains.

de Groot *et al.* (2006), on the basis of the data drawn from various case histories they considered, worked out the diagram displayed in Fig. 8.10 in which $T_{CHAR,DRAIN}$ is the characteristic drainage time, T is the wave period, and D_r is the relative density (or the density index).   The characteristic drainage time is defined by

$$T_{CHAR,DRAIN} = \frac{d^2}{c_v} \tag{8.1}$$

in which d is the thickness of the subsoil layer, measured from the seabed to the impermeable base, and c_v is the coefficient of consolidation, Eq. 3.26. The quantity c_v can, for a completely saturated soil, be taken as (see, for example, Section 3.3.3)

$$c_v = \frac{Gk}{\gamma} \tag{8.2}$$

in which G is the shear modulus, k the coefficient of permeability, and γ the specific weight of sea water.

The physical meaning of the ratio $T_{CHAR,DRAIN}/T$ on the horizontal axis in Fig. 8.10 is as follows. The numerator, the characteristic drainage time, represents the time scale over which the water is drained out of the subsoil. Hence, the ratio can be interpreted as the ratio of the drainage time and the wave period. The larger this ratio, the larger the time required for the water to be drained out of the subsoil by the action of the considered wave loading, and the more susceptible the soil will be to the residual liquefaction. The significance of the parameter $T_{CHAR,DRAIN}/T$ has been recognized by several researchers, e.g., Zen introduced the parameter, termed the drainage coefficient, which is actually the inverse of the ratio $T_{CHAR,DRAIN}/T$, i.e., $C = T/T_{CHAR,DRAIN}$ (Zen, 1993, cited by Tomi et al., 2009).

The diagram given in Fig. 8.10 can be used as a means for the first screening to assess whether or not there is liquefaction potential under the structure. If this exercise indicates a liquefaction potential, then the assessment can be made, adopting the approximate mathematical model used in Example 16 below.

8.2.2   Assessment of liquefaction potential

We will, in the following paragraphs, illustrate how an assessment exercise can be carried out. We will do this by reference to two case studies, Example 16 and Example 17.

Example 16 *A numerical example for the assessment of liquefaction potential at a caisson structure*

We consider the structure studied in Example 15 (Chapter 7) in which a berth at a port container terminal is constructed from precast concrete caissons. The caisson dimensions are 21 m (height) × 41.7 m (length) × 17.5 m (width). The caissons cast on land are towed to a location near the site and placed temporarily on the seabed before they are moved to the construction site. The caissons are, at their temporary positions, exposed to normally incident storm waves, described in Example 15 (Chapter 7). The seabed soil is loose and silty sand, and therefore susceptible to liquefaction.

The soil, water and wave properties are given in Example 15 (Chapter 7). The vertical load on the soil due to the weight of the caisson (including the ballast water) minus the buoyancy force on the caisson is given as $Q = 5.23 \times 10^4$ kN.

In Example 15 (Chapter 7), we have assessed whether or not there is liquefaction potential in front of the caisson under the given storm waves, and found that the liquefaction condition around the nodal section over an area $O(6\,\mathrm{m}) < x < O(15\,\mathrm{m})$ is reached within 12 minutes at the mudline, and within 26 minutes across the entire soil depth. We have also found that the liquefaction will spread out towards the antinode at the offshore face of the caisson rather rapidly, within 40 minutes at longest.

The $H = 3\,\mathrm{m}$ storm waves hit the caissons shortly after they were placed on the seabed. Hence, the relative density of the subsoil under the caissons remains practically unchanged.

The waves will generate cyclic overturning moments, leading to rocking motion of the caissons.

The question is: does this rocking motion generate liquefaction in the subsoil under the caissons?

First screening

We resort to the diagram in Fig. 8.10. The characteristic drainage time, Eq. 8.1,

$$T_{CHAR,DRAIN} = \frac{d^2}{c_v} = \frac{10^2}{0.024} \simeq 4 \times 10^3 \text{ s} \tag{8.3}$$

and therefore the ratio $T_{CHAR,DRAIN}/T$

$$\frac{T_{CHAR,DRAIN}}{T} = \frac{4 \times 10^3}{5.3} \simeq 800. \tag{8.4}$$

The relative density, on the other hand, is given as $D_r = 0.18$. For these values of $T_{CHAR,DRAIN}/T$ and D_r, the diagram in Fig. 8.10 indicates that there is a large potential of liquefaction flow failure.

Detailed calculations

For the detailed calculations, we adopt an approximate mathematical model. The underlying principle in the model is to consider the bed pressure (caused by the rocking motion, which is induced by the overturning moment) analogous to the bed pressure caused by an ordinary standing wave, and on this premise, adapt the mathematical model developed for standing waves (Section 7.2) for the present situation.

(a)

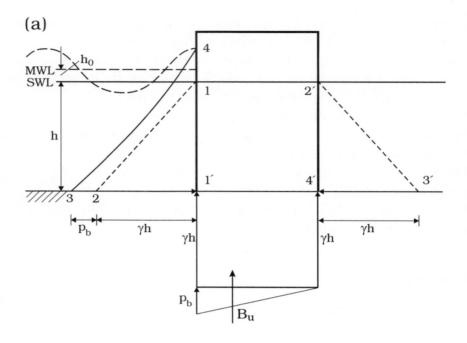

(b)

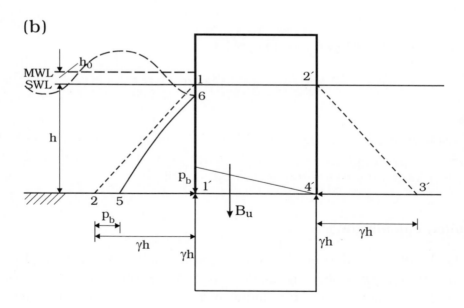

Figure 8.11:   Schematic description of forces on the caisson in the numerical example.
Figure adapted from Shore Protection Manual (1977).

Forces on the caisson

Fig. 8.11 schematically describes forces on the caisson, Fig. 8.11a corresponding to the instant that the wave crest is at the offshore face of the structure, and Fig. 8.11b to that with the wave trough at the offshore face of the structure (Shore Protection Manual, 1977). In the figures, SWL stands for Still Water Level, and MWL for Mean Water Level. The difference between SWL and MWL is grossly exaggerated in the figure for illustration purposes. The latter gave a somewhat distorted picture of the pressure distribution, particularly in Fig. 8.11b, contrary to the familiar picture from the linear wave theory with the excess pressure decreasing with the depth, measured from MWL, where SWL and MWL coincide. We note that this decrease is small for shallow water waves ($h/L < 1/20$, Dean and Dalrymple, 1984), but not so small otherwise.

The quantity p_b in the figure is the amplitude of the pressure on the bed due to the standing wave, given by (Eq. 2.82)

$$p_b = \gamma \frac{(2H_i)}{2} \frac{1}{\cosh(\frac{2\pi h}{L})}. \tag{8.5}$$

In Fig. 8.11a, the forces are:

1. Wave force in the horizontal direction with the onshore-directed force corresponding to the area (41′34), and the offshore-directed force, corresponding to the area (2′3′4′2′); and

2. Uplift force (upward directed), B_u.

In Fig. 8.11b, on the other hand:

1. Wave force in the horizontal direction with the onshore-directed force corresponding to the area (61′56), and the offshore-directed horizontal hydrostatic force, corresponding to the area (2′3′4′2′); and

2. Uplift force (downward directed), B_u.

In Fig. 8.11, the quantity h_0, the difference between MWL and SWL, is found to be $h_0 = 0.165$ m for $2H_i = 3$ m high standing waves (Shore Protection Manual, 1977, p. 7–142) with H_i being the incident wave height. The weight of the caisson and the reaction forces corresponding to the hydrostatic situation are not shown in the figure to keep the figure relatively simple. The hydrostatic forces (namely the weight of the caisson, the hydrostatic reaction forces, and the hydrostatic uplift force) are all dropped in the following analysis because we are interested in the forces and corresponding overturning moments in excess of the hydrostatic situation.

Overturning moment on the caisson

The forces in excess of hydrostatic situation will generate overturning moments; the onshore-directed horizontal wave force and the upward-directed uplift force will generate an overturning moment in the clockwise direction, and the offshore-directed horizontal wave force and the downward-directed uplift force will generate an overturning moment in the anti-clockwise direction. With this cyclic overturning moment, the caisson will execute a rocking motion. Assuming that the caisson rocks about the bottom center point, the overturning moments corresponding to the instants illustrated in Figs 8.11a and b have been calculated and are given in the following paragraphs. Note that (1) in the calculations, the following sign convention has been adopted: plus sign for the clockwise rotation, and minus sign for the anti-clockwise rotation; and (2) the overturning moment values are calculated from the charts given in Shore Protection Manual (1977, p. 7–144, Figs 7–71).

When the wave crest is at the offshore face of the caisson:

1. Overturning moment due to the onshore-directed horizontal wave force corresponding to the area $(41'34) = +10.280 \times 10^3$ kN m per m,

2. Overturning moment due to the offshore-directed horizontal hydrostatic force corresponding to the area $(2'3'4'2') = -9.085 \times 10^3$ kN m per m; and

3. Overturning moment due to the uplift force $= +56.9$ kNm per m.

When the wave trough is at the offshore face of the caisson, on the other hand:

1. Overturning moment due to the onshore-directed horizontal wave force corresponding to the area $(61'56) = +7.833 \times 10^3$ kN m per m,

2. Overturning moment due to the offshore-directed horizontal hydrostatic force corresponding to the area $(2'3'4'2') = -9.085 \times 10^3$ kN m per m; and

3. Overturning moment due to the uplift force $= -56.9$ kNm per m.

From the above values, the total overturning moment corresponding to the instant when the wave crest is at the offshore face of the caisson is

$$M_{crest} = +10280 - 9085 + 56.9 \simeq 1252 \text{ kNm per m}$$

and that when the wave trough is at the offshore face of the caisson is

$$M_{trough} = +7833 - 9085 - 56.9 \simeq -1309 \text{ kNm per m.}$$

Taking the average of the above two values, the amplitude of the cyclic overturning moment can, to a first approximation, be taken as

$$M = \frac{1}{2}(M_{crest} + |M_{trough}|) = 1281 \text{ kNm per m.} \tag{8.6}$$

Now, this generates a pressure distribution on the surface of the subsoil as sketched in Fig. 8.12a (with the wave crest at the offshore face of the caisson), and in Fig. 8.12b (with the wave trough is at the offshore face of the caisson) with an amplitude of

$$P = \frac{6M}{B^2} = 25.1 \text{ kN/m}^2 \tag{8.7}$$

where P is the amplitude of the bed pressure at the two sides of the caisson, and B the caisson width. Therefore, with the rocking motion of the caisson, the subsoil will actually be subject to a cyclic pressure loading with the amplitude 25.1 kN/m². This is a cyclic loading analogous to that caused by an ordinary standing wave around the nodal areas (Fig. 7.15), discussed in Section 7.2.

Implementation of the mathematical model

In order to assess whether or not the pore-water pressure would potentially build up and eventually lead to liquefaction under the previously described cyclic forcing, an approximate analysis can be developed. The following paragraphs will summarize this analysis. This will be achieved in four steps.

1) Cyclic shear box tests. Shear stress ratio versus number of cycles to cause liquefaction

Traditionally the onset of liquefaction is described in terms of the shear stress ratio, τ/σ_0', and the number of cycles to cause liquefaction, N_ℓ, from a simple shear box test where a horizontal, cyclic shear stress load is applied to the soil (Peacock and Seed, 1968, and Alba, Seed and Chan, 1976), as

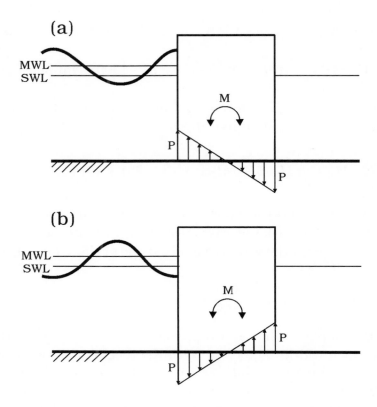

Figure 8.12:   Overturning moments. (a): Wave crest is at the offshore face of the caisson. (b): Wave trough is at the offshore face of the caisson.

discussed in details in Section 3.2.1. Here, τ is the amplitude of the shear stress in the soil, and σ'_0 the initial effective stress. The data obtained in the study of Alba *et al.* (1976) is displayed in Fig. 3.21. The figure represents the results of a set of experiments conducted for several values of the soil relative density D_r. Each curve (corresponding to a particular value of D_r) represents the onset of liquefaction, and divides the $(\tau/\sigma'_0, N_\ell)$ plane in two areas. The area above the curve corresponds to the liquefaction regime, and that below the curve to the "no liquefaction" regime. The larger the value of the shear stress ratio, the more susceptible the soil to liquefaction.

   2) Standing wave case.   Bed-pressure ratio versus number of cycles to cause liquefaction

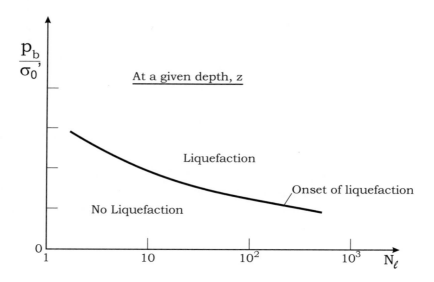

Figure 8.13:   Ratio of the amplitude of the bed pressure to the initial effective stress versus the number of cycles to cause liquefaction. At the nodal section, and at a given depth z.

Now, in the case of a soil exposed to a standing wave, the amplitude of the shear stress in the soil, τ, *at the nodal section* is given by

$$\tau = p_b F(z) \tag{8.8}$$

in which p_b is the amplitude of the bed pressure, given in Eq. 8.5, and $F(z)$ is a given function of z, see Eqs 2.81 and 2.88. Therefore, for a given depth, z, the diagram in Fig. 3.21 can alternatively be plotted in terms of the bed-pressure ratio, replacing the quantity τ with the bed pressure p_b according to Eq. 8.8, as schematically illustrated in Fig. 8.13 in which σ_0' is to be calculated from Eq. 3.6. Clearly, the larger the value of the bed-pressure ratio, p_b/σ_0', the more susceptible the soil to liquefaction.

Now, for the present case, the wave height is given as $2H_i = 3\,\mathrm{m}$, the water depth $h = 17.7\,\mathrm{m}$, and the wave length $L = 43\,\mathrm{m}$. Therefore p_b, the amplitude of the bed pressure, given in Eq. 8.5, will be

$$p_b = \gamma \frac{(2H_i)}{2} \frac{1}{\cosh(\frac{2\pi h}{L})} = 9.81 \frac{3}{2} \frac{1}{\cosh(\frac{2\pi \times 17.7}{43})} = 2.23\,\mathrm{kN/m^2} \tag{8.9}$$

and therefore the bed-pressure ratio, p_b/σ_0', can be calculated from

$$\frac{p_b}{\sigma_0'} = \frac{p_b}{\gamma' z(\frac{1+2k_0}{3})} \tag{8.10}$$

for various values of the depth z. The calculated values of the bed-pressure ratio, p_b/σ_0', from the preceding two equations are tabulated in Table 8.1, column 2 for different z values.

Now, the liquefaction calculations presented in Example 15 (Chapter 7) indicate that the soil is liquefied for this wave. Hence, the values of p_b/σ_0' depicted in Table 8.1, column 2 (under the title standing wave) should fall in the "liquefaction" area of Fig. 8.13 for each and every z.

Table 8.1.   Bed pressure ratios.

Depth z (m)	Standing wave p_b/σ_0'	Rocking caisson[1] P/σ_0'
(1)	(2)	(3)
0.25	1.48	0.51
0.50	0.74	0.49
0.75	0.49	0.48
1.0	0.37	0.47
2.0	0.19	0.42
3.0	0.12	0.38
4.0	0.093	0.35
5.0	0.074	0.32
10.0	0.037	0.23

[1]Here, σ_0' is, as a conservative approach, calculated just below the centerline.

3) Rocking caisson. Bed-pressure ratio versus number of cycles to cause liquefaction

We have found from the force and moment analysis in the preceding subsection that, with the rocking motion of the caisson, the subsoil will actually be subject to a cyclic pressure loading with the amplitude 25.1 kN/m². We have already stressed that this loading is analogous to that experienced by the soil around the nodal areas under an ordinary standing wave. It is remarkable to notice that the bed-pressure amplitude in the rocking caisson case,

$25.1\,\mathrm{kN/m^2}$, is an order of magnitude larger than that of the standing wave, $2.23\,\mathrm{kN/m^2}$ (cf. Fig. 8.14a and Fig. 8.14b), and moreover this pressure distribution is applied over an even smaller x-extent, namely $17.5\,\mathrm{m}$, the width of the caisson. Therefore, this cyclic pressure distribution appears to have a large potential to generate liquefaction. However, whether or not liquefaction occurs depends on the bed-pressure ratio P/σ_0', rather than P alone. We will investigate this in the following paragraph.

In the case of no surcharge, σ_0' is calculated by Eq. 3.6, $\sigma_0' = \gamma'z\frac{1+2k_0}{3}$, the initial mean normal effective stress. However, in the case of a surcharge (the present situation, with the caisson sitting on the bed), the right-hand side of Eq. 3.6 is replaced with (Sumer $et\ al.$, 2010)

$$\sigma_0' = \left(\gamma'z + \frac{Q}{B_cL_c}\right)\frac{1+2k_0}{3} \tag{8.11}$$

representing the initial mean normal effective stress in the soil under the caisson (at the centerline) in which Q is the vertical load on the soil due to the weight of the caisson minus the buoyancy force on the caisson. The quantities L_c and B_c are the length and the width of the caisson, respectively. See also the discussion in Section 3.6 (Chapter 3). This is the expression for σ_0' for shallow soil depths. Otherwise (not-so-shallow soil depths), the above criterion should be corrected by introducing a factor, α, related to the spreading of the loaded area with the soil depth:

$$\sigma_0' = (\gamma'z + \alpha\frac{Q}{B_cL_c})\frac{1+2k_0}{3}. \tag{8.12}$$

The factor α may, to a first approximation, be taken from the contour diagram of the increase in vertical stress below a strip footing (Fig. 3.51). From Fig. 3.51, the value of α corresponding to the subsoil just below the centerline, can be taken as unity (as a conservative value). With this value of α, the bed-pressure ratio values are calculated for the same depths z, and depicted in Table 8.1, column 3 under the title rocking caisson.

4) Liquefaction in the subsoil underneath the rocking caisson?

As mentioned previously, all the values in Table 8.1, column 2 correspond to the liquefaction regime. Comparison of the values of the rocking caisson case (Table 8.1, column 3) with those in Table 8.1, column 2 for the same depth, z, indicates that liquefaction should occur in the case of the rocking

(a)

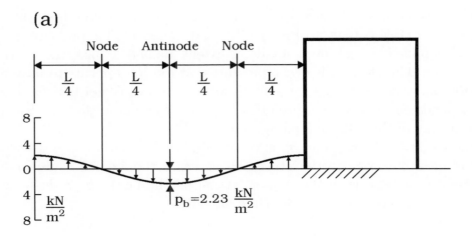

(b)

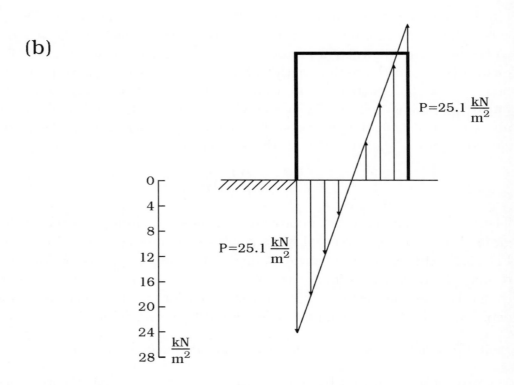

Figure 8.14:   (a): Pressure distribution on the bed in front of the caisson. (b): pressure distribution on the bed underneath the caisson, generated by the rocking motion of the caisson due to the same wave.

caisson for depths $z \gtrsim 0.75$ m (just below the centerline of the caisson), as the bed-pressure ratios for the case of the rocking caisson are larger than (or approximately equal to, for $z = 0.75$ m) those for the case of the standing wave for these depths.

This is what would occur in the subsoil just below the middle section of the caisson. For the offshore and onshore sides of the caisson, the factor α can be taken as $\alpha = 0.5$ (from Fig. 3.51) for shallow soil depths. An exercise similar to that presented in Table 8.1 indicates that the soil is liquefied for depths $z \gtrsim 0.4$ m.

The analyses carried out in Examples 15 and 16 show that there are two zones of liquefaction, one in front of the caisson, Zone 1 (Example 15), and the other underneath the caisson, Zone 2 (Example 16), Fig. 8.15. Zone 1 spreads out towards the two neighbouring antinodal sections (the onshore antinodal section being the offshore face of the caisson) rather fast, as discussed in Example 15. Likewise, Zone 2 also spreads out offshore (as well as onshore), and the two zones should eventually merge. With this, the caisson will fail.

Clearly, the implication of this failure can be catastrophic; the caisson will be displaced seaward and, at the same time, sink (cf., Example 14 in Chapter 6 for the amount of sinking and the time scale). When this happens, the displaced and sunk caisson will have to be recovered. This would involve very large costs. Therefore, this risk should not be overlooked.

Seaward tilting

Seaward tilting of failed caissons has been noted by several researchers, e.g., Oumeraci *et al.* (2001), Zhang, Lee and Leung (2009), Puzrin *et al.* (2010). See Fig. 8.9, reproduced from Puzrin *et al.* (2010). There are several theories to explain this.

Oumeraci *et al.* (2001) suggested that the net forces in the seaward direction may be larger than the opposite direction.

Zhang *et al.* (2009) addressed this problem in their study of subsoil liquefaction under a caisson breakwater subjected to impulsive loads in a $100\,g$ centrifuge experiment in which the wave loading was simulated, using an in-flight wave actuator system. Two kinds of wave loads were simulated: (1) load that was applied only in the landward direction, and (2) load that was applied in both the seaward direction and the landward direction in an alternating manner.

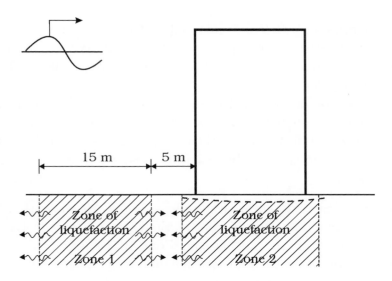

Figure 8.15:   Zones of liquefaction in the example. Zone 1: generated by the 3 m standing wave in front of the caisson; Zone 2: generated by the rocking motion of the structure due to the same wave. Zone 1 and Zone 2 eventually merge; see the text.

Of particular interest is the second type of wave loading. In this exper-iment, the seaward-directed cycle had an amplitude appreciably lower than the landward-directed one (representing typical loading induced by a break-ing wave). Yet, the model caisson was, upon the liquefaction of the soil, tilted in the seaward direction. Zhang *et al.* (2009) interpreted this as follows. The large landward-directed half cycle induces unloading of the offshore-side toe of the breakwater, allowing inflow of water into the sand, causing further loosening of the soil therein. As the caisson settles back into the soil, it embeds itself deeper into the sand at the offshore-side toe, causing a seaward tilt.

Clearly, in Zhang *et al.*'s (2009) experiment, the incoming waves were not simulated. Instead, the action of the waves was simulated, using the previ-ously mentioned actuator. In reality, however, the incoming waves may also induce liquefaction in front of the structure, as in Example 15 (Chapter 7), Zone 1 in Fig. 8.15.

As mentioned in Section 7.2, in a physical modelling study of liquefaction in front of a breakwater with a bedding layer, Tomi *et al.* (2009) observed seabed liquefaction in front of the breakwater, with the breakwater structure undergoing a substantial amount of settlement, and seaward tilting.

Combined with Zhang *et al.*'s (2009) explanation, the presence of the liquefaction zone in front of the structure would further substantiate the explanation for the generally observed seaward-tilting mechanism.

Example 17 *Pressure buildup under an offshore gravity-base oil tank*

We will continue to study the subject with another example, taken from Rahman *et al.* (1977). The latter authors developed a procedure to analyse the pore pressure response under an offshore oil tank during a storm, and subsequently they illustrated the procedure using the Ekofisk oil tank as a case study.

The Ekofisk oil tank has a shape of a vertical circular cylinder; see Fig. 8.16, which is reproduced from Rahman *et al.* (1977). No data was included in Rahman *et al.* (1977) for the sand size, except that the soil type is given as sand with a relative density of $D_r = 0.85$, and the coefficient of permeability, k_r and k_z, and the coefficient of volume compressibility m_v values as indicated in Fig. 8.16.

Lee and Focht (1975), in an independent study, pointed out that only the upper sand layer was of serious concern to the foundation stability, and reported that the sand in this upper sand layer was composed of at least 90% quartz particles, with various other rock fragments and shells making up the rest. Lee an Focht (1975) gave the sand size as $d_{50} = 0.11$ mm.

The design storm in the application example in Rahman *et al.* (1977) is simulated in three consecutive "sea states", the first representing the state just before the storm, the second the storm, and the third the state just after the storm. The wave height reaches the peak value $H = 25$ m (with a period of $T = 13.5$ s) within a period of 3.2 hours.

As mentioned previously (Section 8.1.2), Rahman *et al.* (1977) found that the shear stress induced by the subsurface pressure and the cyclic buoyancy forces are very small (compared with those induced by the wave-induced forces and the resulting overturning moment), and therefore they ignored these effects in the calculations.

The procedure, as implemented for Ekofisk oil tank, may be summarized as follows.

(1) Calculate the horizontal forces and the overturning moment, using the MacCamy and Fuchs (1954) solution for potential flow around the cylinder;

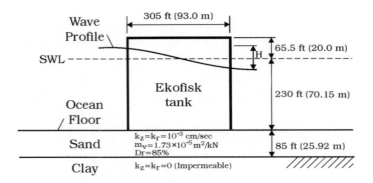

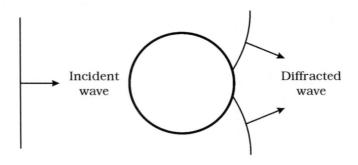

Figure 8.16:   Ekofisk tank and soil profile. Top: longitudinal section in the offshore–onshore direction. Bottom: plan view. Adapted from Rahman *et al.* (1977).

see Sumer and Fredsøe (1997, Chapter 6) for a detailed account of MacCamy and Fuchs' (1954) study.

(2) Treat these wave forces as static forces for the purpose of stress evaluation, and evaluate the amplitudes of cyclic shear stress, τ, using the corresponding amplitudes of the wave forces, utilizing stress distribution theory (stress analysis technique). It may be noted that the stresses in the Rahman *et al.* (1977) analysis were tuned so that a good match between the calculations and the physical model experiments was obtained.

(3) Determine the shear stress ratios, τ/σ'_{v0}, for all the node points of the computational mesh in which σ'_{v0} is the initial vertical effective stress.

(4) With the evaluated values of τ/σ'_{v0}, find the number of cycles to cause liquefaction, N_ℓ, for each and every point of the mesh, reading off the values of N_ℓ from a graph similar to that in Fig. 3.21, specially prepared for the Ekofisk sand with $D_r = 0.85$.

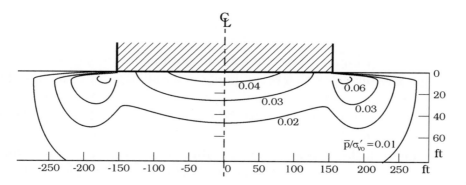

Figure 8.17:   Contour plot of the accumulated excess pore pressure normalized by the initial vertical effective stress just before the peak of the storm, calculated by Rahman *et al.* (1977) for Ekofisk oil tank.

(5) Find the pore pressure generated per unit time and per unit volume of soil (including the pores) from

$$f = \frac{\sigma'_{v0}}{N_\ell T} \frac{1}{\theta \pi} \frac{1}{\sin^{2\theta-1}(\frac{\pi}{2}\frac{\bar{p}}{\sigma'_{v0}}) \cos(\frac{\pi}{2}\frac{\bar{p}}{\sigma'_{v0}})} \tag{8.13}$$

at any time t. Here, T is the wave period, θ is a constant, the average value of which is $\theta = 0.7$, and $\bar{p}$ is the period-averaged pore pressure existing at that time (cf., the above equation with Eq. 3.31).

(6) Then, advance the solution for $\bar{p}$ corresponding to time $t + dt$ by solving the differential equation for the accumulated excess pore pressure $\bar{p}$

$$\frac{1}{r}\frac{\partial}{\partial r}\left(r\frac{k_r}{\gamma}\frac{\partial \bar{p}}{\partial r}\right) + \frac{\partial}{\partial z}\left(\frac{k_z}{\gamma}\frac{\partial \bar{p}}{\partial z}\right) = m_v\left(\frac{\partial \bar{p}}{\partial t} - f\right) \tag{8.14}$$

in which k_r and k_z are the coefficients of permeability in radial and vertical directions, respectively, and m_v is the coefficient of volume compressibility, and related to the coefficient of consolidation by $c_v = k/(\gamma m_v)$ (Lambe and Whitman, 1969, p. 407) (cf., Eq. 8.14 with Eq. 3.28).

Fig. 8.17 displays the contour plot of the calculated $\bar{p}$, normalized by the initial vertical effective stress, $\bar{p}/\sigma'_{v0}$, the pore-pressure ratio, corresponding to time $t = 2.8$ hours (just before the peak of the storm). In Rahman *et al.* (1977), two more contour plots of pore pressure are given, one corresponding to time $t = 3.2$ hours (the peak storm), and the other to time $t = 6$ hours (the end of the storm).

The pattern in Fig. 8.17 resembles that obtained in Sumer *et al.*'s (2008) small-scale experiments with a rocking model foundation, see Fig. 8.8. This is with the exception, however, of small "pockets" at the edges of the structure in Fig. 8.17, which could not be captured in Sumer *et al.*'s (2008) experiments.

Clearly, the pore-pressure ratios displayed in Fig. 8.17 are far too small to cause liquefaction, the maximum value being $\bar{p}/\sigma'_{v0} = O(0.1)$, small compared with unity. Rahman *et al.*'s calculations corresponding to the peak storm, on the other hand, give a maximum value of $\bar{p}/\sigma'_{v0} = 0.2$, still too small to cause liquefaction.

Rahman *et al.* (1977, Fig. 19) repeated their calculations for *undrained situation*. In this case, it was found that the pore-pressure ratio, $\bar{p}/\sigma'_{v0}$, at the center of the tank (just underneath the tank) reached a value of about 35%, and $\bar{p}/\sigma'_{v0}$ at the edge just underneath the structure reached the liquefaction level (100%), i.e., $\bar{p}/\sigma'_{v0} = 1$.

It may be noted that the problem of buildup of pore pressure under the Ekofisk oil tank was addressed by previous authors, Bjerrum (1973), and Lee and Focht (1975), using simplified analyses. As pointed out by Rahman *et al.* (1977), Bjerrum's (1973) analysis ignores both the distributions of stresses in the soil, and the effect of pore-pressure dissipation is considered in an approximate way.

It is interesting to note the following remark in Lee and Focht's (1975) paper under the heading Postscript:

> "the tank had been in place 13 months, and during this time, it has been subjected to numerous major storms, of which at least three produced waves measuring over 21 m crest to trough with one approaching the design wave of 24 m. The performance of the tank and its sand foundation had been entirely satisfactory."

Finally, Li and Jeng (2008) developed a 3-D formulation of buildup of pore pressure around the head of a breakwater, based on the extension of the 1-D formulation (Eq. 3.28) to 3-D, in much the same way as in Sumer and Cheng (1999) where the 1-D formulation was extended to 2-D. Li and Jeng (2008) numerically solved the 3-D differential equation for the accumulated pore pressure, and studied the liquefaction potential. They obtained an increase in the accumulated pore pressure, as much as a factor of 1.5, near the breakwater head with respect to the far-field value.

8.3 Momentary Liquefaction Around a Breakwater

A great many works have been devoted to the investigation of the phase-resolved pore-water pressure (with no residual component) and soil stresses around a breakwater, Mynett and Mei (1982), Mase, Sakai and Sakamoto (1994), Kortenhaus, Oumeraci, Kohlhase and Klammer (1994), Mizutani, McDougal and Mostafa (1996), Mostafa and Mizutani (1997), Mizutani and Mostafa (1996), Mizutani, Mostafa and Iwata (1998), Mostafa, Mizutani and Iwata (1999), Jeng, Cha, Lin and Hu (2001), Ulker, Rahman and Guddati (2009, 2010 a and 2010 b, 2012) (with Ulker *et al.*, 2009 and 2012, with the focus on breaking-wave induced seabed response), and Jeng, Ye and Liu (2012).

With the pore pressure, and the soil stresses fully determined, whether or not momentary liquefaction occurs can, in principle, be determined from

$$\sigma'_z \geq \sigma'_{v0} \tag{8.15}$$

in which σ'_z is the vertical effective stress, and σ'_{v0} is the initial vertical effective stress, similar to the previously studied cases (Chapter 4; Chapter 5, Section 5.7; Chapter 6, Section 6.4; and Chapter 7, Section 7.3).

We will describe the momentary liquefaction by reference to the work of Jeng *et al.* (2012). These authors developed a numerical solution, and presented the method describing the way in which the momentary liquefaction is handled for a composite breakwater (Fig. 8.18). They used an integrated model which combines (1) volume-averaged Reynolds-averaged Navier Stokes (VARANS) equations for the wave motion in the water as well as in the porous structure (the rubble mound in the present case), and (2) the dynamic Biot equations for the seabed. A one-way coupling method was developed to integrate the VARANS equations with the dynamic Biot equations. The dynamic Biot equations are essentially an extension of the Biot equations (Chapter 2) where the accelerations of the pore water and soil particles are considered while the displacement of pore water relative to soil particles is ignored. The authors verified their model against a set of experimental data from various sources.

The seabed soil properties in the numerical example (the composite breakwater displayed in Fig. 8.18) were taken as follows: the shear modulus $G = 10^5$ kN/m^2, Poisson's coefficient $\nu = 0.33$, the coefficient of permeability $k = 10^{-4}$

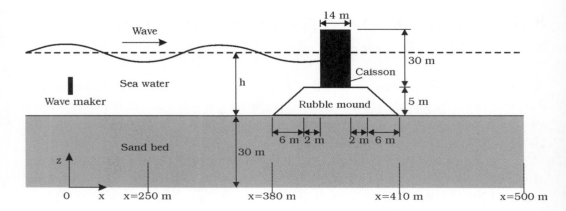

Figure 8.18:   Definition sketch of wave–seabed interaction around composite breakwater studied by Jeng *et al.* (2012).

m/s, the porosity $n = 0.25$, the grain size $d_{50} = 0.5$ mm, and the degree of saturation $S_r = 0.98$. The corresponding figures for the rubble-mound properties were: the shear modulus $G = 5 \times 10^5$ kN/m², Poisson's coefficient $\nu = 0.33$, the coefficient of permeability $k = 0.2$ m/s, the porosity $n = 0.35$, the grain size $d_{50} = 400$ mm, and the degree of saturation $S_r = 0.98$. The wave properties, on the other hand, were as follows: the wave height $H = 3$ m, the wave period $T = 10$ s, and the water depth $h = 20$ m.

Fig. 8.19, adapted from Jeng *et al.* (2012), presents the contour plot of the vertical pore-water pressure gradient. The top panel corresponds to the phase when the wave trough is at the offshore face of the breakwater, and the bottom panel when the wave crest is at the offshore face of the breakwater.

The areas marked (+) in the diagrams correspond to an upward-directed pressure gradient while the areas marked (−) correspond to a downward-directed pressure gradient force. Clearly, when the force induced by the upward-directed pressure gradient exceeds the submerged weight of the soil, the momentary liquefaction will occur, in the same manner as described in Chapter 4. Hence, the (+) areas represent potential liquefaction zones. Note that the liquefaction-prone zones coincide with the troughs of the standing wave, as expected.

Jeng *et al.* (2012) investigated where in the soil the liquefaction criterion is satisfied, and plotted the results in terms of liquefaction and no liquefaction

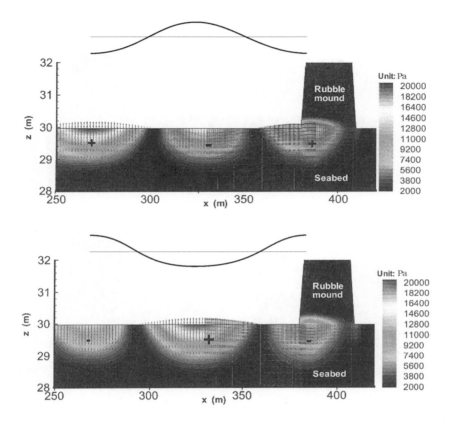

Figure 8.19: Upward/downward-directed pressure-gradient forces in the soil. Top: when the wave trough is at the offshore face of the breakwater. Bottom: when the wave crest is at the offshore face of the breakwater. "+" sign: upward-directed pressure gradient force. "−" sign: opposite. Free-surface elevations are elevated for illustration purposes. Adapted from Jeng *et al.* (2012).

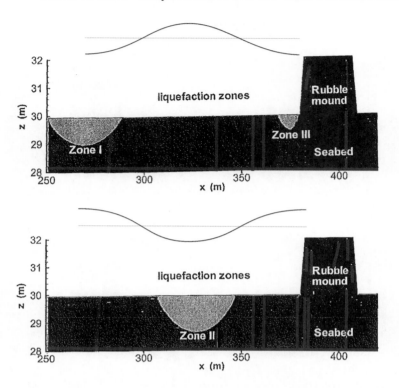

Figure 8.20:   Momentary liquefaction zones corresponding to the phases illustrated in the previous figure.   Free-surface elevations are elevated for illustration purposes.   Adapted from Jeng *et al.* (2012).

zones (Fig. 8.20). As seen, there are three liquefaction zones, Zone I, Zone II and Zone III. It should be noted that although there is an area underneath the breakwater where the pressure gradient in the subsoil is positive (Fig. 8.19, top panel), a liquefaction-prone area, this area is apparently not liquefied because of the surcharge imposed by the breakwater weight. Obviously, of the three liquefaction zones, Zone III will pose a large threat to the stability of the structure. Incidentally, Zen and Yamazaki (1993) noted the same kind of threat at the toe of a rubble mound, leading to enhanced scour and even sand slide failure (along a surface sliding) near the toe. Zen and Yamazaki (1993) related this to the observations of extensive local scour at the edge of a detached breakwater foundation where a large number of concrete blocks, which had formed the breakwater on Niigata coast, were found widely and deeply spread in the seabed.

Li and Jeng (2008), cited earlier in conjunction with Example 17, developed a numerical solution for the 3-D Biot equations under oblique wave attack, and examined potential for momentary liquefaction. These studies showed that, for the most part, the soil in the neighbourhood of the head of the breakwater is more susceptible for liquefaction than any other location.

Finally, Michallet *et al.* (2012 a) report momentary liquefaction observed in a laboratory experiment where a single wave consisting of a trough and followed by a crest breaks on a vertical wall placed on a 1:20 slope. Michallet *et al.* (2012 a) measured the free-surface elevation at the wall and pore-water pressures just below, at four depths, and demonstrated that the upward-directed pressure-gradient forces were generated near the surface of the sediment bed, which were large enough for the sediment to undergo momentary liquefaction. They used in the experiments 0.64 mm grain size light-weight material with a density of 1.18. The bed was initially unsaturated with an estimated amount of air content of $C_{gas} \simeq 4\%$ (see Section 4.5, Chapter 4 for the definition of C_{gas}). The authors observed strong motions in the bed down to 15 cm depth. These motions were recorded, using Particle Image Velocimetry (PIV). In addition to particle velocity, the strain tensor modulus, the vorticity and the divergence were obtained. In a parallel study, Michallet *et al.* (2012 b) carried out tests under periodic waves, with the results similar to the case of the single wave (Michallet *et al.*, 2012 a).

8.4   References

1. Alba, P.D., Seed, H.B. and Chan, C.K. (1976): Sand liquefaction in large-scale simple shear tests. Journal of Geotechnical Engineering Division, ASCE, vol. 102, No. GT9, 909–927.

2. Bjerrum, L. (1973): Geotechnical problems involved in foundations of structures in the North Sea. Géotechnique, vol. 23, No. 3, 319–358.

3. Chaney, R.C. and Fang, H.Y. (1991): Liquefaction in the coastal environment: An analysis of case histories. Marine Geotechnology, vol. 10, No. 3–4, 343–370.

4. Coastal Engineering Manual (2006): Types and functions of coastal structures. Part VI, Chapter 2, Engineer Manual EM 1110-2-1100,

Update: 1 June, 2006. Headquarters, U.S. Army Corps of Engineers, Washington, D.C.

5. Dean, R.G. and Dalrymple, R.A. (1984): Water Wave Mechanics for Engineers and Scientists. Prentice Hall, Englewood Cliffs, New Jersey 07632, xii+353 p.

6. de Groot, M.B., Kudella, M., Meijers, P. and Oumeraci, H. (2006): Liquefaction Phenomena underneath Marine Gravity Structures Subjected to Wave Loads. Journal of Waterway, Port, Coastal and Ocean Engineering, ASCE, vol. 132, No. 4, 325–335.

7. de Groot, M.B. and Meijers, P. (2004): Wave induced liquefaction underneath gravity structures. Proceedings of the International Conference on Cyclic Behaviour of Soils and Liquefaction Phenomena, Bochum, Germany, Balkema, Rotterdam, 399–406.

8. Jeng, D.S., Cha, D.H., Lin, Y.S. and Hu, P.S. (2001): Wave-induced pore pressure around a composite breakwater. Ocean Engineering, vol. 28, No. 10, 1413–1432.

9. Jeng, D.S., Ye, J.H. and Liu, P.L.F. (2012): An integrated model for the wave-induced seabed response around marine structures: Model verifications and applications. Coastal Engineering, in print.

10. Kortenhaus, A., Oumeraci, H., Kohlhase, S. and Klammer, P. (1994): Wave-induced uplift loading of caisson breakwaters. Proceedings of the 24th International Conference on Coastal Engineering, ASCE, Kobe, Japan, 1298–1311.

11. Kudella, M., Oumeraci, H., de Groot, M.B. and Meijers, P. (2006): Large-scale experiments on pore pressure generation underneath a caisson breakwater. Journal of Waterway, Port, Coastal and Ocean Engineering, ASCE, vol. 132, No. 4, 310–324.

12. Lambe, T.W. and Whitman, R.V. (1969): Soil Mechanics. John Wiley and Sons, Inc., New York, 553 p.

13. Lee, K.L. and Focht, J.A. (1975): Liquefaction potential at Ekofisk Tank in North Sea. Journal of Geotechnical Engineering Division, ASCE, vol. 101, No. GT1, 1–18.

14. Li, J. and Jeng, D.S. (2008): Response of a porous seabed around breakwater heads. Ocean Engineering, vol. 35, No. 8–9, 864–886.

15. MacCamy, R.C. and Fuchs, R.A. (1954): Wave forces on piles: A diffraction theory. U.S. Army Corps of Engineers, Beach Erosion Board, Technical Memo No. 69, 17 p.

16. Mase, H., Sakai, T. and Sakamoto, M. (1994): Wave-induced porewater pressures and effective stresses around breakwater. Ocean Engineering, vol. 21, No. 4, 361–379.

17. Michallet, H., Rameliarison, V., Berni, C., Bergonzoli, M., Barnoud, J.-M. and Barthelemy, E. (2012 a): Physical modelling of sand liquefaction under wave breaking on a vertical wall. 33rd Conference on Coastal Engineering, Santander, Spain, ICCE 2012, July 1–6, 2012.

18. Michallet, H., Catalano, E., Berni, C., Chareyre, B., Rameliarison, V. and Barthelemy, E. (2012 b): Physical and numerical modelling of sand liquefaction in waves interacting with a vertical wall. 6th International Conference on Scour and Erosion (ICSE6), August 27–31, 2012, Paris, 679–686.

19. Mizutani, N., McDougal, W. and Mostafa, A. M. (1996): BEM-FEM combined analysis of non-linear interaction between wave and submerged breakwater. Proceedings of the 25th International Conference on Coastal Engineering, ASCE, FL, USA, 2377–2390.

20. Mizutani, N. and Mostafa, A.M. (1996): Nonlinear wave-induced seabed instability around coastal structures. Coastal Engineering Journal, vol. 40, No. 2, 131–160.

21. Mizutani, N., Mostafa, A.M. and Iwata, K. (1998): Non-linear regular wave, submerged breakwater and seabed dynamic interaction. Coastal Engineering, vol. 33, No. 2–3, 177–202.

22. Mostafa, A.M. and Mizutani, N. (1997): Numerical analysis of dynamic interaction between non-linear waves and permeable toe over sand seabed in front of a seawall. Proceedings of the 7th International Offshore and Polar Engineering Conference, ISOPE 1997, HI, USA, vol. 3, 823–830.

23. Mostafa, A. M., Mizutani, N. and Iwata, K. (1999): Nonlinear wave, composite breakwater, and seabed dynamic interaction. Journal of Waterway, Port, Coastal, and Ocean Engineering, vol. 125, No. 2, 88–97.

24. Mynett, A.E. and Mei, C.C. (1982): Wave-induced stresses in a saturated poro-elastic sea bed beneath a rectangular caisson. Géotechnique, vol. 32, No. 3, 235–247.

25. Oumeraci, H. (1994): Review and analysis of vertical breakwater failures: Lessons learned. Coastal Engineering, vol. 22, No. 1–2, 3–29.

26. Oumeraci, H., Kortenhaus, A., Allsop, W., de Groot, M., Croch, R., Vrijling, H. and Voortman, H. (2001): Probabilistic Design Tools for Vertical Breakwaters. Balkema, Lisse, The Netherlands.

27. Peacock, W.H. and Seed, H.B. (1968): Sand liquefaction under cyclic loading simple shear conditions. Journal of Soil Mechanics and Foundations Engineering, ASCE, vol. 94, No. SM3, 689–708.

28. Puzrin, A.M., Alonso, E.E. and Pinyol, N.M. (2010): Caisson failure induced by liquefaction: Barcelona Harbour, Spain. In: Geomechanics of Failures. Springer Science+Business Media B.V., Springer, Dordrecht, Heidelberg, London, New York, 85–148.

29. Rahman, M.S., Seed, H.B. and Booker, J.R. (1977): Pore pressure development under offshore gravity structures. Journal of Geotechnical Engineering Division, ASCE, vol. 103, No. GT12, 1419–1436.

30. Shore Protection Manual (1977): Shore Protection Manual, Vol. II, U.S. Army Coastal Engineering Research Center, Department of the Army Corps of Engineers, Third Edition.

31. Sumer, B.M. and Cheng, N.-S. (1999): A random-walk model for pore pressure accumulation in marine soils. Proceedings of the 9th International Offshore and Polar Engineering Conference, ISOPE-99, Brest, France, 30 May–4 June, 1999, vol. 1, 521–526.

32. Sumer, B.M., Dixen, F.H. and Fredsøe, J. (2010): Cover stones on liquefiable soil bed under waves. Coastal Engineering, vol. 57, No. 9, 864–873.

33. Sumer, B.M. and Fredsøe, J. (1997): Hydrodynamics Around Cylindrical Structures, World Scientific, Singapore, 530 p. Second edition 2006.

34. Sumer, B.M., Fredsøe, J., Christensen, S. and Lind, M.T. (1999): Sinking/floatation of pipelines and other objects in liquefied soil under waves. Coastal Engineering, vol. 38, No. 2, 53–90.

35. Sumer, S.K., Sumer, B.M., Dixen, F.H. and Fredsøe, J. (2008): Pore pressure buildup in the subsoil under a caisson breakwater. Proceedings of the 18th International Offshore (Ocean) and Polar Engineering Conference, ISOPE 2008, Vancouver, BC, Canada, July 6–11, 664–671.

36. Tomi, Y., Zen, K., Chen, G., Kasama, K. and Yahiro, Y. (2009): Effect of relative density on the wave-induced liquefaction in seabed around a breakwater. Proceedings of the ASME 28th International Conference on Ocean, Offshore and Arctic Engineering, OMAE2009, May 31–June 5, 2009, Honolulu, Hawaii, Paper number: OMAE2009-79601.

37. Ulker, M., Rahman, M.S. and Guddati, M.N. (2009): Breaking wave-induced dynamic response of rubble mound and seabed under a caisson breakwater. Proceedings of the 28th International Conference on Ocean, Offshore Mechanics and Arctic Engineering, OMAE 2009, May 31–June 05, 2009, Honolulu, HI, USA.

38. Ulker, M., Rahman, M.S. and Guddati, M.N. (2010 a): Wave-induced dynamic response and instability of seabed around caisson breakwater. Ocean Engineering, vol. 37, No. 17–18, 1522–1545.

39. Ulker, M., Rahman, M.S. and Guddati, M.N. (2010 b): Standing wave-induced dynamic response and instability of seabed under a caisson breakwater. Proceedings of the 29th International Conference on Ocean, Offshore Mechanics and Arctic Engineering, OMAE 2010, June 6–11, 2010, Shanghai, China.

40. Ulker, M., Rahman, M.S. and Guddati, M.N. (2012): Breaking wave-induced response and instability of seabed around caisson breakwater. International Journal for Numerical and Analytical Methods in Geomechanics, vol. 36, No. 3, 362–390.

41. Zen, K. (1993): Study on the wave-induced liquefaction in seabed. Technical Note of the Port and Harbour Research Institute Ministry of Transport, Japan, No. 755 (in Japanese).

42. Zen, K., Umehara, Y. and Finn, W.D.L. (1986): A case study of the wave-induced liquefaction of sand layers under the damaged breakwater. Proceedings of the 3rd Canadian Conference on Marine Geotechnical Engineering. St. Johns's, NL, Canada, 505–519.

43. Zen, K. and Yamazaki, H. (1993): Wave-induced liquefaction in a permeable seabed. Report of the Port and Harbour Research Institute, vol. 31, No. 5, 155–192.

44. Zhang, X.Y., Lee, F.H. and Leung, F. (2009): Response of caisson breakwater subjected to repeated impulsive loading. Géotechnique, vol. 59, No. 1, 3–16.

Chapter 9

Stability of Rock Berms in Liquefied Soil

There are basically three kinds of protection measures for pipelines: (1) the pipeline may be laid in a trench; (2) it may be covered with a stone protection layer; or (3) it may be covered with a protective mattress (see, e.g., Sumer and Fredsøe, 2002).

This chapter is concerned with the first kind of protection.

A viable option for this kind of protection is to install a rock berm over the pipeline (Fig. 9.1a). There are two scenarios related to this option. The trench and the rock berm are left open. However, in this case, the trench may be backfilled due to sediment transport (Fig. 9.1b), the natural backfilling; or, following the installation of the rock berm, the trench is backfilled intentionally with the *in-situ* sediment. The sediment in both cases (being in the loose state because the backfilling processes involve slow sedimentation) may be susceptible to liquefaction under waves. With the liquefaction of the sediment, internal waves emerge at the interface between the liquefied sediment and the water column (Chapter 3, Section 3.1.3), and consequently the liquefied sediment experiences an orbital motion. With this, the rock berm will be exposed to the motion of *liquefied soil*.

Rock berms may also be exposed to the motion of liquefied soil when pipelines (laid not in a trench but directly on the seabed) are buried under migrating sand waves. Here, too, the sediment may be in the loose state, and therefore it may be susceptible to liquefaction under waves, with the berm structure exposed to the motion of the liquefied soil.

The conventional design strategy for a rock berm exposed to the water motion requires the stability of the top layer. For this, the Shields criterion is used; namely, the Shields parameter calculated for the top-layer stones must be smaller than the critical value of the Shields parameter corresponding to the initiation of motion (see, e.g., Sumer and Fredsøe, 2002). Another approach is to accept a certain amount of damage to the protection structure, an approach similar to that used in the design of rubble-mound breakwaters. This approach has been adopted by Klomp and Tonda (1995) and Tørum et al. (2008) to study the stability of rock berms.

Although a substantial amount of knowledge on the stability of rock berms had accumulated over the years, this is not the case when these structures are exposed to the motion of liquefied soil.

Sumer et al. (2011) have conducted laboratory experiments to study the detailed mechanism of stability of such rock berms exposed to the orbital motion of liquefied soil. The present chapter essentially summarizes the results of this latter work.

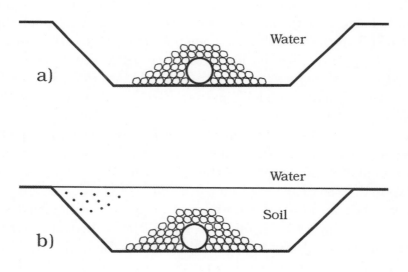

Figure 9.1:   (a): Installation of a rock berm over a pipeline. (b): The trench is backfilled (naturally or intentionally).

9.1 Sequence of Liquefaction with the Berm Present

In Sumer *et al.*'s (2011) study, the experiments were carried out in a wave flume, the same test facility described earlier in Section 3 (see Fig. 3.1). Fig. 9.2 illustrates the test set-up with the model berm. The water depth in the tests was 40 cm. The soil pit had a depth of 16.5 cm, and a length of 84 cm. In the majority of the tests, a trapezoidal-shaped berm was used, Fig. 9.2.

The soil was liquefied by waves (with $H = 17$ cm wave height and $T = 1.6$ s wave period), and the rock berm was subject to the orbital motion of the liquefied soil. The soil was silt with $d_{50} = 0.075$ mm. Various berm materials were used, angular and round stones sizes 0.74–2.5 cm, plastic balls the size 3.6 cm, brass prisms (with hexagonal cross section) sized (height and width) 2.5 cm, and steel circular cylinders sized (height and diameter) 1.0 cm. In the tests, the standard pore-water pressure and water-surface elevation (at the same section as the pressure measurements) were made.

Sumer *et al.*'s (2011) experiments showed that the sequence of liquefaction and compaction process in the present case (in the presence of the model berm) occurred in much the same way as in the "undisturbed" case, described in Section 3.1 (Chapter 3). Sumer *et al.*'s (2011) pore-water measurements indicated that the behaviour of the pressure time series was exactly the same as that in the undisturbed case.

With the soil liquefied, the water column and the liquefied soil formed a two-layered system of liquids, in the same way as in the undisturbed case (Fig. 3.9).

Figure 9.3 presents the data obtained in Sumer *et al.*'s (2011) experiments where, a, is the orbital motion of liquefied soil at two sections, $x = 32$ cm, Section I (see the inset in Fig. 9.3) (square symbols), and $x = 50$ cm, and Section II, at the center of the berm structure (crosses). Fig. 9.3 also includes the orbital-amplitude data corresponding to the undisturbed case, with no berm, for comparison (filled circles). The a value for water particles obtained from the small-amplitude wave theory for a rigid bottom (potential theory) is also plotted in the figure as a reference line (dashed line). It should be noted that the amplitudes were measured for the stage where the soil was in the liquefied state across the entire soil depth.

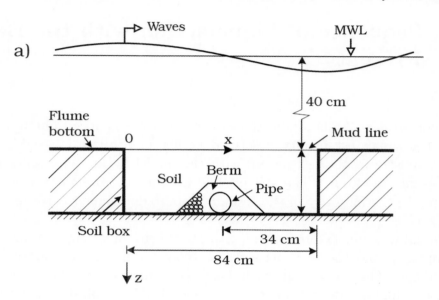

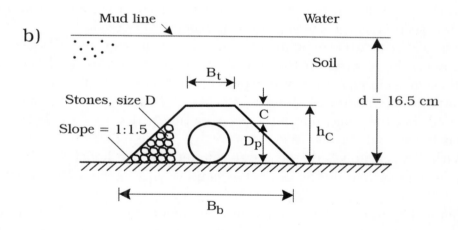

Figure 9.2:   Test set-up in Sumer *et al.* (2011). The figure is not to scale.

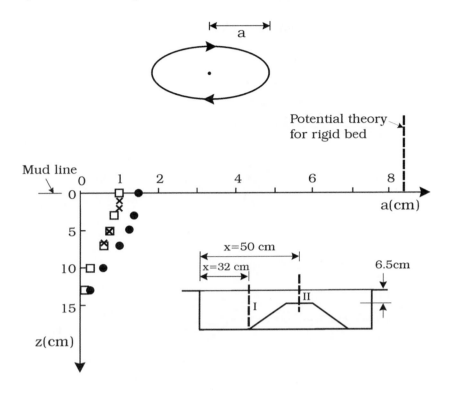

Figure 9.3: Amplitude of liquefied soil as function of depth. Square symbols: liquefied soil at x = 32 cm. Crosses: liquefied soil at x = 50 cm. Filled symbols: liquefied soil, undisturbed case. Sumer *et al.* (2011).

Figure 9.3 shows that the amplitude a is practically the same for both Sections I and II. However, it is clear from the figure that a is smaller than that experienced in the undisturbed case. This is essentially due to the fact that the presence of the berm structure "obstructs" the motion of the liquefied soil, and therefore the amplitude of the orbital motion of the liquefied soil is reduced.

Sumer *et al.* (2011) found that the dissipation and compaction stage of the sequence of the soil behaviour is also qualitatively much the same as in the undisturbed case. Fig. 9.4 illustrates how the compaction front progresses with (Fig. 9.4b) and without (Fig. 9.4a) the presence of the berm structure.

In Fig. 9.4, $t = 0$ coincides with the instant where the first wave arrives at the pressure measurement section. The soil is liquefied first at the surface ($z = 0$) within $t = 0 - 6$ s, and the liquefaction spreads down rapidly over

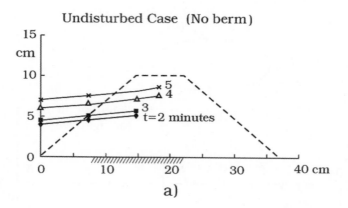

a)

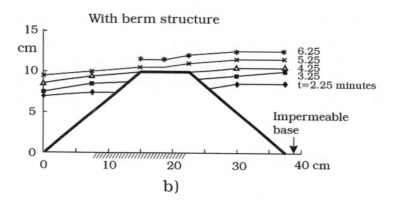

b)

Figure 9.4:   Compaction front progresses (a) in the absence of the berm, and (b) in the presence of the berm. Sumer *et al.* (2011).

the time period from $t = 6\,\mathrm{s}$ to $18\,\mathrm{s}$. Subsequently, the compaction process begins where the soil grains come in contact with each other. The compaction process first begins at the impermeable base and gradually progresses in the upward direction, just as in the undisturbed case, as described in greater details in Chapter 3 (Section 3.1.4). The slight tilt at the compaction front in Fig. 9.4a, the undisturbed case, is attributed to the fact that the liquefaction front reaches the impermeable base slightly earlier in the central areas, and therefore the compaction front will travel to slightly higher elevations in these areas. Sumer *et al.* (2011) also note that the physics of the compaction process implies that the relative density of the compacted soil is expected to

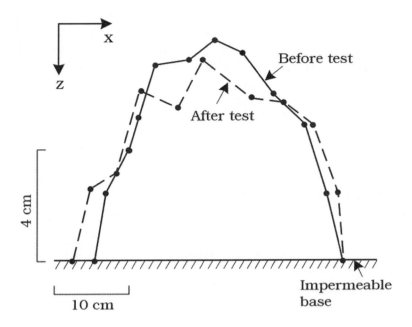

Figure 9.5: Damage to the berm across the cross-shore direction. Sumer *et al.* (2011).

be similar to that in the undisturbed case, measured in their earlier study Sumer *et al.* (2006 a) as $D_r = 0.78$ for similar wave and soil conditions.

9.2 Stability of Berm Structure

9.2.1 General description

Figure 9.5 presents the damage to the berm in one test in Sumer *et al.*'s (2011) experiments, the solid line representing the cross-section of the berm structure before the test, and the dashed line that after the berm structure was exposed to the orbital motion of the liquefied soil.

Figure 9.6, on the other hand, gives the damage across the spanwise length of the berm structure for the same test.

As seen, the damage to the berm is fairly substantial. The crest of the berm experiences displacements of $O(1\,\text{cm})$. The fact that the damage is uneven in the spanwise direction (Fig. 9.6) may be attributed to (1) the unevenness of the construction of the berm structure, and (2) the nonuniform shape and size of the stones. Although not shown in the paper, Sumer *et al.* (2011) note that the photographs of the berm structure before and after the test showed that the stones during their displacement underwent not only translation but also rotation.

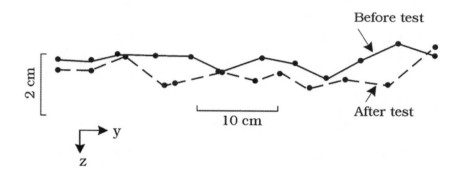

Figure 9.6:   Damage to the berm across the spanwise direction. Sumer *et al.* (2011).

The displacement of stones (i.e., the damage) is due to the hydrodynamic forces on individual stones. There are two kinds of forces: (1) the agitating forces, the in-line force and the lift force, caused by the orbital motion of the liquefied soil, similar to those in the case when an ordinary berm structure is exposed to wave-induced flow in water; and (2) the stabilizing forces, the friction force and the submerged weight of stones. Obviously the stones were displaced because the resultant agitating force exceeded the resultant stabilizing force, as will be discussed in detail later.

Although the experimental constraints did not allow Sumer *et al.* (2011) to monitor the upward movement of the compaction front until it arrived at the level of $z \cong 7\,\text{cm}$, Fig. 9.4b shows that the stones above this level were exposed to the liquefied soil for at least $O(2\text{ minutes})$ (the stones just at the top of the berm are exposed to the liquefied soil even longer, for $O(5\text{ minutes})$), and this indicates clearly that the stones were exposed to the orbital motion of the liquefied soil over a time long enough, $O(2-5\text{ minutes})$, so that they can be displaced under the forces induced by the orbital motion of the liquefied soil.

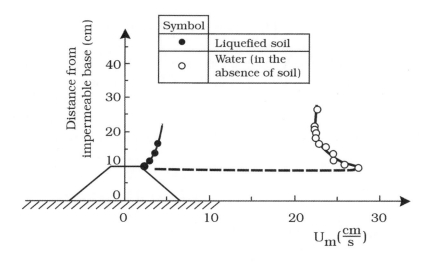

Figure 9.7: Amplitude of velocity as function of vertical distance. Comparison between the water case and the liquefied-soil case under exactly the same wave conditions. Sumer *et al.* (2011).

9.2.2 Comparison with berm exposed to water motion

Sumer *et al.* (2011) carried out two special tests, to compare the response of the berm structure in two cases: (1) when it is exposed to water; and (2) when it is exposed to liquefied soil.

In the first test, the berm structure was placed in position (Fig. 9.2a) without the soil, and it was exposed to water in the test. In the second test, the soil was placed in the sediment box, and therefore the berm structure was, upon liquefaction, exposed to the liquefied soil. The berm material was round stones sized 2.5 cm.

The berm structure was exactly the same in both experiments. The wave climate was also exactly the same in the two tests, with the wave height $H = 17\,\text{cm}$, the wave period $T = 1.6\,\text{s}$, and the water depth $h = 40\,\text{cm}$. The amplitude of the orbital velocity of water particles in the first test, and that of liquefied-soil particles in the second test are plotted in Fig. 9.7 as a function of the vertical distance.

Although the orbital velocity in the water test is a factor of seven larger than in the case of the liquefied soil (Fig. 9.7), stones did not move at all in the water experiment, and the berm structure remained completely intact, whereas a fairly substantial amount of damage occurred (up to as much as $O(2\,\text{cm})$ displacement in the vertical direction) in the case of the liquefied soil.

Sumer *et al.* (2011) checked for the incipient motion in the water case, and found that the Shields parameter was $\theta = 0.027$, smaller than $\theta_{cr} = 0.045$, the critical value for the initiation of the motion (see e.g. Sumer and Fredsøe, 2002, Fig. 1.2), and therefore the stones should not move in the water case, as revealed in the test. Here, the Shields parameter is defined by

$$\theta = \frac{U_{fm,top}^2}{g(s-1)D_{50}} \tag{9.1}$$

in which $U_{fm,top}$ is the maximum value of the friction velocity at the top of the berm, g acceleration due to gravity, s the specific gravity of the stones, and D_{50} the stone size.

The question is why the same stones under the same setting and under the same wave climate (with even a factor of seven smaller orbital-velocity magnitude) move in the case of the liquefied soil. Sumer *et al.* (2011) listed the following factors contributing to the "earlier" incipient stone motion in the case of the liquefied soil:

1. The submerged weight of the stones in water is $g(\rho_s - \rho)\frac{\pi D^3}{6}$ while it is $g(\rho_s - \rho_{liq})\frac{\pi D^3}{6}$ in liquefied soil, meaning that the stones are lighter in the liquefied soil by a factor of 2.3, considering $\rho_{liq}/\rho = 1.93$ (Sumer *et al.*, 2006b) and $\rho_s/\rho = 2.65$.

2. This also implies that the weight and therefore the friction force (the force opposing to the motion of the stones) is a factor of 2 less in the case of the liquefied soil.

3. Sumer *et al.* (2011) pointed out that, no study is available investigating forces on objects exposed to an oscillatory motion (or steady current, for that matter) of a liquefied soil. They argued that one can, to a first approximation, assume that the stones experience an in-line force in the liquefied soil case, similar to the familiar Morison force (Sumer and Fredsøe, 1997). This force consists of two components, the drag component and the inertia component. The "composition" of these forces (the inertia and drag forces) is unknown as the liquefied soil behaves not as a Newtonian fluid. Making an analogy to the steady motion of a body in a fluid, Sumer *et al.* (1999) determined the resistance force and therefore the drag coefficient from an experiment where a "marine" object (a circular cylinder, or a sphere or a cube) falls in a liquefied soil, described in Chapter 6 (Section 6.3). The latter data showed

that the drag coefficient is eight orders of magnitude larger than the water values (see the discussion in Section 6.3 under *Viscosity of Liquefied Soil*), indicating that the drag force in the liquefied soil is expected to be larger in the liquefied soil.

4. Regarding the inertia component of the in-line force, this is composed of two parts: the apparent mass force and the Froude–Krylov force (Sumer and Fredsøe, 1997). The Froude–Krylov force should be increased by a factor of 1.93 in the case of the liquefied soil simply because the density of the liquefied soil is increased by a factor of 1.93. The second part, the apparent-mass force in the case of the liquefied soil, is unknown. Nevertheless, Sumer *et al.* (2011) argued that this, too, is expected to increase because of the increase in the density of the liquefied soil.

From the above considerations, the agitating forces on the stones will be increased while the resistance force will be decreased in the case of the liquefied soil, resulting in an early threshold of the stone motion, as revealed by Sumer *et al.*'s (2011) observations.

9.2.3 Incipient stone motion

As demonstrated in the preceding paragraphs, stones of a berm structure which cannot be moved in water can easily be moved in liquefied soil under the same waves. The degree of the stone movement (i.e., the damage) in the latter case could be fairly substantial (e.g., Figs 9.5 and 9.6), and therefore the berm needs to be designed to ensure its stability if and when the soil is liquefied. The question is: what are the parameters which govern the initiation of stone motion in liquefied soil?

This sub-section attempts to address this question with the purpose of establishing design guidelines to determine the incipient motion.

On dimensional grounds, the initiation of stone motion is governed by three parameters, the so-called mobility parameter,

$$\Psi = \frac{U_m^2}{g(\frac{\rho_s}{\rho_{liq}} - 1)D} \tag{9.2}$$

the stone Reynolds number,

$$\text{Re}_D = \frac{DU_m}{\nu'} \tag{9.3}$$

and the Keulegan–Carpenter number,

$$KC = \frac{U_m T}{D}. \tag{9.4}$$

Here, U_m is the amplitude of the orbital velocity of liquefied soil *in the undisturbed case at the level of the top of the berm*. The quantity ρ_{liq} is the density of the liquefied soil. The kinematic viscosity of the liquefied soil, ν', in Eq. 9.3 was calculated in Sumer *et al.*'s (2011) study from

$$\nu' = \frac{2}{2 - 3c}\nu \tag{9.5}$$

where c is the solid concentration (volume concentration) of the liquefied soil (Cheng, 1997). This definition of the kinematic viscosity, different from the actual kinematic viscosity of the liquefied soil ν_{liq} (discussed in Section 6.3, the subsection *Viscosity of Liquefied Soil*) is maintained as in Sumer *et al.*'s (2011) original publication, to avoid confusion.

The mobility parameter, Ψ, can be interpreted as the ratio of the agitating force and the stabilizing force (i.e., the friction force, which is equal to $\mu(W - F_L)$ in which μ is the friction coefficient, W the submerged weight of the stone and F_L the lift force). Sumer *et al.* (2011) note that, in the above formulation, U_m is adopted in favour of the friction velocity U_{fm} for simplicity. The above analysis can be considered as the extension of the classic Shields analysis to the incipient sediment motion in liquefied soil in waves, with the friction velocity replaced by U_m.

Fig. 9.8 shows the data plotted Ψ as function of KC for the range of the Reynolds number experimented with, i.e., $40 \leq \mathrm{Re}_D \leq 240$, reproduced from Sumer *et al.* (2011). The filled symbols represent the experiments where the berm stones moved whereas the empty symbols represent those where they did not move. The nondimensional quantities in the figure are based on U_m values which are calculated from $U_m = \frac{2\pi a}{T}$ in which a values are taken from Fig. 9.3 for the undisturbed case. The specific gravity of the liquefied soil, ρ_{liq}/ρ, is taken as 1.93 (see Chapter 5, Section 5.3, and also Sumer *et al.*, 2006b) whereas the concentration c in Eq. 9.5 is calculated from Eq. 5.17,

$$\frac{\rho_{liq}}{\rho} = (1 - c) + c\frac{\rho_s}{\rho}. \tag{9.6}$$

The specific gravity of stones used in the experiments is given in the legend in Fig. 9.8.

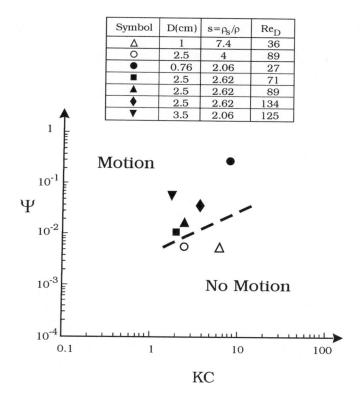

Symbol	D(cm)	s=ρ_s/ρ	Re$_D$
△	1	7.4	36
○	2.5	4	89
●	0.76	2.06	27
■	2.5	2.62	71
▲	2.5	2.62	89
◆	2.5	2.62	134
▼	3.5	2.06	125

Figure 9.8: Incipient stone motion in liquefied soil. Filled symbols: motion. Empty symbols: no motion. For definitions of KC and Re$_D$, see the text. Sumer *et al.* (2011).

As Sumer *et al.* (2011) pointed out, the size of the data in Fig. 9.8 is too small to resolve the KC dependence of the critical mobility number for the range of the Keulegan–Carpenter number encountered in practice, namely $KC < O(40)$. However, the diagram can, to a first approximation, be used when KC remains within the range tested in the experiments, $1 \lesssim KC \lesssim 10$. Sumer *et al.* (2011) also note that the mobility-number data did not give any marked trend when plotted as function of the Reynolds number. The real-life situations may involve Reynolds numbers large compared with the Reynolds numbers in Fig. 9.8. Therefore, caution must be exercised when extending the results in Fig. 9.8 to prototype conditions. See the discussion below under *Remarks on Practical Application*.

9.2.4   Degree of damage in Sumer *et al.*'s (2011) tests

Fig. 9.9 shows the damage in terms of the vertical displacement of the top stones, as a function of the cover thickness, C. The quantity S in the figure represents the average stone displacement in the vertical direction, calculated from

$$S = \left(\frac{\sum_{i=1}^{N}(\Delta z)^2}{N} \right)^{1/2} \tag{9.7}$$

in which Δz is the vertical component of stone displacement, and N the sample size. The root-mean-square value is preferred in favour of the ordinary arithmetic mean to include in the average the negative values of the displacement, which occurred in a few cases. The average in the preceding equation is taken over the entire spanwise length of the berm structure where N was in the range $N = 9 - 12$.

Fig. 9.9 shows that the damage, S, increases with increasing C, the cover thickness (see Fig. 9.2b for the definition sketch for C). This is because the top stones of the berm structure experience larger and larger orbital velocities of the liquefied soil with increasing C (Fig. 9.3), and therefore the damage should also increase with C.

Sumer *et al.* (2011) note the following for the case of breakwaters and regular rock berms. In breakwater tests and in tests related to stability of rock berms over pipelines, another definition of damage is given by

$$S_1 = \frac{A}{D_{n50}^2} \tag{9.8}$$

in which A is the area in cross section of the damaged part at the top of the breakwater or the rock berm, Tørum *et al.* (2008). The latter authors recommend that the damage S_1 should not exceed 50 for mounds covering pipelines, 2–3 m wide on top of the mound and with $D_{n50} = 0.05 - 0.10$ m. For larger stones, the exceeding limit for S_1 should be lower.

Now, returning to Fig. 9.9, the trend in the figure is such that $S \to 0$, as $C \to 0$. This is because when $C = 0$, from Fig. 9.2, h_C becomes $h_C = 5$ cm; this corresponds to a distance of $z = 16.5 - 5.0 = 11.5$ cm, and the orbital amplitude (or alternatively the orbital velocity) corresponding to this value of z is, from Fig. 9.3, apparently very small. Hence, the damage will be practically nil.

The damage S seems to attain a constant value as C approaches the mudline ($C \to (d - D_p) = 16.5 - 5.0 = 11.5$ cm), the total depth of the

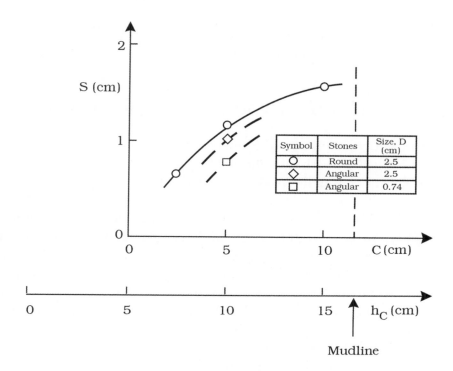

Figure 9.9: Damage as function of cover-layer thickness. S is the average stone displacement in the vertical direction, representing the damage. See Fig. 9.2 for definition sketch for other quantities. Sumer *et al.* (2011).

soil. The figure shows that the maximum percentage damage is $S/h_C = (1.6\,\mathrm{cm})/(15\,\mathrm{cm}) = O(0.1)$.

For the same value of the cover layer thickness ($C = 5\,\mathrm{cm}$), the damage S decreases when the stone shape is changed from round to angular. This can be explained by larger interlocking between the angular stones. In the angular stone case, the damage is smaller for smaller stones ($D = 0.74\,\mathrm{cm}$, square symbol in Fig. 9.9) than for larger stones ($D = 2.5\,\mathrm{cm}$, diamond symbol), and this can be explained by the sheer size of the stones; the smaller the stone size, the smaller the agitating forces, and therefore the smaller the displacement of the stone.

Sumer *et al.* (2011) also tested a pipe with a density intentionally maintained at a very small value, namely $\rho_{pipe}/\rho = 1.2$. With the soil liquefied, the pipe apparently broke through the berm structure, and was observed to float to the surface of the soil. Clearly, with the pipe floating, the cover structure fails, and the result could be catastrophic.

9.3   Remarks on Practical Application

In order to apply the results of Sumer *et al.*'s (2011) study in a real-life problem, one needs to estimate the maximum value of the orbital velocity of liquefied soil in the undisturbed case at the level of the top of the berm structure, U_m. This is to calculate the mobility parameter, Eq. 9.2, as well as the Keulegan–Carpenter number, Eq. 9.4. To this end, the potential theory given in Sassa *et al.* (2001) who implemented Lamb's theory for wave propagation in a two-layer fluid for the present two-layered system (see the discussion in Chapter 3, Section 3.1.3) may be used. However, comparison of the results from the latter theory and Sumer *et al.*'s (2011) experimental results indicate that provisions need to be made in a typical field implementation to accommodate the finite offshore–onshore extent of the liquefied soil to reduce the velocities predicted by the theory.

With the above analysis, the procedure to implement Sumer *et al.*'s (2011) results in a field situation may be summarized as follows. (1) Calculate the maximum value of the orbital velocity of liquefied soil, U_m, in the undisturbed case at the level corresponding to the top layer of the berm structure. (2) Calculate the mobility parameter (Eq. 9.2). For the latter, use the information given in Chapter 5 (Section 5.3, or Sumer *et al.*, 2006b) to assess the density of the liquefied soil, ρ_{liq}. (3) Calculate the Keulegan–Carpenter number from Eq. 9.4. (4) Calculate the Reynolds number from Eqs 9.3 and 9.5. (5) Use Fig. 9.8 (to a first approximation), to check whether or not the designed stone size is large enough for the stone stability.

Clearly, the prototype values of the Reynolds number will be large compared with those corresponding to Sumer *et al.*'s (2011) tests. Sumer *et al.* (2011) note that the way in which the Reynolds number influences the end result is unclear at the present time. Sarpkaya's (1975) water experiments for *free spheres* show that the force coefficients (the drag coefficient and the inertia coefficient) do not depend on the Reynolds number. Whether or not this result can be extended to liquefied soils is unknown. Therefore caution should be observed when using the present results for field conditions. In this context, Sumer *et al.* (2011) also refer to the *steady-case* drag coefficient versus the Reynolds number variation for liquefied soils obtained for various geometrical shapes (circular cylinders, spheres, and cubes) reported in Sumer *et al.* (1999), and reproduced in Chapter 6, Section 6.3. The latter study showed that the drag coefficient for very small Reynolds numbers varies significantly with the Reynolds number whereas, for large Reynolds numbers, no significant variation is observed (Fig. 6.10).

If some damage is allowed to the berm structure, then Ψ can be allowed to have values larger than the critical value for threshold (in Fig. 9.9). In this case, although it is at best suggestive, the model results given in Fig. 9.9 can be scaled up to make an assessment of the degree of damage. For this, (1) the damage S on the vertical axis in Fig. 9.9 should be scaled with h_C to get the percentage damage, and (2) the cover-layer height, h_C, on the horizontal axis should be scaled with d, the soil depth.

9.4 References

1. Cheng, N.-S. (1997): Effect of concentration on settling velocity of sediment particles. ASCE J. Hydraulic Engineering, vol. 123, No. 8, 728–731.

2. Klomp, W.H.G. and Tonda, P.L. (1995): Pipeline cover stability. Proceedings of the 5th International Offshore and Polar Engineering Conference, The Hague, The Netherlands, 11–16. June, 1995, vol. II, 15–22.

3. Sarpkaya, T. (1975): Forces on cylinders and spheres in a sinusoidally oscillating fluid. Transactions of the ASME, Journal of Applied Mechanics, March 1975, 32–37.

4. Sassa, S., Sekiguchi, H. and Miyamoto, J. (2001): Analysis of progressive liquefaction as a moving-boundary problem. Géotechnique, vol. 51, No. 10, 847–857.

5. Sumer, B.M., Dixen F.H. and Fredsøe, J. (2011): Stability of submerged rock berms exposed to motion of liquefied soil in waves. Ocean Engineering. vol. 38, No. 7, 849–859.

6. Sumer, B.M., Fredsøe, J., Christensen, S. and Lind, M. T. (1999): Sinking/Floatation of pipelines and other objects in liquefied soil under waves. Coastal Engineering, vol. 38, No. 2, 53–90.

7. Sumer, B.M. and Fredsøe, J. (1997): Hydrodynamics Around Cylindrical Structures. World Scientific, Singapore, First edition: 1997, Second/revised edition: 2006.

8. Sumer, B.M. and Fredsøe, J. (2002): The Mechanics of Scour in the Marine Environment. World Scientific, Singapore.

9. Sumer, B.M. Hatipoglu, F., Fredsøe, J. and Sumer, S.K. (2006a): The sequence of soil behaviour during wave-induced liquefaction. Sedimentology, vol. 53, 611–629.

10. Sumer, B.M., Hatipoglu, F., Fredsøe, J. and Hansen, N.-E. O. (2006b): Critical floatation density of pipelines in soils liquefied by waves and density of liquefied soils. Journal of Waterway, Port, Coastal and Ocean Engineering, ASCE, vol. 132, No. 4, 252–265.

11. Tørum, A., Kuester, C. and Arntsen, Ø. (2008): Stability against waves and currents of rubble mounds over pipelines. NTNU, Department of Civil and Transport Engineering. Report: BAT/MB-R1/2008. Paper presented at ICCE 2010; the full reference: Tørum, A., Arntsen, Ø.A. and Kuester, C.: Stability against waves and currents of gravel rubble mounds over pipelines and flat gravel beds. Proceedings, Paper No. 268.

Chapter 10

Impact of Seismic-Induced Liquefaction

Earthquakes are an open, direct threat to marine structures (such as quay walls, piers, dolphins, breakwaters, buried pipelines, sheet-piled structures, containers, silos, warehouses, storage tanks located in coastal areas, etc.) when structures are located at or near the epicenter. The structure in this case will be exposed to the devastating shaking effect of the seismic action, and the result can be catastrophic.

Earthquakes may also be a threat to marine structures in an indirect way, through the shaking of the supporting soil. The stability and integrity of structures will be at risk if the soil fails due to liquefaction as a result of the shaking of the soil (Chapter 1, Section 1.3). This kind of failure also can be catastrophic, as observed, for example, in the 1995 Japan Kobe earthquake, and the 1999 Turkey Kocaeli earthquake.

Liquefaction-induced damage to marine structures has been documented quite extensively in the literature: Wyllie *et al.* (1986) (Chile); Iai and Kameoka (1993) (Japan); Iai *et al.* (1994) (Japan); Hall (1995) (USA); Sugano *et al.* (1999) (Taiwan); Boulanger *et al.* (2000) (Turkey); Sumer *et al.* (2002) (Turkey); and Katopodi and Iosifidou (2004) (Greece), to give but a few examples.

A partial list of well-documented case histories can be found in PIANC (2001).

The questions that design engineers face in the case of liquefaction failure are mainly: (1) can the soil be liquefied under a given "design" earthquake? (2) If the soil is liquefiable, how extensive will the damage be to the structure?

(3) Is this damage acceptable (i.e., is it within the limit of damage criteria)? (4) If not, what will the damage (if any) be when some form of "remediation" is implemented? (5) Is the latter damage (if any) within the limit of damage criteria?, etc.

A substantial amount of knowledge on the seismic design of marine structures has accumulated over the past 40–50 years, which has led to excellent treatments on the general subject "seismic design guidelines for marine structures", the most important of which are (1) European Committee for Standardization (CEN), 1994, Eurocode 8: Design Provisions for Earthquake Resistance of Structures; (2) ASCE, 1998, Seismic Guidelines for Ports; and (3) PIANC, 2001, Seismic Design Guidelines for Port Structures.

The aforementioned publications also covered (to some degree) liquefaction design guidelines as well.

The focus of the present chapter is seismic-induced liquefaction and its implications for marine structures. The chapter is organized as follows. The next section presents a general review of seismic-induced liquefaction including the basic concepts, description of the physical process of soil liquefaction under seismic loading and a general overview. The following section gives a detailed review of the existing codes/guidelines regarding seismic-induced liquefaction and its implications for marine structures. This section is followed by two detailed reviews, namely, the Japanese experience of earthquake-induced liquefaction damage on marine structures, and the 1999 Turkey Kocaeli earthquake and liquefaction damage on marine structures. In these reviews, many, well-documented case histories of seismic-induced liquefaction damage are summarized/illustrated; and recommendations which draw on the lessons learned are given. The following sections continue to review two other issues central to marine structures, namely, assessment of liquefaction-induced lateral ground deformations; and tsunamis and their implications.

All information in the present chapter is essentially extracted from a recent publication, Sumer *et al.* (2007), which appeared in a two-volume ASCE *Journal of Waterway, Port, Coastal and Ocean Engineering* special issue on liquefaction around marine structures, edited by the author. The section titles in the following treatment are, except with a few slight changes, maintained as in the original publication. The contributors to the original publication according to the sections were as follows:

- Seismic-Induced Liquefaction by Professor Atilla Ansal;

- Review of the Existing Codes/Guidelines with Special Reference to Marine Structures by Dr. Niels-Erik Ottesen Hansen and Mr. Jesper Damgaard;

- Japanese Experience of Earthquake-Induced Liquefaction Damage on Marine Structures by Professor Kouki Zen;

- Turkey Kocaeli Earthquake and Liquefaction Damage on Marine Structures by Professor Ali Riza Gunbak, Professor Yalcin Yuksel, Dr. Niels-Erik Ottesen Hansen, Professor Andrzej Sawicki, and the author;

- Assessment of Liquefaction-Induced Lateral Ground Deformations by Professor K. Onder Cetin; and

- Tsunamis and Their Impacts by Professor Ahmet C. Yalciner and Professor Costas Synolakis.

In the original publication (Sumer *et al.*, 2007), it was remarked that, with the previously mentioned contributions and new set of information, data and recommendations, their work (Sumer *et al.*, 2007) and the guidelines CEN, ASCE, and PIANC form a complementary source of information on earthquake-induced liquefaction with special reference to marine structures.

10.1   Seismic-Induced Liquefaction

Liquefaction potential depends on the nature of ground shaking and material susceptibility to liquefaction. For potential liquefiability, saturation is an additional necessary condition besides material susceptibility. The cyclic loading induced by seismic excitation represents an ideal loading type for initiation of soil liquefaction.

Liquefaction may be defined as the transformation of a granular material from a solid to a liquefied state as a consequence of increased pore water pressure and reduced effective stress, as described in greater detail in Chapter 3, in conjunction with the wave-induced liquefaction. The tendency of granular materials to decrease in volume when subjected to cyclic shear deformations leads to a positive increase in pore-water pressure resulting in a decrease of effective stress within the soil mass (Chapter 3, Sections 3.1.1 and 3.1.2). The change of state occurs most readily in loose to moderately dense granular soils, such as silty sands and sands and gravels capped by, or containing seams of impermeable sediments. As liquefaction occurs soil stratum softens, allowing large cyclic deformations to occur. In loose materials, softening is also accompanied by a loss of shear strength that may lead to large shear deformations or even flow failure under moderate to high shear stresses; such stress conditions can develop beneath a foundation or in a sloping ground. In moderately dense to dense materials, liquefaction leads to transient soften- ing and increased cyclic shear strains, but a tendency to dilate during shear inhibits major strength loss and large ground deformations. A condition of cyclic mobility or cyclic liquefaction may develop following liquefaction of moderately dense materials (Youd *et al.* 2001).

The term liquefaction has different meanings with respect to various soil conditions. According to Ishihara (1996), the following definitions apply to cohesionless soils: for loose sand, the initial liquefaction is the state of softening in which indefinitely large deformation is produced suddenly with near complete loss of strength during or immediately following the 100% pore-water pressure buildup. For medium dense to dense sand, a state of softening-limited liquefaction, cycling softening, or cycling mobility is also produced with the 100% pore-water pressure buildup accompanied by about 5% double amplitude axial strain but the deformation thereafter does not grow indefinitely large and complete loss of strength does not take place in the sample even after the onset of initial liquefaction. In silty sands or sandy silts, the plasticity of fines has a determinant role in liquefiability (Ishihara and Koseki, 1989). Silty soils with nonplastic fines like many tailings materials are as easily liquefiable as clean sands.

Cohesive fines, as in fluvial deposits, generally increase the cyclic resis- tance of silty soils. The previous definitions of liquefaction for sands are usually applicable to slightly cohesive silty soils also. See the discussion in Section 3.5.

Even though soil liquefaction has been observed in history, intense inves- tigations to assess liquefaction susceptibility have been initiated after the two major events of 1964, the Niigata and the Alaska earthquakes. Both of these earthquakes have produced devastating effects due to liquefaction and attracted the attention of researchers. Since then significant efforts have been made to determine the factors affecting liquefaction susceptibility based on laboratory and field tests.

Two variables are required for the assessment of liquefaction resistance of soils: (1) the seismic demand on a soil layer, expressed in terms of cyclic stress ratio (CSR); and (2) the capacity of the soil to resist liquefaction, expressed in terms of the cyclic resistance ratio (CRR). Triggering of liquefaction is generally represented through a series of relationships between the CSR required to produce 5% double-amplitude axial strain (the assumed onset of liquefaction or cyclic mobility) and the number of cycles N of a uniform, constant amplitude cycling loading. The CSR is defined as the ratio of the maximum cyclically applied shear stress to the effective normal stress acting at the beginning of shaking on the plane where shear stress is applied. The cyclic or dynamic strength is defined as the CSR value at $N = 10$ or 20 cycles. The parameters affecting the liquefaction potential of loose, saturated granular soils that have been investigated in a detailed manner and can be summarized as: relative density, confining pressure, fines content, grain characteristics, plasticity of fines, method of sample preparation (because of the resulting soil structure), and the degree of saturation, as discussed in conjunction with the wave-induced liquefaction in Chapter 3. Other factors include prior seismic straining, the coefficient of earth pressure at rest, k_0, the overconsolidation ratio of the soil deposit, and increased time under pressure.

Quantitative assessment of the likelihood of triggering or initiation of liquefaction is the necessary first step for most projects involving potential seismic-induced liquefaction. There are two general types of approaches available for this: (1) use of laboratory testing of undisturbed samples; and (2) use of empirical relationships based on correlation of observed field behavior with various $in\text{-}situ$ index tests.

A detailed approach to determine liquefaction potential of saturated sand deposits requires cyclic tests, preferably, on undisturbed samples. However, one of the major challenges in assessing the liquefaction susceptibility of the soil layers is the limited capability of obtaining undisturbed specimens to be tested in the laboratory. In the early stages, in order to explain the mechanism of liquefaction, extensive experimental studies have been conducted on reconstituted sand samples (Seed and Idriss, 1971; Martin *et al.* 1975; Mulilis *et al.* 1977; Castro and Poulos 1977). In these studies it was observed that the method of sample preparation strongly affects the liquefaction resistance of laboratory-prepared specimens obtained from remolded samples, and there are some difficulties about sample preparation for silty sands in a wide range of gradation and density. In addition, another problem associated with reconstituted samples is the lack of the $in\text{-}situ$ stress history, which leads to an underestimation of the liquefaction resistance, as can be seen in Fig. 10.1. The challenge in the case of laboratory-prepared specimens is not

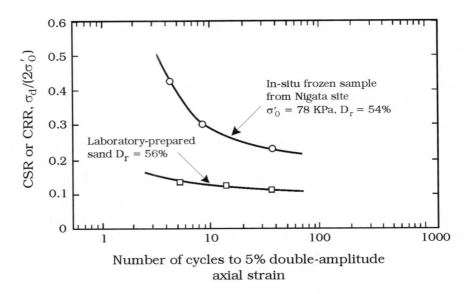

Figure 10.1:   Liquefaction resistance of freshly frozen sand sample obtained from Niigata. Taken from Sumer *et al.* (2007). Original reference: Yoshimi *et al.* (1989).

only in terms of the amplitudes but also in the flatness of the liquefaction resistance curve which makes it very difficult to choose the cyclic shear stress amplitude corresponding to a specified number of cycles in terms of initial liquefaction or in terms of deformation criteria, as can be observed in the liquefaction resistance curve obtained on the laboratory-prepared sample given in Fig. 10.1.

    In recent years, in light of previous findings, the studies of the liquefaction phenomenon of undisturbed sandy, silty soils have received increasing attention to eliminate the effects of the factors mentioned earlier. The use of laboratory testing is complicated by difficulties associated with sample disturbance during both sampling and reconsolidation. It is also difficult and expensive to perform high-quality cyclic simple shear testing. On the other hand, cyclic triaxial testing poorly represents the loading conditions of principal interest for most seismic problems. Both sets of problems can be improved, to some extent, by the use of appropriate sampling techniques, and subsequent testing in a high-quality cyclic simple shear or torsional shear apparatus. The difficulty and cost of these delicate techniques, however, places their use beyond the budget and scope of most engineering studies.

Various advanced undisturbed soil-sampling techniques have been developed by Hatanaka *et al.* (1988), Goto *et al.* (1987), and Yoshimi *et al.* (1989). In these techniques, undisturbed soil samples are obtained by *in-situ* freezing. Hatanaka *et al.* (1988) have shown that liquefaction resistance of reconstituted samples are approximately 50% less than that of undisturbed samples even though they have the same density (cf. Fig. 10.1). In some studies, frozen Shelby tube samples (Ishihara, 1985) and block sampling (Ishihara and Silver, 1977) have been used instead of *in-situ* freezing. In these investigations, it was observed that the reconstituted sand samples also have lower liquefaction resistance compared to undisturbed sand specimens.

Thus, a more empirical approach based on the *in-situ* penetration test results gained popularity in engineering practice as well as in engineering codes. As summarized in a recent state-of-the-art paper (Youd *et al.*, 1997, 2001), four *in-situ* testing methods have now reached a level of sufficient maturity to represent viable tools for this purpose, and these are (1) standard penetration test (SPT); (2) cone penetration test (CPT); (3) measurement of *in-situ* shear wave velocity V_s; and (4) the Becker penetration test. The oldest and still the most widely used of these is the SPT. Assessment methods to determine the liquefaction susceptibility were developed based on the penetration test results coupled with the field observations during major earthquakes. However, even though this has resolved the engineering problem of determining the liquefaction susceptibility of the encountered soil layers, it was always necessary to make some simplifying assumptions to establish the methodology.

The potential of liquefaction is assessed with the aid of liquefaction charts, which are based on observations of liquefaction occurrence and nonoccurrence during past earthquakes. The simplified procedure proposed by Seed *et al.* (1984, 1985) is based on the relationship of SPT N values, corrected for both effective overburden stress and energy, equipment and procedural factors (affecting SPT testing) to $N_{1,60}$ values (Appendix C), versus intensity of cyclic loading, expressed as magnitude-weighted equivalent cyclic stress ratio (CSR_{eq}). The correlation between corrected $N_{1,60}$ values and the intensity of cyclic loading required to trigger liquefaction is also a function of fines content (Seed *et al.*, 2001). CSR_{eq} is estimated from the simplified method of Seed and Idriss (1971). The resulting $N_{1,60}$ values used with modified 5% or less fines content curve of Seed *et al.* (1985) to evaluate liquefaction resistance CRR. The CRR curve represents limiting conditions that determine whether liquefaction will occur. As the curve is valid only for earthquakes with a magnitude of 7.5, the magnitude scaling factor is applied to adjust to the other magnitudes to calculate the corresponding factor of safety (Youd *et al.* 2001).

A factor of safety smaller than one at any depth indicates liquefaction susceptibility at that depth. However, to assess the effect of liquefaction on the ground surface, the variation of the factor of safety with depth needs to be evaluated to determine the possible impact of liquefaction for the engineering structures on the ground surface.

In the context of SPT and CPT methods briefly reviewed in Appendix C, the subject liquefaction assessment is further discussed in Section C.2 (Appendix C).

The data sets used to derive the empirical relationships utilize the observed field evidence of liquefaction as one parameter. But in most cases the liquefaction observations are limited by the liquefaction manifestations observed in the ground surface. It is still necessary to make some simplifications to estimate the depth of liquefaction occurrence in order to derive a correlation between liquefaction potential and penetration resistance.

There are still very few pore pressure records obtained during major earthquakes indicating liquefaction and concerning the change of liquefaction potential with depth based on actual piezometer measurements.

To decide whether liquefaction will or will not inflict damage on the ground surface, the thickness of the liquefiable layer can be compared with the thickness of the surface crust using the criteria given by Ishihara (1985). Iwasaki *et al.* (1982) quantified the severity of possible liquefaction at any site by introducing a factor called the liquefaction potential index. This index gives the liquefiable zone majority in the top 20 m depths of soil deposits through the integration of a function of factor of safety with depth.

10.2   Review of Existing Codes/Guidelines

There are many aspects that need to be considered when reviewing Codes of Practices. There are damage and collapse requirements, compliance criteria, and so on. This particular review focuses on how liquefaction aspects in general are treated.

Earthquake engineering is extensively treated in national and international codes of practices for countries, which are located in the high-intensity earthquake zones (Zone 3 and above, with the earthquake intensity

0.25–0.35 g, see Fig. 1.9). Load systems, damage criteria, and state-of-the-art methods of analyses are defined. Geophysical aspects are also covered as state-of-the-art. The liquefaction problems in marine structures are not so prominently treated, however. They are mostly mentioned and cautioned against. Generally it is specified that the liquefaction potential shall be investigated by certain methods from the literature, in addition to simple rules of thumb for calculations and remediation.

For the European area the Eurocode complex is the governing code. It is denoted Eurocode 8, Design Provision for Earthquake Resistance of Structures (CEN, 1997). The liquefaction aspects are treated in Part 5 Foundations. Definitions are presented and methods to evaluate susceptibility to liquefaction are recommended. For susceptibility, the traditional field test methods of SPT and CPT are specified. In addition, the so-called Field Correlation Approach is mentioned, where future "design" earthquakes are quantified, based on historical earthquakes, which have caused liquefaction.

With respect to remediation, the code specifically mentions the traditional ground improvement methods, compaction and drainage, or deep foundations transferring foundation loads to non-liquefiable strata. It also specifies that densification of soils in connection with cyclic load and liquefaction shall be considered.

The specified liquefaction analyses are not integrated with the other dynamic load effects in the complete analysis of structures. The liquefaction aspects are treated in much less detail than aspects such as earthquake design provisions, repair, and strengthening.

Turkey (a country which lies geographically both in Europe and Asia) is one of a few European countries which lies in Zones 2–3 (Fig. 1.9). Turkey has a detailed code of practice for earthquake engineering. The Turkish Code of Practice for earthquake disaster mainly focuses on evaluations of structures on land. Marine structures are not covered specifically; however, methods for structures such as retaining walls etc. on land can also be applied to marine structures.

The Turkish code is very detailed in presenting analysis methods for the earthquakes, both equivalent loads spectral methods and time-domain methods. It is also very explicit concerning structural details and foundation details for land structures. However, it is somewhat limited with respect to treating liquefaction. It specifies that the "liquefaction potential" shall be investigated for the soils and conditions indicated in Table 10.1 by "appropriate analytical methods" based on *in-situ* laboratory tests. It is to be noted that the code specifies that materials such as soft clay and silty clay also may be liquefiable (Table 10.1). Remedial actions are not specified in the code.

Added after the publication of Sumer *et al.* (2007): A new Code of Practice specifically designed for coastal, port, railway, and airport structures has been published in 2007 with revision in 2008; see the reference Ministry of Transport, the Republic of Turkey (2008).

Table 10.1. Turkish Code of Practice requirements for investigating liquefaction potential.

Soil group	Description of soil groups	Stand. penet. for N	Relat. density D_r	Unconfined compress strength (kPa)	Shear wave velocity (m/s)
D	Soft deep alluvial layers	—			<200
	Loose sand	<10	<0.35		<200
	Soft clay, silty clay	<8		<100	<200

Moving from Europe to Asia, although located in Zone 0 (0.00–0.05 g, see Fig. 1.9), India has a code of practice for earthquake-resistant design of structures (Indian Standards Institution, 1986). It covers civil works in general and not so much marine structures. Structures such as, for instance, dams and bridges, similar to marine structures, are covered for general earthquake analysis as state-of-the-art.

The rules deal with liquefaction as follows:

- Submerged loose sands and soils falling under classification SP with standard penetration values less than the values specified in Table 10.2, shakings caused by earthquake may cause liquefaction or excessive total and differential settlements. In important projects this aspect of the problem needs to be investigated and appropriate methods of compaction or stabilization adopted to achieve suitable N. Alternatively, deep pile foundations may be provided and taken to depths well into the layer, which are not likely to liquefy. Marine clays and other sensitive clays are also known to liquefy due to collapse of soil skeleton/texture and will need special treatment according to site conditions.

- The piles should be designed for lateral loads, neglecting lateral resistance of soil layers liable to liquefy; and

- Desirable field values of N are given in Table 10.2.

Table 10.2. Soils susceptible to liquefaction in Indian Code of Practice.

Zone	Depth below ground level (m)	N values	Remarks
III, IV, and V	Up to 5	15	For values of depth
	10	25	between 5 and 10 m
			linear interpolation
			is recommended
I and II	Up to 5	10	
(for important	10	20	
structures only)			

Japan, one of the few countries in the world with very high earthquake intensity (located in Zone 5 with the earthquake intensity in the range 0.45–0.55 g, see Fig. 1.9) has several codes of practice. A detailed account is given of the latter in Section 10.3.

Reviewing the codes of practices for the European and Asian areas, it may be concluded that, generally, they address earthquake engineering as state-of-art. Liquefaction is mentioned in less detail (except in the Japanese codes). This is not because it has been considered unimportant. But the tendency has been to refer to literature. Some recommendations, however, are given, but they are not treated in great detail. Marine structures are not considered in any detail.

There are also "guidelines", issued by organizations/societies, which are very detailed. An example is the "Seismic Guidelines for Ports" published by ASCE (1998). In this publication, liquefaction is defined clearly and mitigation measures are listed (Chapter 4). Very detailed tables are presented with remedial measures. Furthermore, the available design methods, including assessing liquefaction, reduced effective stresses, and excess pore pressure, are reviewed (Chapter 5). For the Soil Structure Interaction (SSI), reference is generally made to the literature. There is a special chapter (Chapter 6) with guidelines for water front structures. This chapter is very explicit, concerning both design and methods of analysis. The methods refer to piled structures, sheet piling, gravity structures, and crib walls.

The partial conclusion reviewing the ASCE ports recommendations are that an "uncoupled approach" can be used if the soil or fill is potentially liquefiable. Therefore, the recommended procedure is:

- Determine the vertical distribution of reduced effective stresses and pore-water pressure buildup;

- Determine settlements caused by the reduced stresses and pore-water pressure buildup; and

- Use the reduced stresses in the traditional geotechnical methodology.

The ASCE Guidelines also recommend coupled SSI models where the simultaneous development of reduced effective stresses and the corresponding response of the structure are considered. The models, however, are highly specialized and difficult to operate.

The conclusion is that the recommendations and references of the ASCE Guidelines of 1998 are very detailed and operative, and extremely comprehensive. The only missing item, however, is how to analyse the dynamics of lateral spreading, the process in which liquefied soil displaces in the lateral direction, e.g., the displacement along a slope; a detailed account of lateral spreading is given in Section 10.5 below. In the latter code, the effect of lateral spreading on pile foundations is mentioned, but the method of analysis is not given.

The ASCE recommendations of 1998 clearly provided inspiration for the publication of another excellent set of Guidelines, "Seismic Design Guideline for Port Structures" (PIANC, 2001). This publication presents the most extensive guidelines on the "market" concerning the handling of the liquefaction problem for marine structures:

- Detailed phenomenological descriptions;

- Case histories;

- Liquefiable soils;

- Field measurement methodology;

- Laboratory tests;

- Criteria for developing liquefaction;

- Method of analysis; and

- Remedial action.

The guidelines cover both simplified and more advanced dynamic analyses. With the simplified methodology it is possible to determine liquefaction for various different earthquakes, provided the layers are horizontal. Examples are given.

Like the ASCE Guidelines it recommends that mathematical models be used for the development of reduced effective stresses, excess pore-water pressure, and deformations (settlements) due to the reduction in effective stresses. With this input, the traditional geotechnical methods with bearing capacity features and $P-Y$ curves can be used for the foundation design (the "uncoupled approach"). The "Guideline" also advocates that coupled models can be used in the soil–structure-interaction analysis.

The applied methods do not cover phenomena such as:

- The liquefiability of clay and silt materials; and

- Analysis of loads caused by lateral spreading.

The US Corps of Engineers and other American engineering design documents often refer to The Seismic Design of Waterfront Retaining Structures (Ebeling and Morrison, 1993) published by the US Naval Civil Engineering Laboratory. The technical report mainly deals with the calculation of active and passive earth pressures on water and flood retaining structures. It provides guidelines for calculation of design loads for partial and full liquefaction of fill. Interestingly it also contains an Appendix on the Westergaard method (Westergaard, 1933, see also PIANC, 2001, p. 314) for calculating dynamic hydrodynamic pressures on retaining walls during earthquakes. It clearly states that for positive wall base accelerations the hydrodynamic pressure is a tensile. This is important because positive base acceleration (i.e., wall moving against the fill and away from the water) can easily be the design case: in this situation both the lateral earth pressures on the fill side and the dynamic hydrodynamic pressure on the water side generate overturning moments.

Another useful engineering guideline is the "Handbook on Liquefaction Remediation of Reclaimed Land" by the Japanese Port & Harbour Research Institute, published in 1997 (PHRI, 1997). The book is based on extensive Japanese experience in dealing with the liquefaction risk of reclaimed land and it is therefore relevant for design of marine structures. The measurement of relevant soil parameters and the prediction of liquefaction probability are described in detail. The initial "screening" of the grain size distribution is recommended, and the book contains gradation curve envelopes for soils that historically have liquefied (see discussion in the next section, and Fig. 10.6). If the soil is within the range of significant liquefaction risk, the subsequent steps are predominantly based on evaluation of the SPT N values.

Mitigation guidance is provided in the sections on remediation of liquefiable soils. The book divides the basic strategies into soil improvement or structural design mitigation. The soil improvement approaches, in turn, are divided into drainage techniques and soil improvement techniques (e.g., compaction, consolidation, cementation, etc.).

In California, the available guidance on liquefaction hazards is quite advanced. The "Guidelines for Evaluating and Mitigating Seismic Hazards in California" published in 1997 (California Division of Mines and Geology, 1997) contains a chapter dedicated to analysis and mitigation of liquefaction hazards. The publication specifies so-called "screening investigations" for liquefaction potential. The screening investigations are aimed at addressing the following issues:

- Presence of potentially liquefiable soils;

- Degree of saturation of potentially liquefiable soils;

- Geometry of potentially liquefiable soils; and

- *In-situ* soil densities.

If, on the basis of the screening investigations, the responsible engineering body considers that the liquefaction risk is low they may forego the quantitative evaluation of liquefaction resistance that would otherwise be required. The quantitative evaluation of liquefaction resistance is described in the guidelines as well as mitigation measures.

The guidelines also contain a chapter on liquefaction-induced lateral spreading. It is recommended that the lateral spreading hazard is assessed mainly on the basis of SPT or CPT measurements because the inevitable disturbance of samples may render laboratory tests misguiding. For empirical quantification of lateral spreading risk, the guidelines refer to Bartlett and Youd (1995).

Building on the guidelines described above, the Naval Facilities Engineering Center in California published the "Seismic Criteria for California Marine Oil Terminals" in 1999 (Ferrito *et al.*, 1999). The document discusses performance goals and design performance limit states for marine facilities. Chapter 4 specifically deals with evaluation of liquefaction hazards. It contains a useful flow chart for evaluation of liquefaction hazards for pile-supported structures. Finally, various mitigation options are presented. Again, they fall into main categories: compaction, cementation/grouting, improved drainage, or replacement.

It is concluded that a series of recent very detailed recommendations for engineering and the analysis of earthquake-induced liquefaction for marine structures exists — particularly the ASCE "Seismic Guidelines for Ports" (1998), and the PIANC "Seismic Design Guidelines for Port Structures" (2001). Following these recommendations, most marine structures can be treated.

Guidelines for two topics, namely (1) liquefaction in silty and clayey soils, and (2) methods of analysis for lateral spreading, have not been adequately included in the existing codes/publications. However, these subjects are discussed in a number of recent publications, notably in Seed *et al.* (2003). Biondi *et al.* (2002) discusses the methods of analysis for lateral spreading. See Section 10.5 below for a detailed discussion of this issue.

10.3 Japanese Experience

10.3.1 Brief history of earthquakes and design codes

Large earthquakes have occurred in Japan almost every three or four years during the past 40 years (see Table 3 in Sumer *et al.*, 2007). They happened mostly along the boundaries between the Pacific and Philippine plates or the North American and Eurasian plates and caused severe damage to port and harbour facilities located at coastal zones.

The 1964 Niigata earthquake was an epoch-making event in terms of the liquefaction-associated disaster; lots of apartment buildings settled down or tilted due to the bearing capacity loss/reduction of foundation. The girders of Showa Bridge fell down from piers due to large permanent ground displacement. A railway embankment was destroyed by flow failure resulting in other relevant damage. The runway of Niigata airport revealed sand boiling, differential settlement and permanent ground displacement. Buried lightweight structures such as manholes and pipelines floated up from subsoil. A foundation pile was broken due to permanent ground displacement after liquefaction.

Port and harbour facilities were not the exception. Many quay walls and other facilities were damaged due to liquefaction. In this earthquake, many varieties of liquefaction-associated damage were presented, as if it were a showcase of a department store on liquefaction disasters. Since then, the significance of the seismic-induced liquefaction problem was widely recognized among researchers and engineers, triggering the subsequent research activities and engineering practice. After five years, a procedure to assess liquefaction potential using the grain size distribution curve and SPT N-value was reported in 1970 based on the results of laboratory shaking table tests and field investigations made after the 1968 Tokach-oki earthquake (Tsuchida, 1970). These advanced outcomes were introduced into the "Supplements for Design Standards for Ports and Harbour Structures" in Japan (MOT, 1971) and the "Technical Standard for Port and Harbour Facilities with Commentary" (MOT, 1979). It may be noted that the procedure to assess liquefaction potential using the grain size distribution and SPT N-values in the latter publication has been revised to account for the increasing number of loading cycles during long-duration earthquakes such as that which occurred in the 2011 off the Pacific Coast of Tohoku (Shinji Sassa, 2012, personal communication).

The 1983 Nihonkai-chubu earthquake urged consolidation of the existing guidelines on liquefaction measures, because most of the heavily damaged quay walls were associated with the liquefaction of fill behind the quay walls, as shown in Fig. 10.2. The research results relevant to measures against liquefaction, such as philosophy of measures, *in-situ* soil investigation, laboratory test, earthquake response analysis, liquefaction assessment, countermeasures, etc., made by many researchers and engineers were compiled and drafted in a Guideline in 1984. An assessment procedure of liquefaction potential using cyclic triaxial tests on undisturbed samples was introduced to supplement the existing procedure. New findings and knowledge on soil dynamics were also

Figure 10.2: Damage to a quay wall at Akita Port (1983 Nihonkai-chubu earthquake). Reproduced from Sumer *et al.* (2007). Courtesy of Professor Kouki Zen of Kyushu University.

included. The Guideline was not disclosed to the public but was reflected in the "Technical Standard for Port and Harbour Facilities with Commentary" published in 1989 (MOT, 1989). Its English version was published in 1991 (MOT, 1991).

In 1993, after the 1993 Hokkaido-nansei-oki earthquake, the "Handbook on Liquefaction Remediation of Reclaimed Land" was published (MOT, 1993) and its English version was published in 1997 (PHRI, 1997). The handbook describes methods for evaluating liquefaction potential and counter measures against liquefaction in reclaimed land. It provides for the state-of-the-art technologies to supplement the existing design code/standard. In 1997, after the 1995 Hyogoken-nanbu earthquake (EDIC, 1995), the handbook was revised, taking into account the lessons and experiences learned from the catastrophic disaster of port and harbour facilities at Kobe Port and its vicinities (Fig. 10.3 as an example). The "Technical Standard for Port and Harbour Facilities with Commentary" was revised in 1999 (MOT, 1999), including the content of the handbook. (In 1995, an official notice was made public from MOT reflecting the lessons learned from the Kobe earthquake.) Levels 1 and 2 earthquake motions were specified for design practice and a performance-based design concept was introduced in measures against

Figure 10.3:   Destroyed container crane on quay wall at Kobe Port (1995 Hyugoken-nanbu earthquake). Reproduced from Sumer *et al.* (2007). Courtesy of Professor Kouki Zen of Kyushu University.

liquefaction. Here, the two levels of earthquake motions are defined as follows: Level 1 is the level of earthquake motions that are likely to occur during the life span of the structure, and Level 2 is the level of earthquake motions associated with infrequent rare events, which typically involve very strong ground shaking.

10.3.2   Current design code/standard for port and harbour facilities

Earthquake motion in the base layer

The earthquake motion in the base layer, used as the input for the calculation of acceleration, velocity, strain, and stress in the objective layers, is specified by the maximum acceleration given in Fig. 10.4. (Here, Gal is a unit widely used in the geotechnical engineering literature for earthquake acceleration; 1000 Gal means 1 g in which g is the acceleration due to gravity.) In Fig. 10.4, the locations for seismic hazard estimation are 190 coastal regions in Japan. The maximum acceleration is calculated until present on

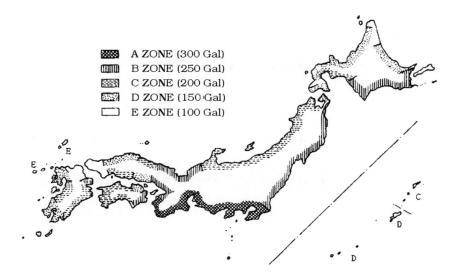

Figure 10.4: Regional divisions in Japan for maximum acceleration used for design (base layer). Taken from Sumer *et al.* (2007). Courtesy of Professor Kouki Zen of Kyushu University.

the basis of the expected maximum acceleration with a return period of 75 years for the coastal region of Japan, using the past earthquake data acquired for the period of 97 years from 1885 to 1981 with reference to the older data from 1200 to 1884. It is noted that the earthquake motion used herein corresponds to the Level 1 earthquake motion. It is also noted that the following description refers to horizontal grounds.

Acceleration waveform in the base layer

The waveforms affect the evaluation of liquefaction because they differ in the predominant frequency contents depending on the characteristics of objective layers. Among the limited number of records on the strong earthquake motion with a large magnitude of approximately 8.0, the two types of recorded waveforms shown in Fig. 10.5 are recommended for use for the liquefaction assessment in design practice. One is the waveform calculated by deconvolution from the wave recorded at the ground surface of Hachinohe Port during the 1968 Tokachi-oki earthquake with the magnitude of 7.9 (referred, S-252 NS Base). The other is the waveform directly recorded at

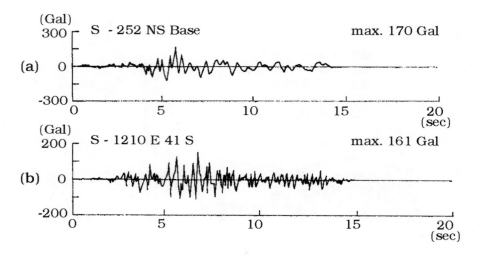

Figure 10.5:   Acceleration wave forms used for earthquake response analysis in design (base layer). Taken from Sumer *et al.* (2007). Courtesy of Professor Kouki Zen of Kyushu University.

the base rock at Ofunato Port during the 1978 Miyagi-ken-oki earthquake with magnitude of 7.4.

Assessment of liquefaction potential

The procedure for liquefaction potential assessment is divided into two. The first step utilizes the grain size distribution of soil and N value by the standard penetration test (SPT N value). If the liquefaction potential by the first step is found to be close to the borderline between liquefaction and non-liquefaction, then the second step using the cyclic triaxial test is introduced to supplement the first step.

(1) Grain size distribution. The grain size distribution curve obtained from a soil sampled *in-situ* is drawn in Fig. 10.6. In this case, either one of the figures in Fig. 10.6 is selected based on the coefficient of uniformity of the soil. If the curve falls outside of the "possibility of liquefaction", the soil is considered non-liquefiable. On the other hand, if it falls inside the "possibility of liquefaction", the liquefaction potential assessment moves onto the next step.

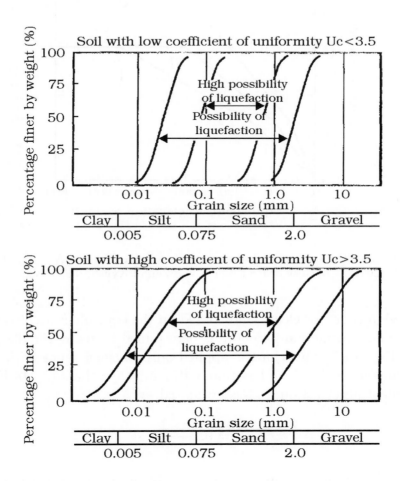

Figure 10.6: Grain size distribution curves with the possibility of liquefaction ranges marked. Taken from Sumer *et al.* (2007). Courtesy of Professor Kouki Zen of Kyushu University.

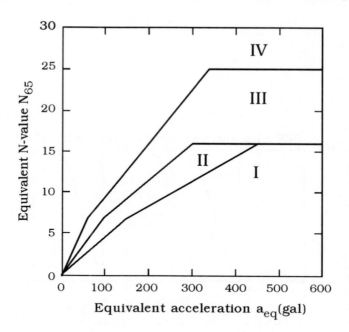

Figure 10.7:   Liquefaction potential assessment based on equivalent N value. Taken from Sumer *et al.* (2007). Courtesy of Professor Kouki Zen of Kyushu University.

(2) Design chart for clean sands based on the equivalent acceleration and N value. The liquefaction potential is determined by using the design chart shown in Fig. 10.7. This chart is applied only to clean sands with the fine-content (fine-grained soils of less than 0.075 mm) less than 5%. In Fig. 10.7, the equivalent acceleration, α_{eq}, is evaluated by

$$\alpha_{eq} = 0.7\frac{\tau_{\max}}{\sigma'_v} g \qquad (10.1)$$

in which $\tau_{\max}$ is the maximum shear stress in the layer calculated with response analysis, and σ'_v the effective overburden pressure. In the response analysis, the waves in the base layer mentioned in the previous section (Fig. 10.5) are utilized.

The equivalent N value, N_{65}, is calculated using the measured SPT N value, N_m, with the following equation:

$$N_{65} = \frac{N_m - 0.019(\sigma'_v - 65)}{0.0041(\sigma'_v - 65) + 1.0} \qquad (10.2)$$

The equivalent N value, N_{65}, refers to the SPT N value corrected for an effective overburden pressure of 65 kPa. The conversion reflects the past design practice in which the assessment of liquefaction potential was performed in the vicinity of the ground water level. Thus, every SPT N value measured along the depth of layer is converted to the equivalent N value. Here, σ'_v is to be inserted in the above equation in kPa.

Being determined from the α_{eq} and N_{65}, the plot falls onto either Zones I, II, III, or IV. The liquefaction potential is assessed by the zone where the plot belongs in Fig. 10.7. The meanings of I to IV are listed as follows:

- Zone I. Possibility of liquefaction is very high. Decide that liquefaction will occur.

- Zone II. Possibility of liquefaction is high. Decide either to determine that liquefaction will occur, or to conduct further evaluation based on cyclic triaxial tests.

- Zone III. Possibility of liquefaction is low. Decide either to determine that liquefaction will not occur, or to conduct further evaluation based on cyclic triaxial tests. When it is necessary to allow a significant safety margin for a structure, decide either to determine that liquefaction will occur, or to conduct further evaluation based on cyclic triaxial tests.

- Zone IV. Possibility of liquefaction is very low. Decide that liquefaction will not occur. Cyclic triaxial tests are not required.

(3) Correction for silty soils. The equivalent N value is corrected for a soil with the fine content more than 5% and/or for a soil with plasticity. For such soils, usually the equivalent N value is increased compared to that for clean sands. The detailed procedure is found elsewhere (MOT, 1999; or PIANC, 2001, p. 202).

(4) Assessment with cyclic triaxial tests. When the assessment of liquefaction possibility cannot be determined, say when the plot falls in Zones II or III, the evaluation can be made by taking advantage of cyclic triaxial tests on undisturbed samples. Liquefaction potential is predicted by using the following procedure.

The time history of the shear stress generated within the soil layer is computed with an earthquake response analysis (where the waves in the base layer mentioned in the previous section, Fig. 10.5, are utilized). The maximum shear stress ratio, L_{max} at the depth of a layer is determined from the effective overburden pressure, σ'_v, and computed maximum shear stress, τ_{max},

$$L_{\max} = \frac{\tau_{\max}}{\sigma'_v}. \tag{10.3}$$

The $L_{\max}$ value is computed for each representative depth of layer and its value is plotted with depth as a distribution.

The cyclic triaxial test is performed on undisturbed/reconstituted samples to obtain the relationship between the cyclic shear stress ratio, τ_ℓ/σ'_v, and the number of load cycles to cause liquefaction, N_ℓ. From this relationship, the liquefaction strength ratio $R_{\max}$ is evaluated by making the corrections as follows

$$R_{\max} = \frac{0.9}{C_k}\frac{1+2k_0}{3}\left(\frac{\tau_\ell}{\sigma'_c}\right)_{N_\ell=20} \tag{10.4}$$

in which σ'_c is the effective confining pressure in the test, namely

$$\sigma'_c = \sigma'_v \frac{1+2k_0}{3}, \tag{10.5}$$

k_0 is the coefficient of lateral earth pressure at rest, and C_k the conversion coefficient for the waveform pattern, say 0.55 for the impact type of waveform pattern and 0.7 for the vibration type. The value of the cyclic shear stress ratio, τ_ℓ/σ'_c, at the number of load cycles, $N_\ell = 20$, is used in the equation.

The safety factor against liquefaction, F_L, is estimated with the following equation

$$F_L = \frac{R_{\max}}{L_{\max}}. \tag{10.6}$$

The liquefaction potential is determined with F_L. If F_L is smaller than 1.0, the soil is expected to liquefy.

If only some portion of a soil layer undergoes liquefaction, the requirement for remedial measures has to be finally assessed by an overall consideration of whether or not damage will be caused in the facilities, taking the sliding failure of foundation, settlement, and horizontal deformation of structures into account. A more sophisticated analysis coupling soil–water–structure interaction and the dynamic earthquake response analysis can be utilized (PIANC, 2001).

Table 10.3. Remedial measures against liquefaction referred to in the text under Section 10.3.3.

Object	Principle	Methods
Soil	Compaction	Sand compaction pile method, vibration method, vibrofloatation method, dynamic compaction method, compaction grouting method, and static densification pile method
	Cementation and solidification	Deep mixing method, premixing method, and chemical grouting method
	Replacement	Replacement method
	Stress history/ overconsolidation	Preloading method
Pore water	Pore-water pressure dissipation	Gravel drain method, piles with a drainage device
	Lowering of ground water level	Deep-well method
	Replacement of pore water	Chemical grouting method
Structure	Shear strain/ restrain	Underground wall method
	Structural reinforcement	Pile foundation, sheet pile, etc.

10.3.3   Mitigations/remediation

When a site is assessed to be liquefiable, usually remedial solutions against liquefaction are applied. Table 10.3 highlights the major remediations currently used at the port and harbour areas in Japan.

Remediation against liquefaction can be classified into three categories by the object of improvement: (a) improvement of soil skeleton, (b) improvement of pore water, and (c) structural reinforcement. Even in each category the principle of improvement is quite different, reflecting the mechanism of liquefaction phenomena.

Detailed descriptions of each method are mentioned elsewhere (JGS, 1998). In practice, firstly, selection of a method is made, and then the advantages and disadvantages of different solutions are compared for a specific project. Sometimes two or more countermeasures are combined.

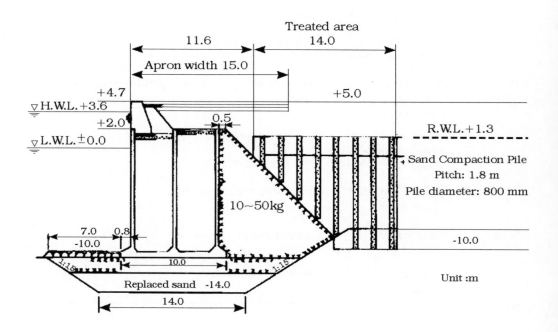

Figure 10.8:   Sand compaction pile method. Taken from Sumer *et al.* (2007). Courtesy of Professor Kouki Zen of Kyushu University.

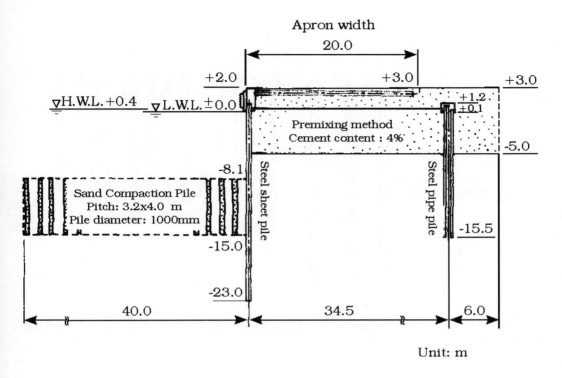

Figure 10.9: Premixing method. Taken from Sumer *et al.* (2007). Courtesy of Professor Kouki Zen of Kyushu University.

Figs 10.8–10.10 show the typical cross-sections of quay walls, respectively, gravity type, sheet pile type, and open-type pier, where remedial solutions are adapted in actual construction projects. The effectiveness of counter measures developed on the basis of the fundamental principles, such as densification (compaction), drainage, and cementation, is indicated in practice by the experience of the past large earthquakes in Japan.

10.3.4 Concluding remarks

The coastal zone, where many port and harbour facilities are densely located, is one of the most liquefiable areas in Japan. After the 1964 Niigata earthquake, many researchers and engineers made efforts to understand the mechanism of liquefaction and to establish design codes/standards to overcome the disaster caused by seismic-induced liquefaction. Almost 50 years since then,

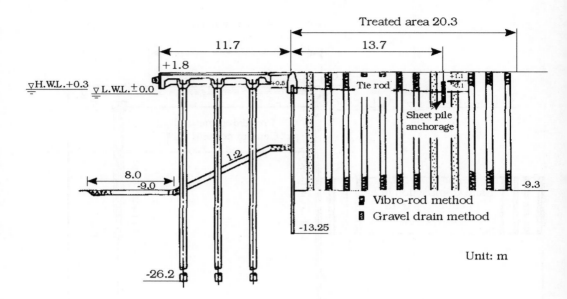

Figure 10.10:   Vibrorod method. Taken from Sumer *et al.* (2007). Courtesy of Professor Kouki Zen of Kyushu University.

the measures against seismic-induced liquefaction have notably progressed in both design and practice. However, liquefaction is still a complicated and thus not fully understood phenomenon. Issues on the most appropriate, optimum, and rational measures against liquefaction require further considerations.

10.4   Turkey Kocaeli Earthquake

10.4.1   Background

In 1999, Turkey experienced two earthquakes: (1) August 17, 1999 Kocaeli earthquake; and (2) November 12, 1999 Duzce earthquake. Both occurred on the North Anatolian Fault in north-western Turkey (Fig. 10.11). The Kocaeli earthquake, which had a magnitude of M_w (the moment magnitude) $= 7.4$ with its epicenter located rather close to the south-east corner of Izmit Bay (Figs 10.12 and 10.13) and lasted 42 s with the largest horizontal acceleration of $0.407\,g$ (Safak *et al.*, 2000), caused extensive damage to marine structures along the coast of the Izmit Bay.

Several papers/reports reported the damage. Boulanger *et al.* (2000) discuss the damage to and the performance of the marine structures in a special volume of the journal *Earthquake Spectra* (2000) dedicated to the this earthquake. Gunbak *et al.* (2000) give a detailed list of the damage over more than 20 marine structures whereas Yuksel *et al.* (2003) elaborate further on the effects of the Kocaeli earthquake on the majority of the marine structures in the region. Sumer *et al.* (2002) give an "inventory" of the damage to marine structures caused by soil liquefaction alone, an exhaustive list. The latter was later reproduced (with some additional material) in Sumer *et al.* (2007).

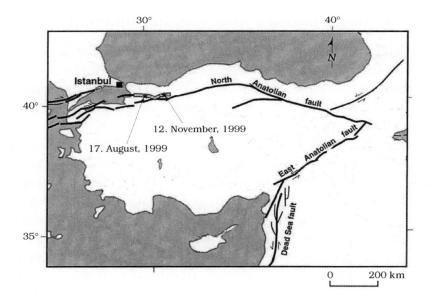

Figure 10.11: The North Anatolian fault. Taken from Sumer *et al.* (2007). Original reference: Lettis *et al.* (2000).

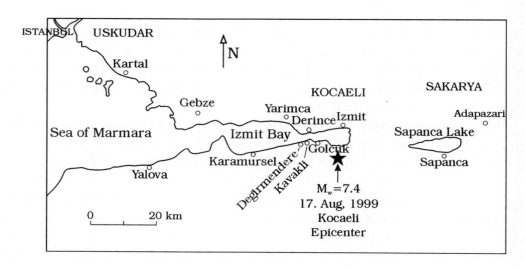

Figure 10.12:   Map of the area stricken by the August 17, 1999 Kocaeli, Turkey Earthquake. Taken from Sumer *et al.* (2007).

10.4.2   Damage caused and lessons learned

In the following paragraphs, we will highlight the impact of liquefaction (the damage inflicted on structures as well as the cases where the structures survived the catastrophic effect of liquefaction). The reader is, however, referred to Sumer *et al.* (2007, Table 6) for a complete documentation of the damage inflicted on the structures.

1. Almost invariably, backfill areas behind quay walls and sheet-piled structures failed due to liquefaction, although, in some cases, the failure in the backfill areas may have been influenced by other factors as well. As discussed in Chapter 3, the liquefaction stage is followed by the densification or compaction stage in which the soil "settles" (Section 3.1.4). The settlement in the backfill areas in the Turkey Kocaeli earthquake varied from $O(10\,\text{cm})$ to $O(1\text{m})$ in which O indicates order-of-magnitude value. The magnitude of the settlement generally decreased with the distance from the epicenter of the earthquake, as anticipated. One of the implications of this kind of failure is that rail foundations for cranes present in the area settled unevenly, leading to tilting of (and eventually damage to) cranes, as revealed clearly in the case of Derince Port (Fig. 10.14; see Fig. 10.13 for the location).

Although the backfill material varied from one case to another, it was typically hydraulically-placed sand from the seabed. In the case of Derince

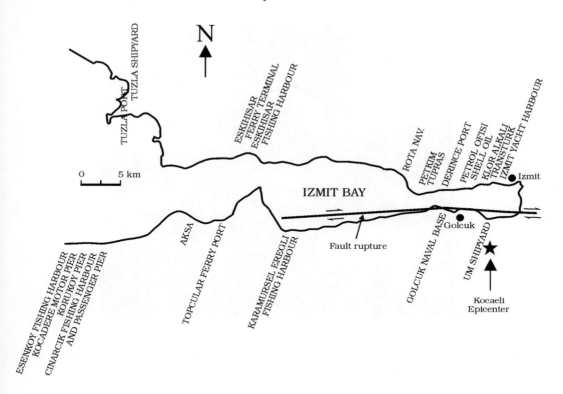

Figure 10.13: Partial layout of coastal structures along the coastline of the Izmit Bay, Turkey. The epicenter of the Kocaeli earthquake is marked on the map. Taken from Sumer *et al.* (2007).

Port, Yuksel *et al.* (2003) report that (1) not much information could be obtained about the backfill material, but it was probably a kind of deltaic sediment, because it had been dredged from a river mouth in the sea and (2) samples taken from liquefied sand were within the liquefiable grain size limits (Yuksel *et al.*, 2003, Fig. 12). A relevant question here is: could the liquefaction failure have been avoided had the backfill material been replaced with a coarser material, a material which is sufficiently permeable so that all pore pressures developed in the backfill would dissipate as rapidly as they developed? Unfortunately, data do not exist in conjunction with the August 17, 1999 earthquake to reveal whether this is the case, and, if so, how coarse this material should be.

2. Quay walls and sheet-piled structures were displaced seaward, the displacements being in the range from $O(10\,\mathrm{cm})$ to $O(1\,\mathrm{m})$ (cf. the ranges of

Figure 10.14:   Rail foundations for cranes settled unevenly, leading to tilting of (and eventually damage to) cranes, Derince Port, Turkey after the 1999 earthquake.

displacement of gravity walls and anchored sheet pile walls at liquefied sites given in PIANC, 2001, p. 325). At this junction, we note the comment by Sumer *et al.* (2002, p. 507) that there is an additional force on a quay wall in the seaward direction in the case of liquefied sand fill, equal to $(1/2)\gamma'h^2(1 - k_0)/3$, in which h is the height of the wall, γ' is the submerged weight of the liquefied sand fill, and k_0 is the coefficient of lateral earth pressure. This additional force obviously helps displace the structure seaward. See also similar considerations in Gerwick (2007, p. 331) in connection with the extensive displacement of the caisson quay walls at Kobe, Japan, caused by by the Great Hanshin earthquake of 1995.

3. An important recommendation which draws on the preceding two "lessons" is that the material to be used as backfill in the coastal structures should be selected properly and compacted densely.

4. Storage tanks near the shoreline tilted due to liquefaction.

5. There are cases where the seabed settled, the settlement being in the range $O(10\,\mathrm{cm}) - O(1\,\mathrm{m})$. However, it is not clear if these settlements

are caused by liquefaction (and therefore by the resulting consolidation/ compaction) or by other processes such as slope instability, surface rupture, etc., or a combination of those processes.

6. There are also cases where structures settled, or they settled and eventually collapsed below water. Again, it is not quite clear if these settlements (and collapses) are caused by liquefaction or by other processes such as slope instability, surface rupture, etc., or a combination of those processes.

7. In addition to the preceding two observations, it may be mentioned that Bahceli-Seymen district at the water front, East of Golcuk Naval Base (Fig. 10.13), slightly West to 29^0 55' (the site quoted as "Liquefaction Zone" in *Earthquake Spectra*, 2000, Fig. 2.6b) experienced heavy liquefaction. Apartment buildings sank in the liquefied soil of $O(20 - 30\,\text{cm})$.

8. The soil samples collected from this area indicated that grain size distributions generally lie in the area marked "possibility-of-liquefaction" in Fig. 10.6. Extensive model runs to make an assessment of liquefaction using the simple mathematical model of Sawicki (Sawicki and Swidzinski, 2007) indicated that the soil in items 4–6 may have experienced liquefaction. The reader is also referred to Sawicki and Swidzinski (2006) for an extensive account of liquefaction susceptibility in the area.

9. Given that the soil had been shaken heavily by the previous earthquakes, one may argue that there was not much "room" for buildup of pore pressure/liquefaction when the August 17, 1999 earthquake struck. One reason why the seabed may have been liquefied by the shaking of the August 17, 1999 earthquake may be that this latest earthquake had a magnitude significantly larger than the previous ones, the amplitude of the ground motion being at least a factor of two larger in the 1999 earthquake ($M_w = 7.4$) than the strongest earthquake that occurred before the 1999 earthquake. (The seismic data for this locality, although limited to recent times, indicate that the strongest earthquake in the area before the 1999 earthquake happened in 1967 with $M_w = 7.1$, Sumer *et al.*, 2002). If the soil has been liquefied before, one may also argue that once sand is completely liquefied, all effects of previously experienced cyclic loading are erased. Then the question is whether or not the 1999 shaking is strong enough to liquefy the soil, which has been through the liquefaction and consolidation/compaction sequence. Indications strongly suggest that it is. It may also be noted that the duration of the earthquake is also an influencing factor.

10. It is to be noted that although a large reclamation area settled in front of 95,000-ton capacity silos in Derince Port TMO facilities, these silos

survived the earthquake. Likewise, a shipyard crane (with a capacity of 510 ton) also survived the earthquake despite the large settlement of the area adjacent to this structure in a shipyard. These structures survived the earthquake largely because of their foundations; both the silos and the crane are supported on piles penetrating into the stiff soil, and therefore avoided any problem caused by liquefaction/weakening of the soil in the top layers. It is also to be noted that a new pier also survived the earthquake whereas the neighbouring old pier did not. This may also be attributed to the same effect, i.e., the piles penetrated into the stiff soil. It is to be noted, however, that there exists a potential problem here of horizontal loading of piles and consequent damage to piles due to liquefaction of layers between the stiff soil and the pile head.

11. The rubble-mound breakwaters in the area largely survived the earthquake. No information is available as to whether these breakwaters are founded on sand or otherwise. This is with the exception, however, that the Eskihisar (Fig. 10.13) breakwater, one of the rubble-mound breakwaters in the area, is founded on sandy coarse gravel, not prone to liquefaction, as the soil information from the neighbouring structures point in this direction.

12. However, some damage occurred to a rubble-mound breakwater, namely the breakwater in Karamursel Eregli Fishing Harbour (Fig. 10.13). The damage was mostly in the form of flattening of the cross-section, sliding of the slope, and intrusion of the lower mound material into the loose sand Yuksel *et al.*, 2004). Yuksel *et al.* (2004) report that the breakwater settled approximately 1.5 m along its entire axis. Their analysis to assess the impact of the earthquake on the structure (using a dynamic finite-difference model) showed that liquefaction/consolidation caused 1.2 m total settlement, largely in agreement with the observed settlement. Soil profile at the breakwater location contains medium-fine sand overlying stiff-hard silty clay. It was observed that, with these structures, a crack of a few centimeters existed along the crest or slightly back at the fill area. This crack may demonstrate an initiation of a slip circle which terminates by an increase of friction immediately after initiation.

13. Liquefaction caused significant problems for other structural components of ports, such as breakwaters, cranes, container terminals, and warehouses. For this reason special caution must be taken if the natural ground contains liquefiable soil, sand, and silt. The data relevant to geology, geomorphology and seismology should be collected and examined with regard to liquefaction.

10.5   Lateral Ground Deformations

10.5.1   Introduction

This section concerns the liquefaction-induced lateral ground deformations. Over the past decade, major advances have occurred in both understanding and practice with regard to engineering treatment of seismic soil liquefaction. Initially, this progress was largely confined to improved ability to assess the likelihood of initiation (or "triggering") of liquefaction in soils. As the years passed, and earthquakes continued to provide lessons and data, researchers and practitioners became increasingly aware of the issue of post-liquefaction strength, and stress-deformation behaviour also began to attract increased attention. Within the confines of this section it was intended to present a brief discussion on existing predictive methods for the assessment of post-liquefaction ground deformation problems, more specifically of lateral ground spreading problem.

When soil liquefaction is accompanied by different forms of ground deformation, then the consequences are destructive to the surrounding environment. During an earthquake shaking, when a subsurface soil sublayer liquefies, the intact surface soil blocks will move down to a gentle slope and/or towards a vertical free face. Therefore, due to soil liquefaction during past earthquake events, large areas and masses of soils were observed to have moved and shifted laterally to new positions, resulting in significant destructive effects for both infrastructures and the overlying surface constructions. Fig. 10.15, taken by Izmit Bay after the 1999 Kocaeli earthquake, presents vivid examples of liquefaction-induced lateral spread. The magnitudes of these liquefaction-induced lateral spreads range from a few centimeters to more than a couple of meters. Fig. 10.16 illustrates a schematic diagram to describe a seismic soil liquefaction-induced lateral spread during an earthquake event, and the associated critical consequences. When a liquefiable soil sublayer exists as an underlying stratum, the overlying soil may slide during an earthquake shaking; even though the ground surface is level or gently sloping a couple of degrees.

10.5.2   Existing lateral-spread predictive models

Several different approaches for predicting the lateral ground deformation magnitudes have been introduced, and from the technical point of view,

(a)   (b)

Figure 10.15:   Liquefaction-induced lateral spreading cases after the 1999 Kocaeli Turkey Earthquake within the Izmit Bay area, from the sites of: (a) Bay Park; and (b) Soccer Field. Reproduced from Sumer *et al.* (2007). Courtesy of Professor K Onder Cetin of METU, Turkey.

they can be categorized as: (1) numerical analyses in the form of finite element and/or finite difference techniques (e.g., Finn *et al.* (1994), Arulanandan *et al.* (2000), and Liao *et al.* (2002)); (2) soft computing techniques (e.g., Wang and Rahman (1999)); (3) simplified analytical methods (e.g., Newmark, 1965, Towhata *et al.*, 1992, Kokusho and Fujita, 2002); and (4) empirical methods developed based on the assessment of either laboratory test data or statistical analyses of lateral spreading case histories (e.g., Hamada *et al.*, 1986, Shamoto *et al.*, 1998, and Youd *et al.*, 2002). Due to their simplicity and ease of use, empirical/semi-empirical and laboratory based methods have been widely used and will be the scope of this section. A discussion of currently available empirical/semi-empirical models (Hamada *et al.*, 1986, Rauch, 1997, and Youd, *et al.*, 2002), and of laboratory based

Figure 10.16: Schematic examples of liquefaction-induced global site instability and/or large displacement horizontal deformations. Reproduced from Sumer *et al.* (2007). Original reference: Seed *et al.* (2003).

methods (Ishihara and Yoshimine, 1992, and Shamoto *et al.*, 1998) will be presented next.

10.5.3   Empirical or semi-empirical models

A number of researchers proposed empirical predictive methods, based on regression analyses of large suites of previous lateral spreading case histories. The simplest of all these models was proposed as part of the pioneering studies by Hamada *et al.* (1986). In his model, the magnitude of horizontal ground deformations at a site composed of potentially liquefiable layers is predicted only in terms of slope and thickness of the liquefied layers

$$D = 0.75H^{0.75}\theta^{0.33} \tag{10.7}$$

in which D is the horizontal displacement (m); θ the slope (%) of ground surface or the base of the liquefied soil, and H the total thickness (m) of the liquefied layers.

In another study, Rauch (1997) followed a different methodology where liquefaction-induced ground deformations were modelled as slides of finite area rather than displacement points where displacement vectors were mapped. Multiple linear regression was used as the tool to estimate model parameters. As a conclusion, an empirical model was proposed for the estimation of liquefaction-induced lateral ground displacements

$$D_R = (613M_w - 13.9R_f - 2.420A_{\max} - 11.4T_d)/1000 \tag{10.8}$$

$$D_S = (50.6Z_{FS\min} - 86.1Z_{liq})/1000 \tag{10.9}$$

$$D_G = (D_R - 2.21)^2 + 0.149 \tag{10.10}$$

$$D = (D_R + D_S + D_G - 2.44)^2 + 0.124 \tag{10.11}$$

in which D is the average horizontal displacement (m); R_f the shortest horizontal distance (km) to fault rupture; M_w the moment magnitude of the earthquake; $A_{\max}$ the peak horizontal soil acceleration (g) at the ground surface; T_d the duration (s) of strong earthquake motions ($>0.05\,g$); $Z_{FS\min}$ the average depth (m) from ground surface to the top of the soil layer with a minimum factor of safety against liquefaction; and Z_{liq} is average depth (m) from ground surface to top liquefied layer.

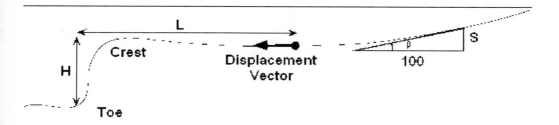

Figure 10.17:   Topography-related descriptive variables. Taken from Sumer *et al.* (2007).

Similarly, starting in early 1990s Bartlett and Youd (1992, 1995) introduced empirical methods for predicting lateral spread displacements at liquefiable sites. The procedure of Youd *et al.* (2002) is a refinement of these early efforts and the new and improved predictive models for either (1) sloping ground conditions, or (2) relatively level ground conditions with a "free face" towards which lateral displacements may occur. These were developed through multilinear regression of a large case history database. The proposed predictive models for the sloping ground and "free face" conditions are given in Eqs 10.12 and 10.13 along with topography-related descriptive variables given in Fig. 10.17:

$$\log D_H = -16.213 + 1.532M - 1.406 \log R^* - 0.012R + 0.338 \log S$$
$$+ 0.540 \log T_{15} + 3.413 \log(100 - F_{15}) - 0.795 \log(D50_{15} + 0.1\,\mathrm{mm})$$
$$(10.12)$$
$$\log D_H = -16.713 + 1.532M - 1.406 \log R^* - 0.012R + 0.592 \log W$$
$$+ 0.540 \log T_{15} + 3.413 \log(100 - F_{15}) - 0.795 \log(D50_{15} + 0.1\,\mathrm{mm})$$
$$(10.13)$$

in which D_H is the horizontal ground displacement in meters predicted by a multiple linear regression model; M the earthquake magnitude (M_w was primarily used whenever reported); R the horizontal distance to the nearest seismic source or to nearest fault rupture (km); $R^* = R + R_o$, and $R_o = 10^{(0.89M-5.64)}$; S the gradient of surface topography or ground slope (%); W the free-face ratio, defined as the height of the free-face divided by its distance to calculation point; T_{15} the thickness of saturated layers with $(N_1)_{60} \lesssim 15$; F_{15} the average fines content (particles $< 0.075\,\mathrm{mm}$) in T_{15} (%); and $D50_{15}$ the average D_{50} in T_{15}.

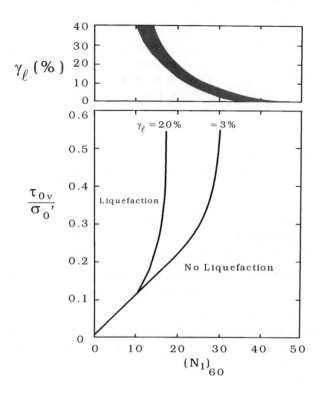

Figure 10.18:   Limiting shear strain potential chart. Taken from Sumer *et al.* (2007). Original reference: Tokimatsu and Seed (1984).

10.5.4   Laboratory-based models

Laboratory based predictive models require the estimation of induced cyclic shear strains which will be further used to estimate the contribution of each potentially liquefiable layer to the overall lateral ground deformation. Employing high-quality cyclic test (triaxial, simple shear or torsional shear test) results, Tokimatsu and Seed (1984), Ishihara and Yoshimine (1992), and Shamoto *et al.* (1998) proposed charts for the estimation of shear strains for cohesionless soils under cyclic loading. Different definitions of cyclic shear strain were adopted by these researchers. Tokimatsu and Seed (1984) used limiting shear strain, γ_ℓ, defined as the strain level that commences following liquefaction but is arrested after a finite displacement, usually as a consequence of dilatancy induced pore-pressure drop and an accompanying increase of effective stress. These deformations are accompanied by

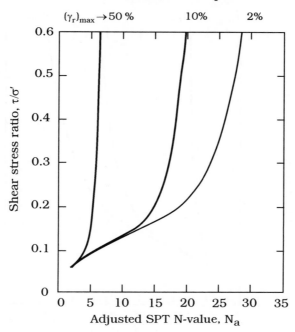

Figure 10.19:   Residual shear strain potential charts. Fines ($<74\,\mu$) $= 0\%$. Taken from Sumer *et al.* (2007). Original reference: Shamoto *et al.* (1998).

transient loss of shear resistance rather than permanent loss of shear strength. Shamoto *et al.* (1998) adopted a residual (permanent) shear strain definition as opposed to Ishihara and Yoshimine's (1992) maximum shear strain definition.

As shown in Figs. 10.18 and 10.19 through to 10.21, for both Tokimatsu and Seed (1984) and Shamoto *et al.* (1998), the capacity and demand terms of shear strain predictive models were chosen as corrected/adjusted standard penetration test blow counts, N_a, fines content (FC) and cyclic stress ratio (or shear stress ratio defined as the ratio of uniform cyclic shear stress, τ, to vertical effective stress, σ') corrected for effective stress conditions, *in-situ* static shear stresses, and duration of earthquake.

However, Ishihara and Yoshimine (1992), still used corrected standard penetration test blow counts (N_1) as the capacity term, preferred factor of safety, FS_L, as the demand term and proposed Fig. 10.22 for the estimation of liquefaction-induced maximum shear strains (γ_{max}). In Fig. 10.22, D_r

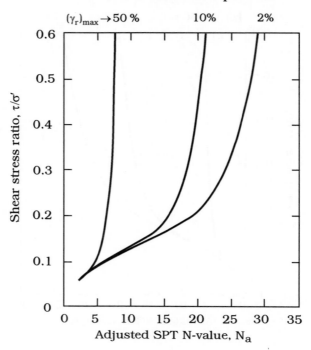

Figure 10.20:   Residual shear strain potential charts. Fines $(<74\,\mu) = 10\%$. Taken from Sumer *et al.* (2007). Original reference: Shamoto *et al.* (1998).

denotes relative density, q_{ct} and ϵ_v denote cone tip resistance and volumetric strain, respectively.

Employing these laboratory-based estimates of liquefaction-induced limiting shear strains coupled with an empirical adjustment factor to relate these laboratory values to observed field behavior, Shamoto *et al.* (1998) proposed a new predictive approach for the estimation of liquefaction-induced ground deformations.

After having determined residual shear strain potential through available methods of Tokimatsu and Seed (1984), or Ishihara and Yoshimine (1992), or Shamoto *et al.* (1998), liquefaction-induced lateral ground deformations can be evaluated by simply multiplying these values by the thickness of the soil substrata which is potentially liquefiable.

As proposed by Shamoto *et al.* (1998), the summation of these lateral deformation values are further multiplied by an empirical factor of 1.0 or 0.16

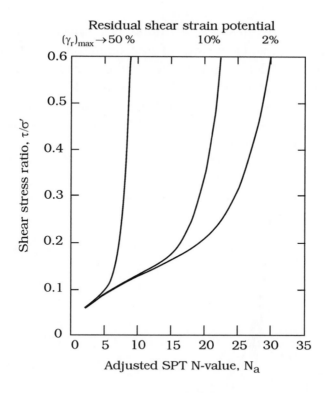

Figure 10.21:   Residual shear strain potential charts. Fines $(<74\,\mu) = 20\%$. Taken from Sumer *et al.* (2007). Original reference: Shamoto *et al.* (1998).

in order to predict lateral displacements at the sites with or without "free face" ground conditions, respectively.

10.5.5   Final remarks

Due to either difficulties (uncertainties) in estimating model input parameters (i.e., estimating the representative CSR, factor of safety against liquefaction, representative standard penetration blow counts, and fines content in the critical strata, etc.) or predictive model inaccuracies (or both), these currently available predictive models, best of their kind, may occasionally produce lateral ground deformation predictions off by factor of more than three (e.g., Cetin *et al.*, 2004). This is also illustrated by Fig. 10.23. Thus, these models, before widely accepted as reliable engineering tools, need to be further calibrated and improved. As a conclusion, their predictions, especially in the range of small to moderately significant lateral ground deformations (lateral displacements of approximately 0.1 to 2.5 m) should be interpreted as an order of magnitude guidance rather than the results of reliable and well-calibrated engineering tools, and for the projects of higher relative importance, higher order tools including numerical simulations calibrated with observations from documented case histories, should be preferred for more reliable assessments.

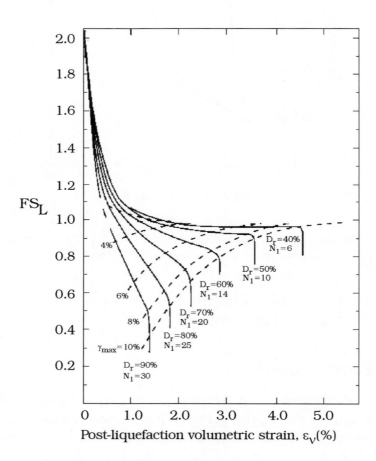

Figure 10.22:   Chart for determination of post-liquefaction volumetric strain as function of FS (factor of safety). Taken from Sumer *et al.* (2007). Original reference: Ishiara and Yoshimine (1992).

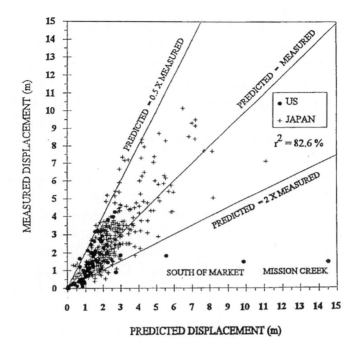

Figure 10.23: Comparison of the predicted and measured lateral spread values. Taken from Sumer *et al.* (2007). Original reference: Bartlett and Youd (1995).

10.6 Tsunamis and their Impacts

10.6.1 Introduction

It is known that tsunami attacks cause erosion and scour on shorelines and at structures. A recent laboratory study (Tonkin *et al.*, 2003) on scour around piles exposed to a tsunami wave showed that very rapid, transient scour at the back of the pile during the drawdown stage of the tsunami is, for the most part, governed by the upward-directed pore-pressure gradients, causing a substantial reduction in the effective stress in the soil and therefore enhanced scour. It is reported in the latter publication that, although transient, this brief but very rapid scour creates the deepest scour holes. This effect was also mentioned by Sumer *et al.* (2002) in a previous publication in connection with the Turkey Kocaeli earthquake.

Although scour around marine structures has been studied quite extensively in the past (Sumer and Fredsøe, 2002), this transient scour caused by
tsunami waves has been brought into light only recently. The mechanism in
which the effective stress in the soil is reduced resembles the mechanism of
the momentary liquefaction under waves, studied in Chapter 4. These pressure gradients are generated during the drawdown stage in the case of the
tsunami, whereas they are generated during the passage of the wave trough
in the case of the momentary liquefaction.

Although not directly related to tsunami waves, pore-water pressure measurements show that sediment at or near the surface of a sloping bed experiences upward-directed pressure gradient forces during the rundown stage
of a solitary wave (Sumer et al., 2011). This force is caused by the delay in
the pore pressure in responding to the fluid loading (infiltration and exfiltration), as already mentioned in Section 1.4 (Chapter 1) and Section 4.4
(Chapter 4). If this upward-directed pressure gradient force is larger than
the submerged weight of the soil column, the soil at these locations will be
liquefied, as revealed by the experiments conducted by Young et al. (2009),
Section 4.4 (Chapter 4).

Because the scour hole caused by this effect is deepest, tsunami-generated
scour should be a design consideration for structures such as bridge piers,
piles, piers, etc. in coastal areas. In general, it is difficult to identify tsunami-
induced liquefaction in field surveys, as flooding from subsequent waves often
erases the liquefaction signature. It is thus harder to differentiate liquefaction
scour from the effects of the tsunami hydrodynamic forces themselves.

Another tsunami-related process concerns forces on quay walls and sheet-
piled structures. Referring to the tsunami experienced in the Izmit Bay triggered by the Turkey Kocaeli Earthquake, Yalciner et al. (2000) reported that
the sea first receded and subsequently rose and flooded the in-land areas. As
mentioned earlier in conjunction with the Turkey Kocaeli Earthquake, backfill areas behind quay walls and sheet-piled structures on the shoreline failed
due to liquefaction. With regard to forces on quay walls and sheet-piled
structures, Sumer et al. (2002) commented that because the water receded
during the tsunami immediately after the earthquake, hydrostatic forces on
quay walls at the waterside decreased (or completely vanished), and therefore quay walls underwent relatively larger, seaward resultant pressure forces
(comprising the hydrostatic pressure force and the pressure force due to the
accumulated excess pore pressure in the liquefied backfill soil). Sumer et al.
(2002) pointed out that this effect may have played a significant role in the

seaward displacement of quay walls and sheet-piled structures, reported to be in the range $O(10\,\text{cm}) - O(1\,\text{m})$ (Section 10.4). Clearly, worst-case design scenarios in which the quay walls and sheet-piled structures are subject to maximum forces when the backfill soil is liquefied should include tsunami attacks, particularly at the stage where the water recedes.

Tsunamis are clearly a colossal threat to coastal structures not only due to the sheer size of the forces that they generate but also due to the previously mentioned processes, which involve soil/structure/wave interaction. For this reason, and because no guidelines currently exist relating flow characteristics to structure performance, a short account of tsunamis and their impacts is included in the present chapter, giving a state-of-the-art review.

Tsunamis are water waves generated by large-scale short-duration energy transfer to the entire water column by earthquakes, coastal and submarine landslides, volcanic eruptions, caldera collapse, or meteor impact. The number of waves and polarity of the initial wave depend on the seafloor motion, and the subsequent evolution over the seafloor terrain is in the classic manner predicted by long wave theory (Synolakis, 2003).

Tsunamis and landslide waves may attain large amplitudes in closed basins or shallow regions (Yalciner *et al.*, 2001, 2002, 2004 and 2005). They are generally classified as long-period waves and now all are referred to as tsunamis. Tsunamis were first described in history 2,500 years ago by Thucydides, Herodotus, Aristotle, and later Strabo (Schonberg, 1997). Thucydides in his *History of the Peloponnesian War* explained the relation between earthquakes, great waves, and topography, observing the frequent earthquakes and a tsunami in 426 B.C. He described the action of the sea as it "subsided from what was then the shore and afterwards swept up again in a huge wave" (Thucydides 247; 3.89). More modern observations and analysis confirm the persistence of leading depression tsunami waves. (Tadepalli and Synolakis, 1996). A leading depression wave may drop the shoreline MWL rapidly, thus creating liquefaction potential.

The word "tsunami" appears to have been communicated outside Japan in the aftermath of the 1896 Great Meiji tsunami which claimed 22,000 people's lives. It translates into harbour wave, quite possible because the more common occurrences of tsunamis in Japan had been as unusual waves in small bays and ports, typically. The English term tidal wave is a translation of the Greek terminology, and it reflects that in the Aegean Sea, tsunamis often manifest themselves along the coast as surges or rapid changes in the water level.

Landslide-generated waves generally have only nearfield impact. They are of a shorter wavelength than tectonic tsunamis, and their initial amplitude depends on the depth, thickness, and initial acceleration of the triggering slide. The amplitude of tectonic tsunamis depends on the length of fault rupture and the slip. Generally, tectonic tsunamis radiate energy in a direction perpendicular to the length of the triggering fault (Ben-Menahem and Rosenman, 1972), whereas landslide tsunamis radiate energy in a radial fashion. The run-up distribution on adjacent shorelines is so different between landslide waves and tsunamis that it allows discrimination of the source (Okal and Synolakis, 2004).

The current paradigm is to model tsunamis with the nonlinear shallow-water (NSW) equations (Synolakis *et al.*, 1997, Synolakis, 2003). Different numerical solution methods of these equations exist (Yeh *et al.*, 1996). The current state of the art for some models such as MOST (Titov and Synolakis, 1998) allows for real-time tsunami inundation forecasting by incorporating real-time data from tsunameters. MOST, TUNAMI-N2, and COMCOT calculate tsunami inundation by computing the wave evolution on dry land and have been validated by comparing their results with exact analytical solutions and laboratory measurements, and results from field surveys. The model MOST is used most often in the United States for developing inundation maps (Borrero *et al.*, 2003). TUNAMI N2 was originally authored by Imamura 1988, Imamura and Goto, 1988, and Imamura and Shuto 1989, for the Tsunami Inundation Modeling Exchange (TIME) program (Goto *et al.*, 1997, Shuto *et al.*, 1990). It is a registered copyright of Professor Imamura, Professor Yalciner, and Professor Synolakis and has been applied to several tsunami events (Imamura, 1996, Yalciner *et al.*, 2000, 2001, 2002, 2004 and Zahibo *et al.*, 2003) and also for computation of resonant oscillations of basins for understanding indirect effects of tsunamis.

10.6.2   Direct effects of tsunamis on coastal structures

As tsunamis approach coastlines or enter bays, they initially evolve in the classic manner predicted by linear long-wave theory (Synolakis and Skjelbreia, 1993). Close to shore, nonlinear effects often become important, as the height increases and wavelength decreases. If breaking occurs, bores often result as first documented in the 1960 Chilean tsunami impact in Hilo, Hawaii, or as seen in the tens of amateur video clips from the beaches of Thailand, following the December 26, 2004 mega-tsunami. Because the offshore wavelength of tsunamis can be tens of kilometers, the resulting bores can penetrate a few kilometers inland.

Direct effects of tsunamis on coastal and marine structures can be extensive and often disastrous. Tsunami waves can (1) move entire structures off their foundations and carry them inland (2) damage buildings through impact with vessels carried from offshore and other debris accumulated as the wave advances inland, (3) undercut foundations and pilings with erosion caused by receding waves, (4) overturn structures by suction of receding or thrust of advancing waves and (5) cause impact of large ships with docks during oil or cargo transfer operations, often causing fires. The damage can be quite unexpected. During the 1998 Hokkaido-Nansei-Oki tsunami, Aonae in Okushiri was consumed by fires triggered after the waves subsided.

In terms of economic impact, Borrero *et al.* (2005) report that the losses resulting from a landslide in the San Pedro Escarpment in California triggering a tsunami and impacting the ports of Los Angeles and Long Beach will range from $7 to $42 billion, in addition to the losses due to structural damage.

Impact forces can cause collapse of coastal structures. The estimation of impact forces and currents is still an art, and far less understood than hydrodynamic evolution and inundation computations. Existing analyses only extend to suggesting methods for calculating forces on piles, impact forces on seawalls and structures, with provisions available for breaking wave loads. No methods exist for calculating debris-impact, beyond the suggestions provided in the Coastal Construction Manual (CCM), which were derived from results from steady flows. A comprehensive discussion maybe found in Synolakis (2003).

There are no existing guidelines for erosion due to tsunamis, although a large amount of data on the erosion and deposition during a tsunami attack has been accumulated (Gelfenbaum and Jaffe, 2003), but they have yet to be translated into standards and guidelines for engineered structures. Although there are a large number of studies of scour around cylindrical piers, for steady flows and combinations of steady flows and waves (Sumer and Fredsøe, 2002), only Tonkin *et al.* (2003) describe a laboratory experiment with erosion from a solitary wave (simulating a tsunami) attacking a circular cylinder (see the discussion in the beginning of this section). One of the results they obtained concluded that the time scale of the tsunami attack is critical in the scouring process. Another interesting result from the study is that the upward-directed pore-pressure gradient in their laboratory test reached as much as 0.5, and the pore-pressure gradient required for full liquefaction was about 0.9, indicating a considerable reduction in the effective stress, as mentioned in the preceding paragraphs.

10.6.3   Indirect effects of earthquakes and tsunamis. Resonant oscillations in enclosed basins

Tsunamis may cause resonant oscillations in lakes, basins, and harbours, as tsunami periods are often in the range of resonant frequencies of large, closed or semi-enclosed water bodies. The process is also referred to as sloshing or seiching. Even small tsunamis can trigger resonance and can cause damage. Resonance can also be triggered directly by the earthquake shaking, as observed during the 1994 Northridge earthquake in the Los Angeles Reservoir in Northridge, California. Further, seiching may persist for several hours after the tsunami's arrival and continue loading, cyclically, foundations already weakened by the tsunami. Therefore, seiching needs to be investigated when evaluating liquefaction potential.

The problem of sloshing is a classical problem of hydrodynamics and was first described in the context of moving atmospheric fronts in Lac Leman in the 19th century (Wilson, 1972). Raichlen (1966) described the problem of harbour resonance and presented a simple analytical method for calculating harbour oscillations.

Most modern harbours have individual interconnected basins and are protected by attached or detached breakwaters. Although elegant semi-analytic methods exist (Lee, 1969) to calculate amplification, complex boundaries and variable depth often require more sophisticated approaches. Lepelletier and Raichlen (1987) introduced a finite-element solution of a nonlinear-dispersive-dissipative model and compared their predictions to laboratory measurements from a narrow constant depth rectangular harbour and a sloping harbour with excellent results, even for transient waves. Zelt and Raichlen (1990) used a Lagrangian formulation to solve a 2-D Boussinesq-type model and calculated the evolution of a solitary wave in a parabolic bay.

It is now standard practice to combine the effects of diffraction and refraction and use the Mild Slope Equation (MSE), a linear, steady-state, depth-averaged elliptic partial differential equation, which, however, assumes that changes in topography are small within one wavelength. The equation has been rederived by Berkhoff (1972) and most recently solved by Tsay and Liu

(1983), Mei (1989), and Chen and Huston (1987). As written by Demirbilek and Panchang (1998) it is

$$\nabla(CC_s)\nabla\eta + (C_s/C)\omega^2\eta = 0 \tag{10.14}$$

in which $C = \omega/k$ is the phase velocity, $C_s = \partial\omega/\partial k$ the group velocity, ω the frequency, k the wave number, and η the wave amplitude. The MSE is not easy to solve numerically, as its solution spectrum covers waves of arbitrary wavelength. In the context of tsunami impact studies, it is impractical to extract waveforms at the entrance of a harbour/basin to initiate solution of the MSE, as the MSE formulation is steady state. Tsunami impact is transient, and while the Lepelletier and Raichlen (1987) formulation is still the "gold" standard, it is too computationally intense for real-time use. Thus, one interesting problem is whether the shallow water (SW) equations can adequately model harbour resonance for tsunami attacks, since presumably the forcing consists of long waves.

Yalciner *et al.* (1996) solved the classic problem of a closed basin and compared the periods of free oscillations with theoretical values. The periods of free oscillations, T_n, inside a closed basin with vertical, solid, smooth, and impermeable boundaries can theoretically be determined as

$$T = \frac{2\ell}{n\sqrt{gh}} \tag{10.15}$$

in which ℓ is the length in the direction of wave, h is the water depth of the basin, and n is integer. Yalciner *et al.* (1996) used a 5,013.33 m square-shaped and 100 m deep closed test basin and forced it with 1 m amplitude, 10 sec period of a single sinusoid, to determine its impulse response, hence the natural frequencies (see Fig. 10.24).

Yalciner and Synolakis (in the paper Sumer *et al.*, 2007) compared theoretical predictions for the resonant frequencies with the nonlinear shallow water (NSW) numerical results for different modes inside the rectangular test basin. This comparison indicates that NSW identifies the resonant frequencies within 1–3%, at a small computational cost. Further comparisons of NSW predictions of resonant frequencies for the Ports of Los Angeles and Long Beach agree with the classic results of Lee (1969) within 3% and the variance may also be due to differences in bathymetric resolution. Comparisons with predictions from CGWAVE (Demirbilek and Pachang, 1998) suggest that the NSW model can calculate amplification factors within 10–20%

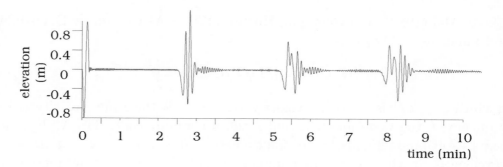

Figure 10.24:   A segment of time history of water surface fluctuation at a point inside the enclosed square flat basin (5,013.3 m dimension and 100 m deep) agitated by a single sinusoidal wave of 1 m amplitude and 10 s period. Taken from Sumer *et al.* (2007).

of the results of CGWAVE. This is perplexing, yet so is the almost ubiquitous ability of NSW theory to adequately model tsunami evolution problems. Although the results do not imply that the NSW model can, in general, adequately model amplification factors, for liquefaction analysis what is needed is the determination of whether resonant excitation may occur.

Yalciner and Synolakis further conjectured that this methodology can be extended to larger water bodies, where presumably NSW may be more appropriate and studied the periods of free oscillations of Izmit Bay (Figs. 10.12 and 10.13), severely impacted by the Izmit tsunami in the 1999 Turkey Kocaeli earthquake. For computing the periods of resonant oscillations of Izmit bay, Yalciner and Synolakis agitated it separately in two normal direction by two different N wave initial impulses (reported in the paper Sumer *et al.*, 2007). Resonant periods were obtained by identifying spectral peaks in the time histories of water surface fluctuations at multiple coastal locations. Results indicated that the resonant periods in the north–south direction have periods ranging from 5.75 to 3.03 min. In the east–west direction the periods range from 16.39 to 3.03 min. Tsunamis with energy at these periods or in the other resonance periods may cause seiching which would impact coastal structures in addition to any direct effects from the causative earthquake and tsunami.

In summary, when evaluating liquefaction potential, it is recommended that the objective be to determine resonance periods and the amplification at resonance. The most likely impact is the interruption of oil or cargo transfer operations, and damage to wharfs and docks from impact with vessels.

Impact forces can be calculated as needed from the previous section.

10.6.4 Final remarks/guidelines for protecting coastal structures

As preliminary survey results from the December 26, 2004 mega-tsunami suggest (Synolakis *et al.*, 2005, Dalrymple and Kriebel, 2005), simple mitigation measures can work, and protecting structures for tsunami attack and associated co-tsunami effects such as liquefaction is not complex.

Since no guidelines currently exist that relate earthquake or shaking measures to flooding parameters, the study of liquefaction potential must be undertaken in the context of comprehensive inundation studies. Specific tasks for preparedness and mitigation studies for marine terminals and other harbour facilities are the following:

1. Identification of offshore hazards, such as submarine landslide sources and submarine faults. In many areas around the world, offshore hazards remain unmapped. For example, in California, only about 50% of the coastline is mapped at resolution to permit hazard evaluation and for the other regions in the world it is much less.

2. Recurrence interval estimates by increasing the coverage and quality of the historic and prehistoric tsunami information. Analysis of near-field and far-field sources and estimation of credible probabilities for the events.

3. Inundation maps reflecting specific scenarios for evacuation planning and for emergency port operations are necessary. Thus, inundation modelling to evaluate the consequences of the generated tsunamis for relevant geologic sources for the localities must also be performed.

4. Determination of 100-year and 500-year recurrence for specific flow depths. The choice of 100 and 500 year intervals is guided by how existing flood hazard maps are produced, much the same as the existing maps do not include tsunami hazards. The same intervals would permit easier evaluation of relative flooding hazards. Also, insurance companies generally evaluate actuarian flooding risk on 100- and 500-year intervals.

5. Evaluation of the strength of critical coastal structures with tsunami force estimates, using wave heights, inundation depths and currents identified by the detailed numerical inundation modelling per the previous discussion.

6. Evaluation of harbour response to determine whether a scenario tsunami could trigger seiching.

10.7   References

1. Arulanandan, K., Li, X.S. and Sivathasan, K. (2000): Numerical simulation of liquefaction-induced deformation. Journal of Geotechnical Engineering, ASCE, vol. 126, No. 7, 657–666.

2. ASCE (American Society of Civil Engineer) (1998): Seismic Guidelines for Ports. Written by the Ports Committee of the Technical Council on Lifetime Earthquake Engineering, Monograph No. 12. Ed. Stuart D. Werner, New York.

3. Bartlett, S.F. and Youd, T.L. (1992): Empirical Analysis of Horizontal Ground Displacement Generated by Liquefaction-induced Lateral Spreads. Technical report NCEER-92-0021, National Center for Earthquake Engineering Research, State University of New York, Buffalo, NY, 5-14–15.

4. Bartlett, S.F. and Youd, T.L. (1995): Empirical prediction of liquefaction-induced lateral spread. Journal of Geotechnical Engineering, vol. 121, No. 4, 316–329.

5. Ben-Menahem, A. and Rosenman, M. (1972): Amplitude patterns of tsunami waves from submarine earthquakes. Journal of Geophysical Research, vol. 77, No. 17, 3097–3128.

6. Berkhoff, J.C.W. (1972): Computation of combined refraction-diffraction. Proceedings of the 13th Conference on Coastal Engineering, ASCE, New York, 471–490.

7. Biondi, G., Cascone, E. and Mangui, M. (2002): Flow and deformation failure of sandy slopes. Soil Dynamics and Earthquake Engineering, vol. 22, No. 9–12, 1103–1104.

8. Borrero, J.C., Cho, S., Moore, J.E, Richardson, H.W. and Synolakis, C.E. (2005): Could it happen here. Civil Engineering (N.Y.), vol. 75, No. 4, 55–67.

9. Borrero, J.C., Yalciner, A.C., Kanoglu, U., Titov, V., McCathy, D. and Synolakis, C.E. (2003): Producing tsunami inundation maps in California. Submarine Landslides and Tsunamis, Kluwer Academic, Dordrecht, 315–329.

10. Boulanger, R., Iai, S., Ansal, A., Cetin, K.O., Idriss, I.M., Sunman, B. and Sunman, K. (2000): Performance of Waterfront Structures. 1999 Kocaeli, Turkey, Earthquake Reconnaissance Report, Chapter 13, Supplement A to Earthquake Spectra, vol. 16, T.L. Youd, J.-P. Bardet, J.D. Bray, eds., 295–310.

11. California Division of Mines and Geology (1997): Guidelines for Evaluating and Mitigating Seismic Hazards in California. Special Publication No. 117, California Department of Conservation, Sacramento, California.

12. Castro, G. and Poulos, S.J. (1977): Factors affecting liquefaction and cyclic mobility. Journal of Geotechnical Engineering, ASCE, vol. 103, No. GT6, 501–506.

13. Cetin, K.O., Youd, T.L., Seed, R.B., Bray, J.D., Durgunoglu, H.T., Lettis, W. and Yilmaz, M.T. (2004): Liquefaction-induced lateral spreading at Izmit Bay during the 1999 Kocaeli (Izmit) Turkey Earthquake. Journal of Geotechnical Geoenvironmental Engineering, ASCE, vol. 130, No. 12, 1300–1313.

14. Chen, H.S. and Houston, J.R. (1987): Calculation of Water Oscillation in Coastal Harbors: CERC-87-2, USACE/WES, Coastal Engineering Research Center, Vicksburg, MS.

15. Dalrymple, R.A. and Kriebel, D.L. (2005): Lessons in engineering from the tsunami in Thailand. The Bridge, National Academy of Engineering, NW, Washington, D.C., vol. 35, No. 2, 1–8.

16. Demirbilek, Z. and Panchang, V. (1998): CGWAVE: A Coastal Surface Water Wave Model of the Mild Slope Equation, Technical Report, Coastal Hydraulic Laboratory, U.S. Army Corps of Engineers, Vickburg, MS.

17. Earthquake Damage Investigation Committee (EDIC), ed. (1995): Study on Damage to Port Structures during Hyogoken-Nambu Earthquake. Technical Note of Port and Harbour Research Institute, (PHRI), Port and Harbor Research Institute, Ministry of Transport, No. 813 (in Japanese).

18. Earthquake Spectra (2000): Earthquake of August Reconnaissance Report. Supplement vol. 16, Earthquake Engineering Research Institute, Oakland, CA.

19. Ebeling, R.M. and Morrison, E.E. (1993): The Seismic Design of Waterfront Retaining Structures. Technical Report No. ITL-92-11, NCEL TR-939, Information Technology Laboratory, Department of Army, Waterways Experiment Station, Corps of Engineers, Vicksburg, MS.

20. European Committee for Standardization (CEN) (1994): Eurocode 8: Design Provisions for Earthquake Resistance of Structures, Part 1-1– General rules for structures (ENV-1998-1-1); Part 5–Formations, retaining structures and geotechnical aspects (ENV 1998-5).

21. European Committee for Standardization (CEN) (1997): Eurocode 7: Geotechnical design. Part 3: Geotechnical design assisted by field tests (ENV-1997-3), Brussels, Belgium.

22. Ferritto, J., Dickenson, S., Priestley, N., Werner, S. and Taylor, C. (1999): Seismic Criteria for California Marine Oil Terminals. Naval Facilities Engineering Center, Port Hueneme, CA.

23. Finn, W.D.L., Ledbetter, R.H. and Wu, G. (1994): Liquefaction in silty soils: design and analysis. Ground failures under seismic conditions. Geotechnical Special Publication No. 44, S. Prakash and P. Dakoulas (eds.), ASCE, Reston, VA, 51–76.

24. Gelfenbaum, G. and Jaffe, B. (2003): Erosion and sedimentation from the July 17, 1998 Papua New Guinea Tsunami. Pure and Applied Geophysics, vol. 160, 1969–2000.

25. Gerwick, B.C. (2007): Construction of Marine and Offshore Structures. Third Edition. CRC Press, Taylor & Francis Group, Boca Raton, London, New York.

26. Goto, C., Ogawa, Y., Shuto, N. and Imamura, F. (1997): Numerical Method of Tsunami Simulation with the Leap-frog Scheme (IUGG/IOC Time Project), IOC Manual, UNESCO, Paris, No. 35.

27. Goto, S., Syamoto,Y. and Tamaoki, S. (1987): Dynamic properties of undisturbed grave samples obtained by *in-situ* freezing method. Proceedings of the 8th Asian Regional Conference on Soil Mechanics and Foundation Engineering, Kyoto, Japan, 233–236.

28. Gunbak, A.R., Muyesser, O. and Yuksel, Y. (2000): Damages recorded at the coastal and port structures around Izmit Bay under the 17th August, 1999 Earthquake, PIANC Buenos Aires Conference, 1–19.

29. Hall, J.F. (ed.) (1995): Northridge Earthquake of January 17, 1994. Reconnaissance Report, vol. 1, Earthquake Spectra, Supplement C to vol. 11.

30. Hamada, M., Yasuda, S., Isoyama, R. and Emoto, K. (1986): Study on Liquefaction Induced Permanent Ground Displacement, Report published for the Association for the Development of Earthquake Prediction, Tokyo.

31. Hatanaka, M., Suzuki, Y., Kawasaki, T. and Masaaki, E. (1988): Cyclic undrained shear properties of high quality undisturbed Tokyo Gravel. Soils and Foundations, vol. 28, No. 4, 57–68.

32. Iai, S. and Kameoka, T. (1993): Finite element analysis of earthquake induced damage to anchored sheet pile quay walls. Soils and Foundations, vol. 33, No. 1, 1, 71–91.

33. Iai, S., Matsunaga, Y., Morita, T., Miyata, M., Sakurai, H., Oishi, H., Ogura, H., Ando, Y., Tanaka, Y and Kato, M. (1994): Effects of remedial measures against liquefaction at 1993 Kushiro-Oki earthquake. Proc. 5th US-Japan Workshop on Earthquake Resistant Design of Lifeline Facilities and Counter Measures against Soil Liquefaction, NCEER-94-0026, National Center for Earthquake Engineering Research, 135–152, Buffalo, NY.

34. Imamura, F. (1988): Numerical simulations of the transoceanic propagation of tsunamis. 6th Congress Asian and Pacific Regional Division, International Association for Hydraulic Research, Kyoto, Japan.

35. Imamura, F. (1996): Review of tsunami simulation with a finite difference method. Long-Wave Runup Models. H. Yeh, P. Liu, and C. Synolakis (eds.), World Scientific, New York, 25–42.

36. Imamura, F. and Goto, C. (1988): Truncation error in numerical tsunami simulation by the finite difference method. Coastal Engineering in Japan, vol. 31, No. 2, 245–263.

37. Imamura, F. and Shuto, N. (1989): Numerical simulation of the 1960 Chilean tsunami. Proceedings of the Japan–China (Taipei) Joint Seminar on Natural Hazard Mitigation, Kyoto, Japan.

38. Indian Standards Institution (1986): Criteria for Earthquake Resistant Design of Structures, Rev. 4. UDC 699-841; 624-042-7. Indian Standards Institution, New Delhi, 2003.

39. Ishihara, K. (1985): Stability of natural deposits during earthquakes. Proceedings of the 11th International Conference on Soil Mechanics and Foundation Engineering, San Francisco, vol. 1, 321–376.

40. Ishihara, K. (1996): Soil Behaviour in Earthquake Engineering. Clarendon Press, Oxford, UK.

41. Ishihara, K. and Koseki, J. (1989): Cyclic shear strength of fines-containing sands, earthquake geotechnical engineering. Proceedings of the, Discussion Session on Influence of Local Conditions on Seismic Response, 12th ICSMFE, Rio de Janeiro, Brasil, 101–106.

42. Ishihara, K. and Silver, M.L. (1977): Large diameter sand sampling to provide specimens for liquefaction testing. Proceedings of the, 9th International Conference on Soil Mechanics and Foundation Engineering, Tokyo, Specialty Session 2, 1–6.

43. Ishihara, K. and Yoshimine, M. (1992): Evaluation of settlements in sand deposits following liquefaction during earthquakes. Soils and Foundations, vol. 32, No. 1, 173–188.

44. Iwasaki, T., Tokida, K., Tatsuoka, F., Watanabe, S., Yasuda, S. and Sato, H. (1982): Microzonation for Soil Liquefaction Potential Using Simplified Methods. Proceedings of the, 3rd International Conference on Microzonation, Seattle, vol. 3, 1319–1330, National Science Foundation.

45. The Japanese Geotechnical Society (JGS), ed. (1998): Remedial Measures against Soil Liquefaction. (ed.). Balkema, Rotterdam, The Netherlands.

46. Katopodi, I. and Iosifidou, K. (2004): Impact of the Lefkada earthquake (14-08-2003) on marine works and coastal regions, Proceedings of the, 7th Panhellenic Geographical Conference, Mytilene, Greece, October 2004, 363–370.

47. Kokusho, T. and Fujita, K. (2002): Site investigations for involvement of water films in lateral flow in liquefied ground. Journal of Geotechnical Engineering, ASCE, vol. 128, No. 11, 917–925.

48. Lee, J.J. (1969): Wave Induced Oscillations in Harbors of Arbitrary Shape. PhD Thesis, California Institute of Technology, Pasadena, CA.

49. Lepelletier, G.T. and Raichlen, F. (1987): Harbor oscillations induced by nonlinear transit waves. Journal of Waterway, Port, Coastal and Ocean Engineering, ASCE, vol. 113, No. 3, 381–400.

50. Lettis, W., Bachhuber, J., Barka, A., Brankman, C., Lettis, C., Somerville, P. and Witter, R. (2000): Geology and seismicity. Earth Spectra, Chapter 1, in 1999 Kocaeli, Turkey, Earthquake Reconnaissance Report, Supplement A to Earthquake Spectra, Vol. 16, T. L. Youd, J.-P., Bardet, J. D. Bray, eds., pp. 19.

51. Liao, T., McGillivray, A., Mayne, P.W., Zavala, G. and Elhakim, A. (2002): Seismic Ground Deformation Modeling, Final report for MAE HD-7a (year 1). Geosystems Engineering/School of Civil & Environmental Engineering, Georgia Institute of Technology, Atlanta, CA.

52. Martin, G.R., Finn, W.D.L. and Seed, H.B. (1975): Fundamentals of liquefaction under cyclic loading. Journal of the Geotechnical Engineering Division, ASCE, vol. 101, No. GT5, 423–438.

53. Mei, C.C. (1989): The Applied Dynamics of Ocean Surface Waves. World Scientific, London.

54. Ministry of Transport, Republic of Turkey (2008): Technical Code of Practice for Seismic Design of Coastal, Port, Railway and Airport Structures. Published in the Official Gazette of the Republic of Turkey, Publication Date: August 18, 2007, with Official Gazette No: 26617, and Revision: December 26, 2008, with Official Gazette No: 27092 (in Turkish).

55. Ministry of Transport (MOT). (1971): Supplements for Design Standards for Port and Harbour Structures. Ports and Harbours Bureau, Japan Ports and Harbours Association, Tokyo (in Japanese).

56. Ministry of Transport (MOT). (1979): Technical Standards for Port and Harbour Facilities with Commentary. Ports and Harbours Bureau, Japan Ports and Harbours Association, Tokyo (in Japanese).

57. Ministry of Transport (MOT). (1989): Technical Standards for Port and Harbour Facilities with Commentary. Ports and Harbours Bureau, Japan Ports and Harbours Association, Tokyo (in Japanese).

58. Ministry of Transport (MOT). (1991): Technical Standards for Port and Harbour Facilities in Japan, New Edition. Ports and Harbours Bureau, The Overseas Coastal Area Development Institute of Japan (OCDI), Tokyo, Japan.

59. Ministry of Transport (MOT). (1993): Handbook on Liquefaction Remediation of Reclaimed Land. Ports and Harbours Bureau, Coastal Development Institute of Technology, Tokyo (in Japanese).

60. Ministry of Transport (MOT). (1999): Technical Standards for Port and Harbour Facilities with Commentary. Ports and Harbours Bureau, Japan Ports and Harbours Association, Tokyo (in Japanese).

61. Mulilis, J.B., Seed, H.B., Chan, C.K., Mitchell, J.K. and Arulanandan, K. (1977): Effects of sample preparation on sand liquefaction. Journal of Geotechnical Engineering Division, ASCE, vol. 103, No. 2, 91–108.

62. Newmark, N. M. (1965): Effects of earthquakes on embankments and dams. Géotechnique, vol. 15, No.2, 139–160.

63. Okal, E.A. and Synolakis, C.E. (2004): Source discriminants for near-field tsunamis. Geophysical Journal International, vol. 158, No. 3, 899–912.

64. PIANC (2001): Seismic Design Guidelines for Port Structures. Working Group No. 34 of the Maritime Navigation Commission, International Navigation Association, Balkema, Rotterdam, The Netherlands.

65. Port and Harbor Research Institute (PHRI), ed. (1997): Handbook on Liquefaction Remediation of Reclaimed Land. Balkema, Rotterdam, The Netherlands.

66. Raichlen, F. (1966): Harbor resonance. In: Coastline and Estuarine Hydrodynamics, (ed. A.T. Ippen), McGraw Hill, NY.

67. Rauch, a. F. (1997): EPOLLS: An Empirical Method for Predicting Surface Displacement due to Liquefaction-induced Lateral Spreading in Earthquakes, PhD thesis, Virginia Polytechnic Institute, VA.

68. Safak, E., Beyen, K., Carver, D., Cranswick, E., Celebi, M., Durukal, E., Ellsworth, W., Erdik, M., Holzer, T., Meremonte, M., Mueller, C., Ozel, O. and Toprak, S. (2000): Recorded main shock and after shock motions. 1999 Kocaeli, Turkey, Earthquake Reconnaissance Report, Supplement A to Earthquake Spectra, vol. 16, Chapter 5. T.L. Youd, J.-P. Bardet, J.D. Bray, eds., 97–112.

69. Sawicki, A. and Swidzinski, W. (2006): A study on liquefaction susceptibility of some soils from the coast of Marmara sea. Bulletin of the Polish Academy of Sciences, Technical Sciences, vol. 54, No. 4, 405–418.

70. Sawicki, A. and Swidzinski, W. (2007): Simple mathematical model for assessment of seismic-induced liquefaction of soils. Journal of Waterway, Port, Coastal and Ocean Engineering, ASCE, vol. 133, No. 1, 50–54.

71. Schonberg, J. (1997): The Spartan Earthquake of 464–65 B.C. Brown Classical Journal, vol. 11. Brown University, Classics Department, U.S.

72. Seed, R.B., Cetin, K.O., Moss, R.E.S., Kammerer, A., Wu, J., Pestana, J., Riemer, M., Sancio, R.B., Bray, J.D., Kayen, R.E. and Faris, A. (2003): Recent advances in soil liquefaction engineering: A unified and consistent framework. White Paper for Keynote Presentation, 26th Annual ASCE Los Angeles Geotechnical Spring Seminar, Long Beach, CA.

73. Seed, R.B., Cetin, K.O., Moss, R.E.S., Kammerer, A.M., Wu, J., Pestana, J.M. and Riemer, M.F. (2001): Recent advances in soil liquefaction engineering and seismic site response evaluation. Proceedings

of the, 4th International Conference Recent Advances in Geotechnical Earthquake Engineering and Soil Dynamics, San Diego, CA.

74. Seed, H.B. and Idriss, I.M. (1971): Simplified procedure for evaluating soil liquefaction potential. Journal of Soil Mechanics and Foundations Division, ASCE, vol. 97, No. SM9, Proceedings Paper 8371, 1249–1273.

75. Seed, H.B., Tokimatsu, K., Harder, L.F. and Chung, R.M. (1984): The Influence of SPT Procedures in Soil Liquefaction Resistance Evaluations. Earthquake Engineering Research Center Report No. UCB/EERC-84/15, University of California at Berkeley, Berkeley, CA.

76. Seed, H.B., Tokimatsu, K., Harder, L.F. and Chung, R.M. (1985): Influence of SPT procedures in soil liquefaction resistance evaluations. Journal of Geotechnical Engineering, ASCE, vol. 111, No. 12, 1425–1445.

77. Shamoto, Y., Zhang, J.M. and Tokimatsu, K. (1998): Methods for evaluating residual post-liquefaction ground settlement and horizontal displacement. Soils and Foundations, vol. 2, No. S2, 69–84.

78. Shuto, N., Goto, C., Imamura, F. (1990): Numerical simulation as a means of warning for near field tsunamis. Coastal Engineering in Japan, vol. 33, No. 2, 173–193.

79. Sugano, T., Kaneko, H. and Yamamoto, S. (1999): Damage to port facilities. The 1999 Ji-Ji Earthquake, Taiwan. Investigation into the Damage to Civil Engineering Structures, Japan Society of Civil Engineers, 5-1–5-7.

80. Sumer, B.M., Ansal, A., Cetin, K.O., Damgaard, J., Gunbak, A.R., Hansen, N.-E.O., Sawicki, A., Synolakis, C.E., Yalciner, A.C., Yuksel, Y. and Zen, K. (2007): Earthquake-induced liquefaction around marine structures. Journal of Waterway, Port, Coastal and Ocean Engineering, ASCE, vol. 133, No. 1, 55–82.

81. Sumer, B.M. and Fredsøe, J (2002): The Mechanics of Scour in the Marine Environment. World Scientific, Singapore, 552 p.

82. Sumer, B.M., Kaya, A. and Hansen, N.-E.O. (2002): Impact of liquefaction on coastal structures in the 1999 Kocaeli, Turkey Earthquake.

Proceedings of the 12th International Offshore and Polar Engineering Conference, KitaKyushu, Japan, May 26–31, 2002, vol. II, 504–511.

83. Sumer, B.M., Sen, M.B., Karagali I., Ceren, B., Fredsøe, J., Sottile, M., Zilioli, L. and Fuhrman, D.R. (2011): Flow and sediment transport induced by a plunging solitary wave, Journal of Geophysical Research, vol. 116, No. C01008, 1–15.

84. Synolakis, C.E. (2003): Tsunamis and seiches. Earthquake Engineering Handbook, W.F. Chow and C. Scawthorn (eds), CRC, Boca Raton, Fla. Press, 9-1–9-90.

85. Synolakis, C.E., Liu, P.L.-F., Yeh, H. and Carrier, G. (1997): Tsunamigenic seafloor deformations. Science, vol. 278, No. 5338, 598–600.

86. Synolakis, C.E., Okal, E.A. and Bernard, E.N. (2005): The megatsunami of December 26, 2004, The Bridge, National Academy of Engineering, vol. 35, No. 2, 35–48.

87. Synolakis, C.E. and Skjelbreia, J.E. (1993): Evolution of maximum amplitude of solitary waves on plane beaches, Journal of Waterway, Port, Coastal and Ocean Engineering, ASCE, vol. 119, No. 3, 323–342.

88. Tadepalli, S. and Synolakis, C.E. (1996): Model for the leading waves of tsunamis. Physical Review Letters, vol. 77, No. 10, 2141–2144.

89. Thucydides (425BC). History of the Peloponnesian War. Translated by Rex Warner. London: Penguin Books, 1972.

90. Titov, V.V. and Synolakis, C.E. (1998): Numerical modeling of tidal wave runup. Journal of Waterway, Port, Coastal and Ocean Engineering, ASCE, vol. 124, No. 4, 157–171.

91. Tokimatsu, K. and Seed, H.B. (1984): Simplified Procedures for the Evaluation of Settlements in Clean Sands. Technical report UCB/EERC-84-16, University of California, Berkeley, Earthquake Engineering Research Center, Berkeley, CA.

92. Tonkin, S., Yeh, H., Kato, F. and Sato, S. (2003): Tsunami scour around a cylinder: An effective stress approach. Journal of Fluid Mechanics, vol. 496, 165–192.

93. Towhata, I., Sasaki, Y., Tokida, K.-I., Matsumoto, H., Tamari, Y. and Yamada, K. (1992): Prediction of permanent displacement of liquefied ground by means of minimum energy principle. Soils and Foundations, vol. 32, No. 3, 97–116.

94. Tsay, T.-K. and Liu, P.L.-F. (1983): A finite element model for wave refraction and diffraction. Applied Ocean Research, vol. 5, No. 1, 30–37.

95. Tsuchida, H. (1970): Prediction of and Measures Against Liquefaction of Sandy Ground. Seminar Report, Port and Harbour Research Institute, 3-1–3-33 (in Japanese).

96. Turkish Code of Practice (1998): Specification for Structures to be built in Disaster Areas. Part III. Earthquake Disaster Prevention, issued 2.9.1977. Official Gazette No. 33098 (Effective from January 1 1998, amended on July 2, 1998, Official Gazette No. 23390).

97. Wang, J.G. and Rahman, m.S. (1999): A neural network model for liquefaction-induced horizontal ground displacement. Soil Dynamics and Earthquake Engineering, vol. 18, No. 8, 555–568.

98. Westergaard, H.M. (1933): Water pressure on dams during earthquakes. Transactions of the American Society of Civil Engineers, vol. 98, No. 1850, 418–472.

99. Wilson, B.W. (1972): Seiches. Advances in Hydroscience, vol. 8, (ed., V.T. Chow), Academic Press, New York.

100. Wyllie, L.A., Abrahamson, N., Bolt, B., Castro, G., Durkin, M.E., Escalante, L., Gates, H.J., Luft, R., McCormic, D., Olson, R.S., Smith, P.D. and Vallenas, J. (1986): The Chile earthquake of March 3, 1985. Earthquake Spectra, vol. 2, No. 2, 293–371.

101. Yalciner A.C., Alpar B., Altinok Y., Ozbay I. and Imamura, F. (2002): Tsunamis in the Sea of Marmara: Historical documents for the past, models for future. Marine Geology, 2002, vol. 190, No. 1–2, 445–463.

102. Yalciner, A.C., Altinok, Y. and Synolakis, C.E. (2000): Tsunami waves in Izmit Bay after the Kocaeli Earthquake. Chapter 13, Kocaeli Earthquake, Earthquake Engineering Research Institute, Oakland, CA.

103. Yalciner, A.C., Pelinovsky, E., Çakiroglu, Y. and Imamura, F. (1996): The properties of resonance due to the geometry of the basins XXV. Tsunamis Impacting on the European Coasts: Modelling, Observation and Warning. Session NH5, General Assembly of European Geophysical Society, The Hague, The Netherlands.

104. Yalciner, A., Pelinovsky, E., Talipova, T., Kurkin, A., Kozelkov, A. and Zaitsev, A. (2004): Tsunamis in the Black Sea. Journal of Geophysical Research, 109, No. C12, C12023.1–C12023.13

105. Yalciner, A.C., Perincek, D., Ersoy, S., Prasetya, G., Rahman, H. and McAdoo, B. (2005): Report on January 21–31, 2005 North Sumatra Survey on December, 26, 2004 Indian Ocean Tsunami by ITST of UNESCO IOC, November 11, 2006.

106. Yalciner, A.C., Synolakis, A.C., Alpar, B., Borrero, J., Altinok, Y., Imamura, F., Tinti, S., Ersoy, S., Kuran, U., Pamukcu, S. and Kanoglu, U. (2001): Field surveys and modeling 1999 Izmit Tsunami. International Tsunami Symposium, ITS 2001, Seattle, WA, Paper 4-6, 557–563.

107. Yeh, H., Liu, P-L.F and Synolakis, C.E. (1996): Long Wave Runup Models. World Scientific, London.

108. Yoshimi, Y., Tokimatsu, K. and Hosaka, Y. (1989): Evaluation of liquefaction resistance of clean sands based on high-quality undisturbed samples. Soils and Foundations, vol. 29, No. 1, 93–104.

109. Youd, T.L., Hansen, C.M. and Bartlett, S.F. (2002): Revised multilinear equations for prediction of lateral spread displacement. Journal of Geotechnical and Geoenvironmental Engineering, ASCE, vol. 128, No. 12, 1007–1017.

110. Youd, T.L., Idriss, I.M., Andrus, R.D., Arango, I., Castro, G., Christian, J.T., Dobry, R., Finn, W.D.L., Harder, L.F. Jr., Hynes, M.E., Ishihara, K., Koester, J.P., Liao, S.S.C., Marcuson, W.F. III., Martin, G.R., Mitchell, J.K., Moriwaki, Y., Power, M.S., Robertson, P.K.,

Seed, R.B. and Stokoe, K.H., II. (1997): Summary paper. Proceedings of NCEER Workshop on Evaluation of Liquefaction Resistance of Soils, NCEER- 97-0022, Buffalo, NY.

111. Youd, T.L., Idriss, I.M., Andrus, R.D., Arango, I., Castro, G., Christian, J.T., Dobry, R., Finn, W.D.L., Harder, L.F. Jr., Hynes, M.E., Ishihara, K., Koester, J.P., Liao, S.S.C., Marcuson, W.F. III., Martin, G.R., Mitchell, J.K., Moriwaki, Y., Power, M.S., Robertson, P.K., Seed, R.B. and Stokoe, K.H., II. (2001): Liquefaction resistance of soils: Summary report from the 1996 NCEER and 1998 NCEER/NSF Workshops on Evaluation of Liquefaction Resistance of Soils. Journal of Geotechnical and Geoenvironmental Engineering, vol. 127, No. 10, 817–833.

112. Young, Y.L., White, J.A., Xiao, H. and Borja, R.I. (2009): Liquefaction potential of coastal slopes induced by solitary waves. Acta Geotechnica, vol. 4, No. 1, 17–34.

113. Yuksel, Y., Alpar, B., Yalciner, A., Cevik, E., Ozguven, O. and Celikoglu, Y. (2003): Effects of the Eastern Marmara Earthquake on the marine structures and coastal areas. Proceedings of the Institution of Civil Engineering (ICE, UK), Water and Maritime Engineering, vol. 156, Issue WM2, 147–163.

114. Yuksel, Y., Cetin, K.O., Ozguven, O., Isik, N.S., Cevik, E. and Sumer, B.M. (2004): Seismic response of a rubble mound breakwater in Turkey. Proceedings of the Institution of Civil Engineering (ICE, UK), Maritime Engineering, vol. 157, Issue MA4, 151–161.

115. Zahibo, N., Pelinovsky, E., Yalciner, A.C., Kurkin, A., Kozelkov, A. and Zaitsev, A. (2003): The 1867 Virgin Island tsunami. Oceanologica Acta, 26, No. 5–6, 609–621.

116. Zelt, J.A. and Raichlen, F. (1990): A Lagrangian model for wave-induced harbor oscillations. Journal of Fluid Mechanics, vol. 213, 203–225.

Chapter 11

Counter Measures

As seen in the previous chapters, liquefaction occurs essentially in loose sands with fine grain size, when pore-water pressures build up and cause the soil to behave as a liquid, the residual liquefaction. This process occurs in waves (Chapters 1, 3, and Chapters 5–8), earthquakes (Chapter 10), and in cases such as shocks, or blasts, and when structures execute rocking motions (Chapters 1 and 8). Liquefaction also occurs momentarily in waves if the seabed contains air or gas (Chapters 1, 4, and 5–8), the momentary liquefaction. Under liquefaction conditions, the seabed soil fails, thus precipitating failure of the supported structure. With the soil liquefied, buried pipelines or outfalls may float to the surface of the seabed (Chapter 5), or they may sink, penetrating to larger depths (Chapter 6), large individual blocks and sea mines may penetrate into the seabed and may eventually disappear (Chapter 6), and gravity structures (caisson breakwaters, offshore platforms, box caissons, etc.) may fail (Chapter 8). Therefore, preventive measures need to be considered in the design of marine structures if and when there is potential for liquefaction.

Gerwick (2007, Chapter 7) gives a comprehensive account of prevention of liquefaction under the general title of seafloor modifications and improvements. In this chapter, we will discuss the prevention measures against liquefaction with reference to the following four cases: buried pipelines, gravity-base structures, hydraulic fills, and sheet-pile cofferdams.

11.1   Buried Pipelines

The burial of pipelines is required when there is a risk of dragging ship anchors, impact of dropped anchors, snagging by trawl boards, or fishing gears. Burial of the pipe also permits the pipe to be designed with less net weight (less coating) (Gerwick 2007, p. 617). The burial also provides the best protection against hydrodynamic loading (drag and lift forces, and flow-induced hydroelastic vibrations) (Sumer and Fredsøe, 1997), and scour (Sumer and Fredsøe, 2002). We note that stone protection and protective mattresses are also viable options for pipeline protection (Sumer and Fredsøe, 2002). The latter protection measures are preferred, however, when the required burial depth is large, or when it is required that the pipeline should be easily accessible, or when the cost of the pipe burial is simply too high.

Figure 11.1:   Pipeline laid in a trench, with the trench backfilled.

The burial of pipelines involves trenching. For example, a cutter suction dredger excavates a trench, and pumps the material to a temporary stockpile area, then the pipeline is placed in the trench, and subsequently, the cutter dredger re-dredges the stockpile, pumps the backfill to the trench, and discharges the material around the pipeline (Fig. 11.1). The backfill material, when placed in the trench, will inevitably be loose, and if the grain size is fine (say, smaller than $O(0.1\,\mathrm{mm})$, or coarser in the case of earthquakes), the backfill will be prone to liquefaction (the residual liquefaction) under storm waves, or earthquakes, due to the buildup of pore-water pressure. The soil may also be subject to momentary liquefaction under waves if it contains air or gas.

Preventive measures need to be considered if there is risk of liquefaction of the backfill soil. These measures include: design of the burial depth, use of select backfill, and surface protection by cover stones.

Design of burial depth

In the case of the residual liquefaction, as a conservative approach, it may be assumed that the liquefaction spreads across the entire depth, and therefore the design of the burial depth may not be an option as a preventive measure, although the methods available to assess liquefaction potential can resolve the penetration depth of liquefaction (Chapter 3, Section 3.2.4). Involving the penetration depth of liquefaction in design will be inconsistent due to various uncertainties related to the soil and the design wave, as well as due to the finite trench depth.

In the case of the momentary liquefaction, the penetration depth of liquefaction can be assessed, using the approach described in Sections 5.7 or 6.4. On this basis, the burial depth of the pipe is designed such that liquefaction will not occur below the top of the pipeline, which will ensure the stability of the pipeline.

The remainder of this section is concerned with the residual-liquefaction scenarios in which two counter measures will be discussed, namely, use of select material, and surface protection by cover stones.

Use of select material

If the native backfill sand does not have the quality to protect the pipe against the risk of liquefaction, it must be replaced with a coarser material, a material sufficiently permeable so that all pore-water pressures accumulated by cyclic loading (due to waves, or earthquakes, or other effects) will dissipate as rapidly as they develop (Seed and Rahman, 1978).

In order to design the select backfill material, the assessment exercise described in Example 4, Section 3.2.4, should be repeated for coarser materials until the grain size is judged to be sufficiently large so that liquefaction does not occur. Once this "critical" grain size is determined, then any material available with grain size larger than this critical grain size can be chosen as the select material.

Surface protection by cover stones

This method is resorted to when a select backfill material is not available. Herbich (1981) notes that the surface protection by cover stones is less expensive than the use of select material.

Fig. 11.2 gives a schematic description in which the surface of the backfill soil is covered by a stone protection. The cover may be one-layer deep, or two-layers deep, etc., extending across the entire width of the trench. Gerwick (2007, p. 246), in the context of strengthening of weak soils, notes that an effective means of consolidating weak soils is by surcharging in which an underwater fill of stones can be placed by bottom-dump barge, and allowed to exert its excess pressure on the soil. Gerwick (2007) comments on the length of time necessary for complete consolidation as six months to a year, or more. The latter can easily be worked out, using the classic consolidation theory (e.g., Lambe and Whitman, 1969) for a given set of soil and cover stones conditions.

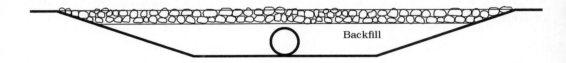

Figure 11.2:   Stone protection over the backfill soil.

It is also interesting to recall the application cited in Example 13 in Chapter 6. Here, stone cover over the backfill is adopted as a special measure to offset upheaval buckling since the required download to prevent upheaval buckling substantially exceeded the self weight of the pipe. As pointed out in Example 13, the project eventually chose to adopt an increase on the required stone-cover height to reduce the likelihood of the pipe buckling even if a liquefaction event occurred.

Sekiguchi *et al.* (2000) in a centrifuge wave flume (simulating field conditions), and Sumer *et al.* (2010) in a standard wave flume studied the influence of cover stones/surcharge on the liquefaction behaviour of soils, see Chapter 3, Section 3.6. Both studies found that the underlying soil is not liquefied when the cover layer is heavy enough to densify the soil to the degree that the resulting soil relative density, D_r, is presumably too large for liquefaction to occur. This is the key mechanism for surface protection by cover stones in which the backfill soil is densified by consolidation achieved by the surcharge provided by the cover stones. A numerical example given at the end of this sub-section addresses the issue of how to design cover stones, ensuring liquefaction-resistant backfill.

It is prudent to use an intermediate filter layer between the cover stones and the bed to minimize the sinking of the cover stones into the soil in extreme-case scenarios where the soil is liquefied, and therefore, the cover stones undergo sinking. Sumer *et al.*'s (2010) experiment with a filter layer, although limited to a single test, demonstrated that the stones are prevented from sinking in the soil, and the downward displacement of the stones (where the stones and the filter layer sink en masse) is greatly reduced. Sumer *et al.* (2010) report:

> "the cover layer (the cover stones plus the filter stones) experienced a downward displacement significantly smaller than that of the stones in the case of no filter, largely because of the considerable reduction in the fall velocity in the liquefied soil of the individual grains of the filter layer compared with that of the individual stones in the case of no filter."

Finally, it may be noted that Sumer *et al.*'s (2010) research also showed that the soil under a stone cover may not be liquefied, even with a single-layer stone cover, when the stones are densely packed so "tight" that they do not allow the soil to undergo sufficiently large periodic expansion and contraction under the action of the cyclic bed pressure, and therefore the soil experiences no significant cyclic shear strains (and thus no significant cyclic shear stresses). The latter implies no significant shaking up of the soil, and hence no liquefaction. The densely-packed cover-stone arrangement in Sumer *et al.*'s (2010) experiment involved rather small porosities ($n = 0.36$), which may not be achievable in practice (reported values for porosity being 0.38–0.40 for quarry stones, 0.47 for cubes, 0.50 for tetrapods and 0.63 for dolos; Shore Protection Manual, 1977), and therefore this latter case appears to be only of academic interest.

Example 18. *Design of cover stones. A numerical example (Sumer* et al., *2010).*

The soil properties are given as follows. The soil depth, $d = 1\,\text{m}$; the grain size, $d_{50} = 0.135\,\text{mm}$; the submerged specific weight, $\gamma' = 9.42\,\text{kN/m}^3$; the shear modulus, $G = 2.23 \times 10^3\,\text{kN/m}^2$; the coefficient of permeability, $k = 8.1 \times 10^{-5}\,\text{m/s}$; the porosity, $n = 0.435$; the degree of saturation, $S_r = 1$; the coefficient of lateral earth pressure, $k_0 = 0.37$; Poisson's ratio, $\nu = 0.3$; the void ratio, $e = 0.77$; the maximum void ratio, $e_{\text{max}} = 0.888$; the minimum void ratio, $e_{\min} = 0.627$; the relative density, $D_r = 0.452$; and the coefficient of consolidation, $c_v = 0.00197\,\mathrm{m}^2/\mathrm{s}$.

This soil is exposed to a progressive wave with the following properties: the wave height, $H = 2\,\mathrm{m}$; the period, $T = 7.9\,\mathrm{s}$; and the water depth, $h = 9\,\mathrm{m}$.

Find the stone size for a three-layer stone cover, which would prevent soil liquefaction. See the inset in Fig. 11.3 for the definition sketch.

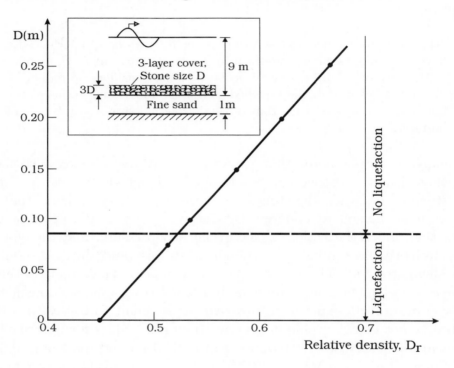

Figure 11.3: Critical stone size for no liquefaction for a three-layer stone cover in the example. Wave height $H = 2\,\mathrm{m}$, wave period $T = 7.9\,\mathrm{s}$ and water depth $h = 9\,\mathrm{m}$. Sumer *et al.* (2010).

We first consider the undisturbed case, i.e., the case without cover stones. The mathematical model described in Section 3.2.4 gives, in this case, that the soil under the given wave is liquefied at the mudline after 11 waves, and that the liquefaction spreads across the entire soil depth after 16 waves (cf., Example 4 in Section 3.2.4).

Now, when stone cover is used as protection against liquefaction, the relative density of the soil, D_r, will change, depending on the surface loading (or the surcharge), p_s. The new relative density will be

$$D_{r,new} = \frac{e_{max} - e_{new}}{e_{max} - e_{min}} \qquad (11.1)$$

in which e_{new} is the new void ratio, achieved by the use of cover stones. In order to determine the new relative density, we need the void ratio e_{new}, which is calculated from the porosity information

$$e_{new} = \frac{n_{new}}{1 - n_{new}} \qquad (11.2)$$

with the porosity determined from

$$n_{new} = \left(\frac{e}{1+e} - \frac{k}{\gamma c_v} p_s \right) \left(1 - \frac{k}{\gamma c_v} p_s \right)^{-1} \qquad (11.3)$$

in which p_s is the surface loading, or the surcharge, corresponding to the cover stones, which is actually the submerged weight of the stones per unit area of the bed, and given by (Section 3.6)

$$p_s = wN - (1 - n)\overline{D_z}N\gamma \qquad (11.4)$$

in which w is the weight of stones per unit area of the bed per stone layer, N is the number of stone layers, and $\overline{D_z}$ is the mean stone height. In the derivation of Eq. 11.3, the expression for the final settlement of the underlying sand due to consolidation

$$\rho_1 = d\, m_v p_s \qquad (11.5)$$

is utilized (Terzaghi, 1948, p. 281). Here, m_v is the coefficient of volume decrease, $m_v = k/(\gamma c_v)$. The quantities in Eqs 11.3 and 11.4, e, k, c_v, and n are the values corresponding to the initial condition before the stone cover is installed.

The relative density of the soil, $D_{r,new}$, is worked out for various values of the stone size for a three-layer stone cover where the porosity of the stone cover is taken as 0.39.

The change in the soil permeability with an increase in the relative density is also taken into account in the calculations as a function of the new void ratio, utilizing the chart in Fig. B.2 in Appendix B.

Subsequently, the mathematical model described in Section 3.2.4 is implemented to see whether or not the soil under the given wave is liquefied, using the liquefaction criterion $\bar{p}/\sigma'_0 > 1$ (Section 3.1.2) in which σ'_0, the initial mean normal effective stress, is calculated from Eqs 3.80 and 3.81 (Section 3.6). The results obtained for various values of the stone size are plotted in Fig. 11.3. The figure shows that the soil is not liquefied when the stone size is larger than about $D = 0.09\,\mathrm{m}$. As a conclusion, the soil can be protected against liquefaction with a three-layer stone cover provided that the stone size is larger than $0.09\,\mathrm{m}$.

As a final note, we add the following. The purpose of this numerical example is to illustrate how to design cover stones, which ensures a liquefaction-resistant backfill, utilizing the mathematical model given in Section 3.2.4. For a final design, the following two issues also need to be checked: (1) stability of stones under extreme current and wave conditions; (2) motion as well as suction removal (or winnowing) of base sediment from between stones by current and wave actions (Sumer, Cokgor, Fredsøe, 2001, Dixen, Sumer, Fredsøe, 2008, and Nielsen, Sumer, Fredsøe, 2012). Also, an intermediate filter layer between the cover stones and the soil may be considered, to minimize the sinking of the cover stones into the soil in extreme flow conditions (see the discussion in the preceding paragraphs for filter).

11.2   Gravity-Base Structures

In this section we will focus on preventive measures for liquefaction risk underneath large, gravity-base structures (such as caisson breakwaters (or vertical-wall breakwaters), offshore platforms, and gravity-base caissons) due to rocking motions of these structures under waves. As discussed in Chapter 8, the seabed soil under a gravity-base structure may undergo liquefaction.

In the case of gravity-base structures with large widths, e.g., caisson breakwaters, seabed liquefaction occurs in two zones, namely in front of the breakwater (Zone 1, Fig. 8.15), and under the breakwater (Zone 2, Fig. 8.15). Preventive measures for the liquefaction in front of the breakwater involve methods similar to those used for hydraulic fills, discussed under hydraulic fills in the next section.

The following paragraphs will summarize special counter measures against liquefaction that occur under these structures.

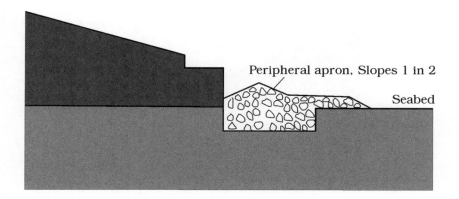

Figure 11.4: Peripheral apron to minimize the pressure buildup at these locations. This is particularly important as these locations are scour sensitive.

Drainage pads and wells

To prevent pore-water pressure from building up, drainage pads and wells are installed underneath the base of the structure. Gerwick (2007, p. 248) notes that such counter measures have been taken for some of the large gravity-base structures in the North Sea. Gerwick also adds that the drainage into the structure (from where the drained water is pumped out) is controlled so that the ambient pore-water pressure is less than the static pressure; therefore, even under storm conditions, pressure accumulation is prevented.

Peripheral apron of graded stones

As seen in Section 8.2.2 (Figs 8.15 and 8.17) and Section 8.3 (Fig. 8.20), the areas at/near the edge of a gravity structure are especially vulnerable to liquefaction, and therefore calls for special measures. To minimize the buildup of pore pressure at these locations (which is particularly important, considering that these locations are scour sensitive), an apron of graded stones (Gerwick, 2007, p. 248) is installed along the periphery of the structure, the peripheral apron (Fig. 11.4). The width of the peripheral apron needs to be designed to ensure that the volume of the "falling apron" is large enough to accommodate the edge scour at the junction between the peripheral apron and the seabed.

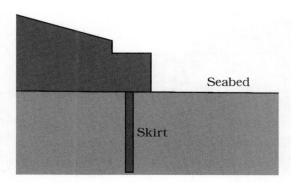

Figure 11.5:   Skirts to prevent the structure from rocking, and therefore to avoid liquefaction due to the rocking motion of the structure.

Skirts

To prevent the structure from rocking, and therefore to avoid liquefaction,
vertical steel "sheets" may be installed to the foundation of the structure, as
sketched in Fig. 11.5.

Consolidation and strengthening of soil by surcharging

There are several methods used for consolidation and strengthening of the
seabed soil prior to the installation of the structure. Of these, surcharging
will be discussed in this section, whereas other methods will be discussed
under hydraulic fills.

As mentioned earlier, an effective method of consolidating weak soils is
by surcharging in which an underwater fill of sand, or rock can be placed on
the seabed and allowed to exert its excess pressure on the soil for a period of
six months to a year or more (Gerwick, 2007, p. 246). Consolidation is even
more effective if the pore water is able to drain through natural or artificial
drains, such as vertical sand drains or wick drains (see next section for wick
drains).

Gerwick (2007, p. 246) notes that, with or without artificial drainage, the
surcharge effect of a fill over a period of six to twelve months will serve to
stabilize the liquefaction-prone seabed to acceptable strength to support the
structure.

The surcharge fill can then be maintained in its place to act as a bedding layer, which will also provide horizontal escape paths for the pore water that is further driven out of the soil (moving upwards through the drains) under the rocking motion of the structure. Alternatively, just before the installation of the structure, the surcharge fill can be spread laterally sideways to a peripheral location where (1) it will provide a peripheral apron described in the preceding paragraphs (Fig. 11.4); and (2) it will lengthen the shear path and act as counter-balancing force against bearing failure (Gerwick, 2007, p. 246).

Other densification methods

These methods are described in greater details in the next section.

11.3 Hydraulic Fills

Hydraulic fills of sand are used to fill areas behind quay walls and sheet-piled structures. They are also used to backfill (1) trenches for burial of pipelines, or (2) trenches for immersed tunnels.

In the case of the hydraulic fills behind quay walls and sheet-piled structures, they are obviously above the water level, and therefore they are not subject to the action of waves. However, they may be subject to seismic actions, and hence prone to liquefaction. In the case of the hydraulic fills used to fill trenches for pipelines or immersed tunnels, they may be subject to both waves and earthquakes, and hence they are prone to liquefaction induced by these actions. Preventive measures in the case of pipeline burials are discussed in the preceding paragraphs (Section 11.1).

Regardless of whether hydraulic fills are onland or underwater, preventive measures are practically the same. The following paragraphs will summarize these measures.

Densification by vibratory compaction

This method is extremely effective for loose sand deposits; Gerwick (2007, p. 246), in this context, quotes Oosterschelde Storm Surge Barrier, where four compactors (which were mounted on a barge) compacted the loose sands to a depth of 50 m below sea level. Several brands of such compactors are available, some of which are able to work underwater (Gerwick, 2007, Fig. 7.13).

Densification by dynamic compaction

In this method, a heavy weight is repeatedly raised and dropped. Gerwick (2007, p. 245) notes that, depending on its mass, density and the distance of fall, this dynamic compaction can consolidate up to as much as 10 m of underwater soils or fills.

In this context, it is interesting to note the following. In order to observe what happens in a test simulating dumping of stones in a real-life situation (Herbich *et al.*, 1984), Sumer *et al.* (2010), in their standard wave flume experiments, carried out a test where stones were released at a distance 15 cm from the bed to form the cover layer over a liquefiable sand bed. Care was taken so that the cover stones had loose packing with a porosity which was the same as that of the other loose-packing experiments, referred to earlier in Section 3.6. This test showed that no liquefaction occurred, although (1) the stones were packed loosely, and (2) they were placed one-layer deep. (Note that liquefaction was observed with a single-layer stone cover when the stones were gently placed on the silt bed). Sumer *et al.* (2010) note that soil in this test was compacted by the falling stones over the entire soil surface, in a way similar to dynamic compaction used in practice for soil improvement, thereby increasing the relative density of the soil, which was presumably too large (also revealed by relative density measurements) for the soil to liquefy.

Densification by surcharging

This method has already been discussed under the previous two sections above, and the reader is referred to these.

Provision for free drainage

Provisions for free drainage enable the accumulated pore pressure to freely escape. To this end, vertical, gravel-filled drains, or wick drains are used. Both of these will also help consolidation. These provisions, much used onland, can also be used underwater if accompanied by a surcharge (Gerwick, 2007, p. 246). Sand drains can be drilled in, or jetted. Wick drains, on the other hand, essentially consist of a central plastic core with multiple, small drainage channels, surrounded by a thin geosynthetic filter jacket (Fig. 11.6). They are typically wide, thin, and very long sheets, for example, 4 inches in width and 1/8 inch in thickness, and they come in rolls. They are installed with the aid of a special installation mandrel; the mandrel, having the wick drain inside (Fig. 11.7), is hydraulically pushed or vibrated into the soil to the desired depth. As the mandrel is withdrawn, the wick drain is left in place within the soil. The water driven out of the soil moves upward through the drains and must have a means of escaping laterally. Therefore, a blanket of coarse sand and gravel should first be installed over the seabed. The latter can also act as part of the surcharge (Gerwick, 2007, p. 246).

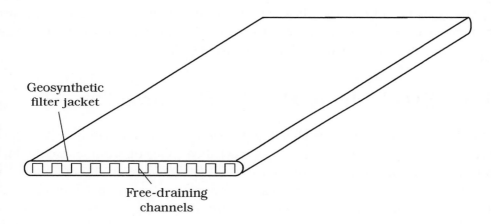

Figure 11.6: Wick drain consists of a central plastic core with multiple, small drainage channels, surrounded by a thin geosynthetic filter jacket.

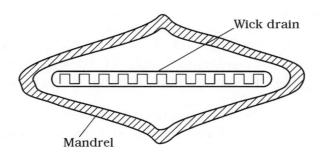

Figure 11.7: Wick drains are installed with the aid of an installation mandrel; the mandrel is hydraulically pushed or vibrated into the soil to the desired depth.

Multiple-pile support with close spacing

Important structures can be designed as pile-supported structures with the piles (closely spaced) penetrating into the deep stiff soil, to avoid any problem caused by liquefaction/weakening of the soil in the top layers. As already mentioned in Section 10.4.2, two large structures, a silo and a crane, survived the 1999 Turkey Kocaeli earthquake, largely because of their pile foundations, designed in this way.

Finally, it is to be noted that the above methods can be used to prevent liquefaction in large areas where the seabed is susceptible to liquefaction, e.g., in front of a breakwater (Zone 1 in Fig. 8.15).

11.4 Steel Sheet Pile Cofferdams

Gerwick (2007, p. 248) notes that liquefaction of silts and sands has recently been recognized in the construction of steel sheet pile cofferdams, particularly when powerful vibratory hammers are used. This is essentially equivalent to the effect created by shocks or blasts, referred to earlier in Chapter 1. As indicated therein, liquefaction may be induced by such effects even in dense soils whereas liquefaction induced by waves or even earthquakes is normally associated with loose soils.

Pre-installation of drainage or pre-drainage of pore water in the construction of such steel sheet pile cofferdams may be necessary.

11.5 References

1. Dixen, F.H., Sumer, B.M. and Fredsøe, J. (2008): Suction removal of sediment from between armour blocks. II. Waves. Journal of Hydraulic Engineering, ASCE, vol. 134, No. 10, 1405–1420.

2. Gerwick, B.C. (2007): Construction of Marine and Offshore Structures. Third Edition. CRC Press, Taylor & Francis Group, Boca Raton, London, New York.

3. Herbich, J.B. (1981): Offshore Pipeline Design Elements. Marcel Dekker, Inc., New York and Basel, 233 p.

4. Herbich, J.B., Schiller, R.E., Jr., Dunlap, W.A. and Watanabe, R.K. (1984): Seafloor Scour. Design Guidelines for Ocean-Founded Structures, Marcell Dekker, Inc., New York, NY, 320 p.

5. Lambe T.W. and Whitman, R.V. (1969): Soil Mechanics. John Wiley and Sons, Inc., 553 p.

6. Nielsen, A.W., Sumer, B.M. and Fredsøe, J. (2012): Experiments on removal of sediment from between armour blocks. Part 3: Breaking waves. To appear in J. Hydraulic Engineering, ASCE.

7. Seed, H.B. and Rahman, M.S. (1978): Wave-induced pore pressure in relation to ocean floor stability of cohesionless soil. Marine Geotechnology, 3, No. 2, 123–150.

8. Sekiguchi, H., Sassa, S., Sugioka, K. and Miyamoto, J. (2000): Wave-induced liquefaction, flow deformation and particle transport in sand beds. Proceedings of the International Conference GeoEng2000, Melbourne, Australia, Paper No. EG-0121.

9. Shore Protection Manual (1977): U.S. Army Coastal Engineering Research Center, vol. II, Department of the U.S. Army Corps of Engineers.

10. Sumer, B.M., Cokgor, S. and Fredsøe, J. (2001): Suction of sediment from between armour blocks. J. Hydraulic Engineering, ASCE, vol. 127, No. 4, 293–306.

11. Sumer, B.M., Dixen, F.H. and Fredsøe, J. (2010): Cover stones on liquefiable soil bed under waves. Coastal Engineering, vol. 57, No. 9, 864–873.

12. Sumer, B.M. and Fredsøe, J. (1997): Hydrodynamics Around Cylindrical Structures, World Scientific, Singapore, 530 p. Second edition 2006.

13. Sumer, B.M. and Fredsøe, J. (2002): The Mechanics of Scour in the Marine Environment. World Scientific, Singapore, 552 p.

14. Terzaghi, K. (1948): Theoretical Soil Mechanics. London: Chapman and Hall, John Wiley and Sons, Inc., NY, 510 p.

Appendix A

Small Amplitude, Linear Waves

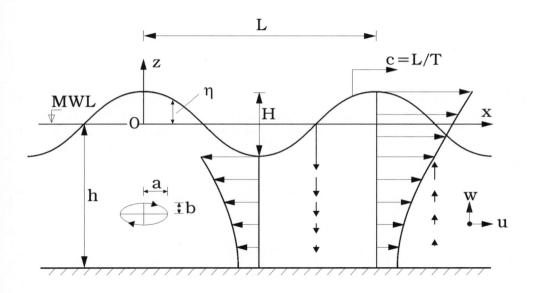

Figure A.1: Definition sketch.

This appendix highlights the small amplitude, linear wave theory. The waves involved in the wave solution are known as small waves, small-amplitude waves, infinitesimal waves, linear waves, sinusoidal waves. These waves are also called first order Stokes waves, simple harmonic waves, or Airy waves. The basic assumption is that the wave height is small compared with the wave length, namely $H/L \ll 1$. A detailed account of linear waves is given in the book of Svendsen (2006). Dean and Dalrymple's book (1984) can also be consulted.

Basic equation:
$$\nabla^2\phi = \phi_{xx} + \phi_{zz} = 0. \tag{A.1}$$

Bed boundary condition:
$$w = \phi_z = 0 \text{ at } z = -h. \tag{A.2}$$

Kinematic, free-surface boundary condition:
$$\left(\frac{\partial\phi}{\partial z}\right)_{z=0} = \frac{\partial\eta}{\partial t}. \tag{A.3}$$

Dynamic, free-surface boundary condition:
$$\left(\frac{\partial\phi}{\partial t} + g\eta\right)_{z=0} = C(t). \tag{A.4}$$

Water surface elevation:
$$\eta = \frac{H}{2}\cos(\omega t - \lambda x). \tag{A.5}$$

Potential function:
$$\phi = -\frac{Hc}{2}\frac{\cosh(\lambda(z+h))}{\sinh(\lambda h)}\sin(\omega t - \lambda x). \tag{A.6}$$

Wave celerity:
$$c = L/T = \omega/\lambda, \tag{A.7}$$
$$\lambda = 2\pi/L = \text{ wave number},$$
$$\omega = 2\pi/T = 2\pi f = \text{ angular wave frequncy},$$
$$f(= 1/T) \text{ being the wave frequency.}$$

Dispersion relation:
$$\omega^2 = g\lambda\tanh(\lambda h) \tag{A.8}$$
or
$$(2\pi f)^2 = g\lambda\tanh(\lambda h), \tag{A.9}$$
$$g \text{ being the acceleration due to gravity.}$$

Horizontal particle velocity:
$$u = \phi_x = \frac{\pi H}{T}\frac{\cosh(\lambda(z+h))}{\sinh(\lambda h)}\cos(\omega t - \lambda x) \tag{A.10}$$

or

$$u = \phi_x = \frac{g\lambda H}{4\pi f} \frac{\cosh(\lambda(z+h))}{\cosh(\lambda h)} \cos(\omega t - \lambda x). \tag{A.11}$$

Vertical particle velocity:

$$w = \phi_z = -\frac{\pi H}{T} \frac{\sinh(\lambda(z+h))}{\sinh(\lambda h)} \sin(\omega t - \lambda x) \tag{A.12}$$

or

$$w = \phi_z = \frac{g\lambda H}{4\pi f} \frac{\sinh(\lambda(z+h))}{\sinh(\lambda h)} \sin(\omega t - \lambda x). \tag{A.13}$$

Horizontal amplitude of particle motion:

$$a = \frac{H}{2} \frac{\cosh(\lambda(z+h))}{\sinh(\lambda h)} \tag{A.14}$$

Vertical amplitude of particle motion:

$$b = \frac{H}{2} \frac{\sinh(\lambda(z+h))}{\sinh(\lambda h)}. \tag{A.15}$$

Pressure:

$$\frac{p}{\rho} = -gz - \phi_t, \tag{A.16}$$

$-gz =$ hydrostatic pressure, and $-\phi_t =$ excess pressure,

ρ being the density of water.

Excess pressure:

$$\frac{p^+}{\rho} = -\phi_t = g\frac{H}{2} \frac{\cosh(\lambda(z+h))}{\cosh(\lambda h)} \cos(\omega t - \lambda x). \tag{A.17}$$

Wave (potential) energy per unit area:

$$E = \frac{1}{L}\left(\int_0^L (\rho g\eta dx)\frac{\eta}{2}\right) = \frac{1}{16}\rho g H^2. \tag{A.18}$$

A.1   References

1. Dean, R.G. and Dalrymple, R.A. (1984): Water Wave Mechanics for Engineers and Scientists. Prentice Hall, Englewood Cliffs, New Jersey 07632, xii+353 p.

2. Svendsen, I.A. (2006): Introduction to Nearshore Hydrodynamics. World Scientific, Singapore, xxii+722 p.

Appendix B

Soil Properties

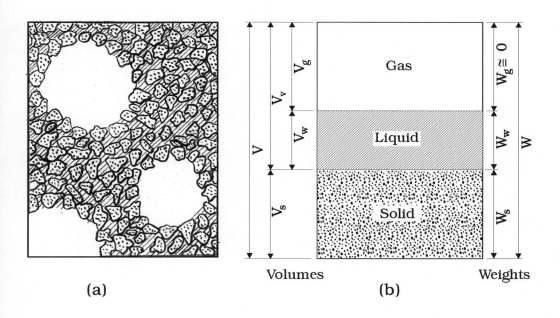

Figure B.1: (a) Close-up picture of unsaturated soil (adapted from Sills *et al.*, 1991). (b) Definition sketch.

Relationships among soil properties

From Lambe and Whitman (1969). See Fig. B.1 for the definition sketch.

(1) *Volume*

413

Porosity:

$$n = \frac{V_v}{V}. \tag{B.1}$$

Void ratio:

$$e = \frac{V_v}{V_s}. \tag{B.2}$$

Degree of saturation:

$$S_r = \frac{V_w}{V_v} \tag{B.3}$$

$$n = \frac{e}{1+e}; \qquad e = \frac{n}{1-n}. \tag{B.4}$$

(2) *Weight*
Water content:

$$w = \frac{W_w}{W_s}. \tag{B.5}$$

(3) *Specific gravity*
Mass:

$$s_t = \frac{\gamma_t}{\gamma_0}, \tag{B.6}$$

Water:

$$s_w = \frac{\gamma}{\gamma_0}, \tag{B.7}$$

Solids:

$$s = \frac{\gamma_s}{\gamma_0}, \tag{B.8}$$

in which γ_0 = unit weight of water at $4°C \approx \gamma$.

(4) *Unit weight*
Total:

$$\gamma_t = \frac{W}{V} = \frac{s + S_r e}{1 + e}\gamma \tag{B.9}$$

$$= \frac{1+w}{1+e}s\gamma.$$

Solids:

$$\gamma_s = \frac{W_s}{V_s}. \tag{B.10}$$

Water:

$$\gamma = \frac{W_w}{V_w}. \tag{B.11}$$

Dry:

$$\gamma_d = \frac{W_s}{V} = \frac{s}{1+e}\gamma$$ (B.12)

$$= \frac{s\gamma}{1+ws/S_r} = \frac{\gamma_t}{1+w}.$$

Submerged (unsaturated soil):

$$\gamma' = \gamma_t - \gamma$$ (B.13)

$$= \frac{s - 1 - e(1 - S_r)}{1+e}\gamma.$$

Submerged (saturated soil):

$$\gamma' = \gamma_t - \gamma$$ (B.14)

$$= \frac{s-1}{1+e}\gamma.$$

Ranges of soil properties

(1) *Soil components* (Lambe and Whitman, 1969, p. 36)

Table B.1. Soil components.

Soil component	Grain size (mm)
Boulder	>300
Cobble	300–150
Coarse gravel	76–20
Fine gravel	20–5
Coarse sand	5–2
Medium sand	2–0.4
Fine sand	0.4–0.074
Silt[1]	<0.074
Clay[2]	<0.074

[1]Particles smaller than 0.074 mm identified by behaviour; slightly or non-plastic regardless of moisture, and exhibit little or no strength when air dried.
[2]Particles smaller than 0.074 mm identified by behaviour; it can be made to exhibit plastic properties within a certain range of moisture and exhibits considerable strength when air dried.

(2) *Void ratio, e and porosity, n* (Lambe and Whitman, 1969, p. 31)

Table B.2.   Void ratio and porosity for granular soils.

Description	e_{max}	e_{min}	n_{max}	n_{min}
Uniform spheres	0.92	0.35	0.476	0.26
Standard Ottawa sand	0.80	0.50	0.44	0.33
Clean uniform sand	1.0	0.40	0.50	0.29
Uniform inorganic silt	1.1	0.40	0.52	0.29
Silty sand	0.90	0.30	0.47	0.23
Fine to coarse sand	0.95	0.20	0.49	0.17
Micaceous sand	1.2	0.40	0.55	0.29
Silty sand and gravel	0.85	0.14	0.46	0.12

(3) *Relative density, D_r* (Lambe and Whitman, 1969, p. 31)
in which the relative density (or density index) is defined as

$$D_r = \frac{e_{max} - e}{e_{max} - e_{min}}.$$ (B.15)

Table B.3.   Soil categories (or soil states) for granular
soils.

Relative density, D_r	Soil category (or soil state)
0.00–0.15	Very loose
0.15–0.35	Loose
0.35–0.65	Medium
0.65–0.85	Dense
0.85–1.00	Very dense

(4) *Specific gravity of soil grains, s* (Lambe and Whitman, 1969, p. 30)
$s = \gamma_s/\gamma_0 \approx \gamma_s/\gamma = 2.65$ for quartz.

(5) *Coefficient of permeability, k* (Lambe and Whitman, 1969, p. 286)
Fig. B.2 presents laboratory permeability test data on a variety of soils.

(6) *Young's modulus, E* (USACE, 1990)

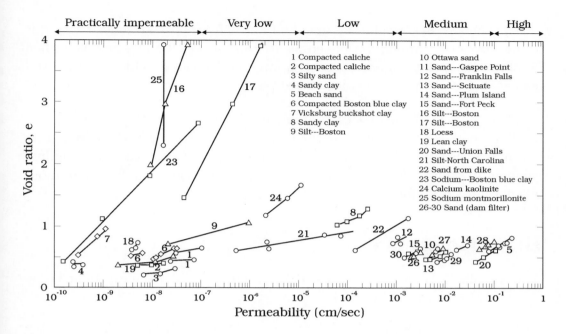

Figure B.2:   Permeability for various kinds of soils. Lambe and Whitman (1969).

Ranges of Young's modulus for sand: $E = 1 - 2.5 \times 10^4 \, \mathrm{kN/m^2}$ (loose sand); $0.25 - 1 \times 10^5 \, \mathrm{kN/m^2}$ (dense sand); $1 - 2 \times 10^5 \, \mathrm{kN/m^2}$ (dense sand and gravel); $0.25 - 2 \times 10^5 \, \mathrm{kN/m^2}$ (silty sand). Ranges of Young's modulus for clay: $0.5 - 5 \times 10^3 \, \mathrm{kN/m^2}$ (very soft clay); $0.5 - 2 \times 10^4 \, \mathrm{kN/m^2}$ (soft clay); $2 - 5 \times 10^4 \, \mathrm{kN/m^2}$ (medium clay); $0.5 - 1 \times 10^5 \, \mathrm{kN/m^2}$ (stiff clay, silty clay); $0.25 - 2 \times 10^5 \, \mathrm{kN/m^2}$ (sandy clay); $1 - 2 \times 10^5 \, \mathrm{kN/m^2}$ (clay shale).

(7) *Shear modulus, G* (Yamamoto *et al.*, 1978)
The shear modulus varies from $G = 4.8 \times 10^2 \, \mathrm{kN/m^2}$ (silt and clay) to $4.8 \times 10^5 \, \mathrm{kN/m^2}$ (dense sand).

(8) *Poisson's ratio, ν* (Lambe and Whitman, 1969, p. 160)
For the early stages of a first loading of a sand (when particle rearrangements are important) ν typically has values of about 0.1 to 0.2. During cyclic loading, however, ν becomes more of a constant with values from 0.3 to 0.4.

(9) *Coefficient of lateral earth pressure, k_0* (Lambe and Whitman, 1969, p. 100)

k_0 typically has a value between 0.4 and 0.5 for a sand deposit which is formed by an accumulation of sediment from above. Lambe and Whitman (1969, p. 100) note, however, that k_0 may well reach a value of 3, if a soil deposit has been heavily loaded in the past.

(10) *Uniformity coefficient* (Lambe and Whitman, 1969, p. 32)
    This coefficient is defined as the ratio of d_{60} to d_{10} where d_{60} is the grain size at which 60% of the soil weight is finer, and d_{10} is the corresponding value at 10% finer. A soil having a uniformity coefficient smaller than about 2 is considered uniform.

B.1   References

1. Lambe T.W. and Whitman, R.V. (1969): Soil Mechanics. John Wiley and Sons, Inc., New York, 553 p.

2. Sills, G., Wheeler, S.J., Thomas, S.D. and Gardner, T.N. (1991): Behaviour of offshore soils containing gas bubbles. Géotechnique, vol. 41, No. 2, 227–241.

3. USACE (1990): Settlement Analysis. Department of the U.S. Army Corps of Engineers, Washington, DC 20314-1000. Engineering Manual 1110-1-1904, 30 September 1990.

4. Yamamoto, T., Koning, H.L., Sellmeijer, H. and van Hijum, E. (1978): On the response of a poro-elastic bed to water waves. Journal of Fluid Mechanic, vol. 87, part 1, 193–206.

Appendix C

In-Situ Relative Density

As discussed in Chapter 3, the relative density (or density index) of the soil, D_r, is one of the key quantities needed in a liquefaction assessment study. Therefore, the *in-situ* value of the relative density of the seabed soil needs to be determined in an accurate manner. This appendix will discuss this issue.

The two most popular methods to determine the *in-situ* value of the relative density are the Standard Penetration Test (SPT) and the Cone Penetration Test (CPT). The first two sections, Sections C.1 and C.2, in the following paragraphs will describe these methods, respectively.

There are situations where the *in-situ* relative density needs to be determined for *disturbed* seabed soils. For instance, in the case of pipeline burial, the pipeline is trenched, and subsequently the trench is backfilled. In order to undertake a backfill liquefaction assessment study, already in the design stage, the initial *in-situ* value of the relative density of the backfill soil obviously needs to be known. Section C.3 will describe the recommended practice for this.

C.1   Standard Penetration Test (SPT)

Penetration test is one of the most widely used field tests for soil investigations, and SPT is the oldest and simplest form of penetration tests. The main idea behind SPT is that a standardized penetrometer is driven or pushed into the ground and the resistance to the penetration is recorded. The latter is correlated with, among others, the relative density of the soil, as will be detailed in the following paragraphs.

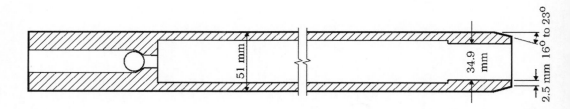

Figure C.1:   SPT sampler. Standard dimensions for SPT sampler, as given in ASTM D
1586-84. Notice that the bottom end (the right end in the figure) has a shape of truncated
cone, the angle being 16°–23°.

Figure C.1 shows a schematic illustration of the standard penetrometer,
a split spoon sampler, for the SPT testing. SPT basically involves driving
the spoon sampler into the ground by dropping a weight, a so-called drop-
hammer (of 140-lb, i.e., 63.6 kg), from a height of 30 in (76 cm) (Lambe and
Whitman, 1969). The way in which the SPT test is carried out is as follows.
The penetrometer is driven into the soil below the bottom of a bore hole to
a depth of $\frac{1}{2}$ ft (15 cm) by means of the drop-hammer. Then the number
of blows to drive the penetrometer 1 ft (30 cm) is recorded (Powrie, 2004).
This number, designated by N, is the SPT blowcount, and represents the
resistance to the penetration. Fig. C.2 gives an example of SPT blowcount
distribution from a site investigation where N (i.e., the number of blows per
ft) are plotted as function of depth.

Table C.1 presents a correlation of SPT blowcount with relative density
for sand (From Terzaghi and Peck, 1967, also reproduced in Lambe and
Whitman, 1969, and Rogers, 2006).

Table C.1.   Standard Penetration Test.

SPT blowcount, N (blows/ft)	Relative density, D_r	Soil category
0–4	0.00–0.15	Very loose
4–10	0.15–0.35	Loose
10–30	0.35–0.65	Medium
30–50	0.65–0.85	Dense
>50	0.85–1.00	Very dense

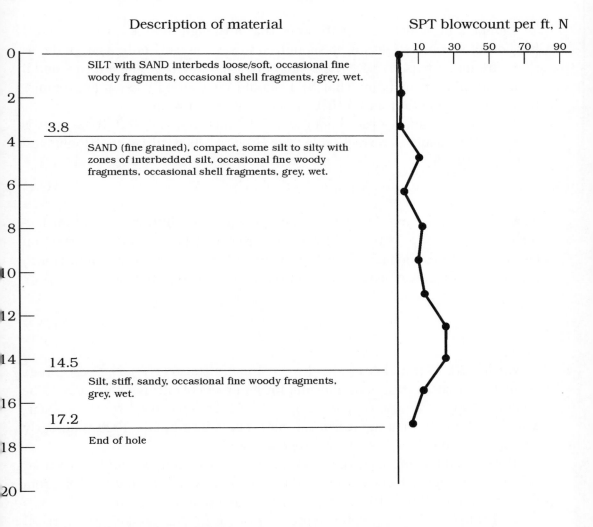

Figure C.2: An example of SPT blowcount distribution from a site investigation.

Lambe and Whitman (1969) notes that although SPT is a very valuable method of soil investigation, it should, however, be used only as a guide, as there are many reasons why the results are approximate. The authors detail their reasoning, concluding that SPT should be used only as an approximation or in conjunction with other methods of exploration.

In order to improve the reliability of the method, the SPT blowcount is corrected by means of correction factors (Rogers, 2006, Idriss and Boulanger, 2008, pp. 70–76):

$$N_{1(60)} = C_N C_E C_B C_R C_S \ N \tag{C.1}$$

in which N is the measured blowcount, C_N the overburden correction factor, C_E the energy correction factor, C_B the bore hole diameter correction factor, C_R the rod length correction factor, C_S the sampling method (liner) correction factor. The energy correction factor is computed as

$$C_E = \frac{ER_m}{60} \tag{C.2}$$

in which ER_m is the Measured delivered Energy Ratio (ER) as a percentage of the theoretical maximum, allowing for hammers of varying efficiency to be accounted for. We note that the original SPT hammer used by H.A. Mohr, who was the first to standardize the SPT procedure in late 1920s and early 1930s, has about 60% efficiency (Rogers, 2006).

Rogers (2006) compiled the recommended corrections for SPT blowcount values, reproduced from Rogers (2006) in Table C.2 (the data taken from Robertson and Wride, 1997, as modified from Skempton, 1986).

In Table C.2, Pa is a reference pressure of 100 kN/m^2, and σ'_{v0} the vertical effective stress (Liao and Whitman, 1986).

The correlation of SPT blowcount with relative density for sand (from Terzaghi and Peck, 1967) given in Table C.1 was modified by Skempton (1986) according to the corrected SPT blowcount $N_{1(60)}$. These results are given in Table C.3, reproduced from Powrie (2004).

The in-situ relative density of sands may be estimated, using the information given in Table C.1 or Table C.3. With the corrected values of the SPT blowcounts, the latter will obviously give a better estimate of the relative density.

Table C.2. Recommended corrections for SPT blowcount values. Compiled by Rogers (2006).

Factor	Equipment variable	Correction factor	Correction
Overburden pressure		C_N	$(Pa/\sigma'_{v0})^{0.5}$ with $C_N \leq 2$
Energy ratio	Donut hammer	C_E	0.5–1.0
	Safety hammer		0.7–1.2
	Autumatic hammer		0.8–1.5
Bore hole diameter	65–115 mm	C_B	1.0
	150 mm		1.05
	200 mm		1.15
Rod length	3–4 m	C_R	0.75
	4–6 m		0.85
	6–10 m		0.95
	10–30 m		1.0
	>30 m		<1.0
Sampling method	Standard sampler	C_S	1.0
	Sampler without liners		1.1–1.3

Table C.3. Standard Penetration Test with corrected blowcount.

Corrected SPT blowcount, $N_{1(60)}$ (blows/ft)	Relative density, D_r	Soil category
		Very loose
3	0.15	
		Loose
8	0.35	
15	0.5	Medium
25	0.65	
		Dense
42	0.85	
		Very dense
58	1.0	

C.2   Cone Penetration Test (CPT)

In this test, a cone at the end of a series of rods is driven at a steady rate of 15–25 mm/s into the soil, and the penetration resistance, measured by a load cell just behind the cone, is recorded continuously. Most cones have a diameter of 35.7 mm, an apex angle of 60°, and a projected area (perpendicular to the direction of penetration) of 1,000 mm². Some versions of the CPT instrument incorporate a pore-water pressure transducer with its filter either on or just behind the cone. Such a device is known as a piezocone. Detailed information on CPT testing is given in Powrie (2004) and Rogers (2006) among others. The subject is covered most comprehensively by the book of Lunne, Robertson and Powell (2002), and the recent review by Lunne (2012).

The penetration resistance is normally expressed as the cone (or tip) resistance, calculated as the force divided by the projected area of the cone, q_c, one of the output parameters from the CPT testing. Others include the sleeve friction and the pore-water pressure. Figure C.3 displays an example of cone resistance distribution from a site investigation where the tip resistance (i.e., q_c with the unit being bar) is plotted as a function of depth.

Observations show that there is, not surprisingly, a correlation of the tip resistance with relative density for sand. Table C.4 provides a guide for the relation between cone resistance and the soil category (and therefore the relative density), taken from Fugro (2004). The latter document notes that the guide applies to unaged, uncemented sands up to about 10 to 15 metres depth. It may be noted that there is a slight difference between the definition of soil categories in the Fugro table (Table C.4) and that given in Table C.1. In the Fugro definition, Very Loose Sand corresponds to $D_r < 0.2$; Loose

Table C.4.   Cone Penetration Test.

Cone resistance, q_c (MPa)	Soil category (See the text for definition of soil categories)
<2	Very loose
2–4	Loose
4–12	Medium
12–20	Dense
>20	Very dense

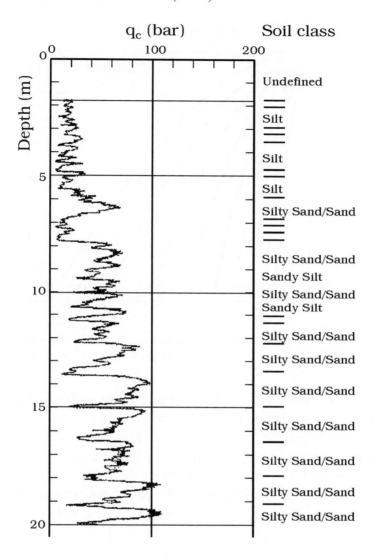

Figure C.3: An example of cone resistance distribution from a site investigation. To convert q_c (bar) to q_c (MPa), divide by 10.

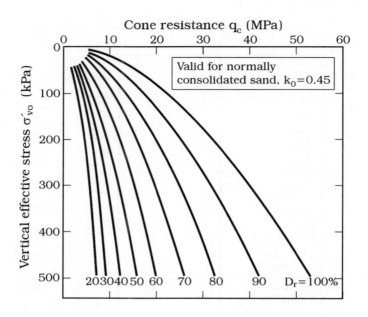

Figure C.4: q_c, σ'_{v0}, D_r relationships for Ticino sand in terms of normally consolidated behaviour. Adapted from Lunne *et al.* (2002). Original reference: Baldi *et al.* (1986).

Sand to $D_r = 0.2 - 0.4$; Medium Dense Sand to $D_r = 0.4 - 0.6$; Dense Sand to $D_r = 0.6 - 0.8$; and Very Dense Sand to $D_r = 0.8 - 1.0$.

More recent correlation between q_c and D_r, which takes account of the effective stress, is given in Figs C.4 and C.5, taken from Lunne *et al.* (2002), for normally consolidated sand and for overconsolidated sand, respectively.

The relationship exhibited in Fig. C.4 can be represented by the following empirical expression

$$D_r = \frac{1}{C_2} \ln \left[\frac{q_c}{C_0(\sigma'_{v0})^{C_1}} \right] \tag{C.3}$$

in which σ'_{v0} is the vertical effective stress, and C_is are coefficients with

$$C_0 = 157, \quad C_1 = 0.55, \quad C_2 = 2.41. \tag{C.4}$$

The relationship in Fig. C.5, on the other hand, can be represented by a similar expression where σ'_{v0} is to be replaced with the mean normal effective stress

$$\sigma'_m = \frac{1}{3}(\sigma'_{v0} + \sigma'_{h0} + \sigma'_{h0}) = \frac{\sigma'_{v0} + 2\sigma'_{h0}}{3} \tag{C.5}$$

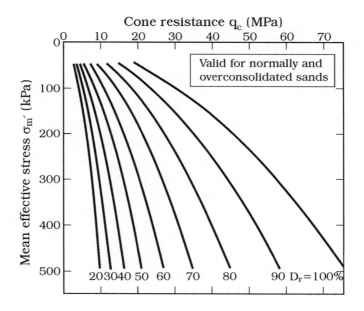

Figure C.5: q_c, σ'_{v0}, D_r relationships for Ticino sand in terms of normally and overconsolidated behaviour. Adapted from Lunne *et al.* (2002). Original reference: Baldi *et al.* (1986).

in which σ'_{h0} is the horizontal effective stress. The C_i coefficients in the latter case (Fig. C.5) are given as

$$C_0 = 181, \quad C_1 = 0.55, \quad C_2 = 2.61. \tag{C.6}$$

From CPT data in cone resistance, *in-situ* relative densities can be obtained, using the information in Table C.4, or in Figs C.4 and C.5, or Eqs C.3–C.6.

Incidentally, correlations between CPT and SPT data have been established so that CPT data can be used in existing SPT-based design approaches. The reader is referred to the review given in Lunne *et al.* (2002, pp. 149–151).

Rogers (2006), in concluding his review of SPT and CPT tests, notes that

"... Cone penetration soundings are being employed with increasing regularity. The engineering geologists should consider employing both techniques whenever possible because each has slight advantages over the other, but they are most powerful when combined. (For example) the SPT allows a firsthand look. . . "

The SPT, or CPT, or combined SPT and CPT, testings are essential to determine the *in-situ* relative density of the soil. This is one of the quantities required for liquefaction assessment studies. Clearly, soil samples also need to be collected to determine the other soil quantities as well as to carry out such an assessment exercise. See Example 4 in Chapter 3 for a complete list of soil quantities required for a numerical assessment of liquefaction potential (see also Table C.6 below for a similar list).

As a final note, SPT and CPT results are directly used in liquefaction triggering analyses in earthquake engineering with the help of charts describing the correlation between the cyclic stress ratio, CSR (the amplitude of the cyclic shear stress divided by the initial effective stress) and $N_{1(60)}$, or the correlation between CSR and the normalized corrected CPT tip resistance, q_{c1N}, similar to the chart given in the bottom panel of Fig. 10.18. The way in which the SPT (or CPT) results are used in these analyses is that (1) CSR is determined for a given set of earthquake parameters at a given depth (e.g., in its simplest and very crude form from Eq. 1.1; however, see the analyses, for example, in Kramer, 1996, PIANC, 2001, and Idriss and Boulanger, 2008); and (2) the SPT (or CPT) number is picked up for this depth from the SPT profile (or CPT profile); and subsequently (3) it is checked if the point with the coordinates $N_{1(60)}$ and CSR (or the coordinates q_{c1N} and CSR) plotted on the aforementioned charts fall into "liquefaction" region. If yes, the soil is expected to be liquefied by that earthquake at that depth of interest.

The subject has been treated extensively in the earthquake engineering literature, e.g. Kramer (1996), PIANC (2001), and Idriss and Boulanger (2008). PIANC (2001, pp. 199–204) also gives a systematic description of the methodology developed and extensively used in Japan, comprising a first step based on grain size distribution and the SPT number, and a second step based on cyclic triaxial test results to be used when the liquefaction potential cannot be determined from the first step.

C.3   *In-Situ* Relative Density of a Backfill

As mentioned previously, in order to carry out a backfill liquefaction assessment study, *already in the design stage*, the initial in-situ value of the relative density of the backfill soil needs to be known. In this case, the recommended course of action may be summarized as follows:

(1) Collect soil samples at locations along the route of the pipeline. (Here we assume that the native soil will be used as backfill.)

(2) Determine classification properties of the collected soil, including the minimum void ratio, e_{min}, and the maximum void ratio, e_{max}.

(3) Use the measured values of e_{min} and e_{max} to select, for example, three soil-density values, representative of a possible range of *in-situ* densities that the soil may end up having in the trench when used as backfill.

(4) Reconstitute three samples at the aforementioned selected densities.

(5) Conduct isotropically consolidated undrained triaxial multi-stage tests on these three samples to determine the soil parameters.

The following paragraphs illustrate this exercise by reference to a soil investigation undertaken along these lines. In this work, three samples were selected, and the samples were reconstituted with initial dry densities of 1.459, 1.525, and 1.595 t/m^3.

Table C.5.   Triaxial test results and derived parameters.

Parameter	Test-1 Soil	Test-2 Soil	Test-3 Soil
Initial dry density (t/m^3)	1.459	1.525	1.595
Friction angle, φ (deg)	38	39	44
Lowest possible confining pressure (kN/m^2)	20	20	20
Permeability, k (m/s)	1.8×10^{-4}	8.1×10^{-5}	8.1×10^{-5}
Young's modulus[1], E (kN/m^2)	4983	5792	18175
Poisson's ratio[2], ν	0.30	0.30	0.30
Coefficient of lateral earth pressure[3], k$_0$	0.38	0.37	0.31

[1]In view of seating problems at low confining pressures, a secant modulus is measured from the triaxial results. The modulus is measured for the first load stage of each test, i.e., the stage where the lowest possible confining pressure is applied.

[2]This is a typical Poisson's ratio for this material type and is not derived from site specific data.

[3]k$_0$ is calculated from $1 - \sin(\varphi)$, the Jaky equation, Eq. 3.8.

The soils corresponding to these three samples are designated Test-1 Soil, Test-2 Soil, and Test-3 Soil. The results of these tests are summarized in Table C.5, which is reproduced from the soil investigation report.

Table C.6, on the other hand, gives a complete list of the soil parameters used in the liquefaction-assessment exercise.

Table C.6.   Soil properties used in the liquefaction assessment.

Parameter	Test-1 Soil	Test-2 Soil	Test-3 Soil
Spec. gravity of soil grains, s	2.70	2.70	2.70
Grain size, d_{50} (mm)	0.135	0.135	0.135
Dry density of soil, ρ_d (t/m^3)	1.459	1.525	1.595
Dry specific weight of soil, $\gamma_d = \rho_d g$ (kN/m^3)	14.313	14.960	15.647
Void ratio, $e = (\gamma/\gamma_d)$s-1	0.85	0.77	0.69
Relative density, D_r	0.146	0.452	0.759
Porosity, n	0.46	0.435	0.41
Young's modulus, E (kN/m^2)	4983	5792	18175
Poisson's ratio, ν	0.30	0.30	0.30
Permeability, k (m/s)	1.8×10^{-4}	8.1×10^{-5}	8.1×10^{-5}
Coeff. of consolidation[1], c_v (m^2/s)	0.123	0.064	0.201
(Total) specific weight of soil, γ_t (kN/m^3)	18.83	19.23	19.68
Submerged specific weight of soil, $\gamma' = \gamma_t - \gamma$ (kN/m^3)	9.02	9.42	9.87

[1]Coefficient of consolidation is determined from Eqs 3.26 and 2.36 with $S_r = 1$ (saturated soil).

C.4   References

1. Baldi, G., Bellotti, R., Ghionna, V., Jamiolkowski, M. and Pasqualini, E. (1986): Interpretation of CPTs and CPTUs; 2nd Part: drained penetration of sands. Proceedings of the Fourth International Geotechnical Seminar, Singapore, 143–156.

2. Fugro (2004): Cone Penetration Testing. Fugro Engineering Services Limited, Wallingford, Oxfordshire, UK, 37 p. and 9 plates.

3. Idriss, I.M. and Boulanger, R.W. (2008): Soil Liquefaction During Earthquakes. Earthquake Engineering Research Institute, Oakland, California, USA. Original Monograph Series MNO-12, 243 p.

4. Kramer, S.L. (1996): Geotechnical Earthquake Engineering. Prentice Hall, New Jersey, USA, xviii+653 pp.

5. Lambe T.W. and Whitman, R.V. (1969): Soil Mechanics. John Wiley and Sons, Inc., New York, 553 p.

6. Liao, S.S.C. and Whitman, R.V. (1986): Overburden correction factors for SPT in sand. Journal of Geotechnical Engineering, vol. 112, No. 3, 373–377.

7. Lunne, T. (2012): The Fourth James K. Mitchell Lecture: The CPT in offshore soil investigations — a historic perspective. Geomechanics and Geoengineering: An International Journal, vol. 7, No. 2, 75–101.

8. Lunne, T., Robertson, P.K. and Powell, J.J.M. (2002): Cone Penetration Testing in Geotechnical Practice. Spon Press, Taylor & Francis Group, London and New York, 312 pp.

9. PIANC (2001): Seismic Design Guidelines for Port Structures. Working Group No. 34 of the maritime Navigation Commission, International Navigation Association (PIANC). A book published by A.A. Balkema Publishers, Lisse/Abingdon/Exton (PA)/Tokyo, XV + 474 p.

10. Powrie, W. (2004): Soil Mechanics. 2nd edition. Spon Press, Taylor and Francis Group, London, x+675p.

11. Robertson, P.K. and Wride, C.E. (1997): Cyclic liquefaction and its evaluation based on the SPT and CPT. In Proceedings of the NCEER Workshop on Evaluation of Liquefaction Resistance of Soils: technical Report NCCEER-97-0022: National Center for Earthquake Engineering Research, Buffalo, NY, 41–87.

12. Rogers, J.D. (2006): Subsurface exploration using the standard penetration test and cone penetrometer test. Environmental & Engineering Geoscience, vol. XII, No. 2, 161–179.

13. Skempton, A.W. (1986): Standard penetration test procedures and the effects in sands of overburden pressure, relative density, particle size, aging and overconsolidation. Géotechnique, vol. 36, No. 3, 425–447.

14. Terzaghi, K. and Peck, R.B. (1967): Soil Mechanics in Engineering Practice, 2nd Edition. John Wiley and Sons, New York. The first edition was published in 1948.

Appendix D

Hsu & Jeng Coefficients

Hsu and Jeng (1994) have developed analytical solutions to the Biot equations (Eqs 2.30, 2.32, and 2.38) for an unsaturated, anisotropic soil of finite depth. For the case of a saturated, isotropic soil exposed to a progressive wave, the solution is given in Section 2.2.1 (Eqs 2.61–2.73) while, for the case of a saturated, isotropic soil exposed to a standing wave, the solution is given in Section 2.2.2 (Eqs 2.76–2.88). The solution includes a number of coefficients. These coefficients are given below.

$$C_i = \frac{D_i}{D_0} \text{ for } i = 1, \ldots, 6 \tag{D.1}$$

in which the D coefficients are given by

$$D_j = C_{j0} + C_{j1}e^{-2\lambda d} + C_{j2}e^{-(\lambda+\delta)d} + C_{j3}e^{-4\lambda d} + C_{j4}e^{-2\delta d} \tag{D.2}$$
$$+ C_{j5}e^{-2(\lambda+\delta)d} + C_{j6}e^{-(3\lambda+\delta)d} + C_{j7}e^{-(4\lambda+2\delta)d} \text{ for } j = 0, 1, \ldots, 6.$$

The coefficients C_{ji} are

$$C_{00} = (\delta - \lambda)^2(\delta - \delta\nu + \lambda\nu)B_1$$

$$C_{01} = -2\delta\left[\left(\lambda^2\nu - \delta^2 + \delta^2\nu\right)^2 + \lambda^4(1 - 2\nu)^2 + 2\lambda^2 d^2(1 - \nu)^2(\delta^2 - \lambda^2)^2\right]$$

$$+ 4\lambda^2 d(\delta^4 - \lambda^4)(1 - 2\nu)(1 - \nu)$$

433

$$C_{02} = -8\delta\lambda^2(1-2\nu)\left[\lambda d(\delta^2-\lambda^2)(1-\nu)-\delta^2(1-\nu)+\lambda^2\nu\right]$$

$$C_{03} = (\delta+\lambda)^2(\delta-\delta\nu-\lambda\nu)B_2$$

$$C_{04} = C_{03}$$

$$C_{05} = C_{01} - 8\lambda^2 d(\delta^4-\lambda^4)(1-\nu)(1-2\nu)$$

$$C_{06} = C_{02} + 16\delta\lambda^3 d(\delta^2-\lambda^2)(1-\nu)(1-2\nu)$$

$$C_{07} = C_{00}$$

$$C_{11} = 2\lambda^2 d(\delta+\lambda)(\delta-\delta\nu-\lambda\nu)B_3$$

$$C_{12} = 4\delta\lambda^3 d(1-2\nu)(\delta^2-\delta^2\nu-\lambda^2\nu)$$

$$C_{15} = 2\lambda^2 d(\delta-\lambda)(\delta-\delta\nu+\lambda\nu)B_4$$

$$C_{20} = (\delta-\lambda)^2(\delta-\delta\nu+\lambda\nu)B_1$$

$$C_{21} = C_{03} + (\delta+\lambda)^2(\delta-\delta\nu-\lambda\nu)B_5$$

$$C_{22} = 4\delta\lambda^2(1-2\nu)[2\delta^2(1-\nu)-2\lambda^2\nu-\lambda d(1-\nu)(\delta^2-\lambda^2)]$$

$$C_{24} = (\delta+\lambda)^2(\delta-\delta\nu-\lambda\nu)B_2$$

$$C_{25} = C_{00} + (\delta-\lambda)^2(\delta-\delta\nu+\lambda\nu)B_6$$

$$C_{31} = 2\lambda^2 d(\delta-\lambda)(\delta-\delta\nu+\lambda\nu)B_3$$

$$C_{33} = C_{14}$$

$$C_{35} = 2\lambda^2 d(\delta+\lambda)(\delta-\delta\nu-\lambda\nu)B_4$$

$$C_{36} = -C_{12}$$

$$C_{37} = C_{10}$$

$$C_{41} = C_{25} - 2C_{00}$$

$$C_{42} = -C_{26}$$

$$C_{43} = -C_{24}$$

$$C_{45} = C_{21} - 2C_{03}$$

$$C_{46} = 4\delta\lambda^2(1-2\nu)\left[2\lambda^2\nu-2\delta^2(1-\nu)-\lambda d(1-\nu)(\delta^2-\lambda^2)\right]$$

$$C_{47} = -C_{20}$$

$$C_{51} = -4\lambda^2 d(1-2\nu)B_3$$

$$C_{52} = -2\lambda^2 d(1-2\nu)(\delta+\lambda)(\delta-\delta\nu-\lambda\nu)$$

$$C_{56} = -2\lambda^2 d(1-2\nu)(\delta-\lambda)(\delta-\delta\nu+\lambda\nu)$$

$$C_{62} = -C_{56}$$

$$C_{64} = C_{53}$$
$$C_{65} = -4\lambda^2 d(1 - 2\nu)B_4$$
$$C_{66} = -C_{52}$$
$$C_{67} = C_{50}$$
$$C_{10}, C_{13}, C_{14}, C_{16}, C_{17} = 0$$
$$C_{23}, C_{26}, C_{27} = 0$$
$$C_{30}, C_{32}, C_{34} = 0$$
$$C_{40}, C_{44} = 0$$
$$C_{50}, C_{53}, C_{54}, C_{55}, C_{57} = 0$$
$$C_{60}, C_{61}, C_{63} = 0$$

in which the B coefficients are given by

$$B_1 = \lambda^2 \nu - (1 - \nu)(\delta^2 + \delta\lambda + \lambda^2)$$
$$B_2 = -\delta^2 + \delta\lambda - \lambda^2 + \delta^2\nu - \delta\lambda\nu + 2\lambda^2\nu$$
$$B_3 = (\delta^3 d - \lambda^2 - \delta\lambda^2 d)(1 - \nu) + \lambda^2\nu$$
$$B_4 = (\delta^3 d + \lambda^2 - \delta\lambda^2 d)(1 - \nu) - \lambda^2\nu$$
$$B_5 = 2\delta\lambda d(\delta - \lambda)(1 - \nu)$$
$$B_6 = 2\delta\lambda d(\delta + \lambda)(1 - \nu).$$

D.1 References

1. Hsu, J.R.S. and Jeng, D.S. (1994): Wave-induced soil response in an unsaturated anisotropic seabed of finite thickness. International Journal for Numerical and Analytical Methods in Geomechanics, vol. 18, No. 11, 785–807.

Appendix E

List of Symbols

The main symbols used in the book are listed below. In some cases, the same symbol is used for more than one quantity. This is to maintain generally accepted conventions in different fields. In most cases, however, their use is restricted to a single chapter. This is marked in the following list by chapter numbers in brackets.

a	amplitude of horizontal component of orbital motion of water particles
a	ground acceleration (Chapter 1)
CC	clay content
C_D	drag coefficient
CPT	Cone Penetration Test
CRR	Cyclic Resistance Ratio, equivalent to: critical value of CSR for liquefaction
CSR	Cyclic Shear stress Ratio
c	concentration
c_v	coefficient of consolidation
D	pipe diameter, stone size
D_r	relative density
d	soil depth
d_{50}	grain size
E	modulus of elasticity (Young's modulus)
e	void ratio
e	depth of sinking (Chapter 6)
$e_x,\ e_y,\ e_z$	soil deformation in x-, y- and z- directions, respectively
Fr	Froude number

437

Gal a unit widely used in geotechnical engineering literature for earthquake accelerations; 1000 Gal means $1\,g$

G shear modulus

g acceleration due to gravity

H wave height, measured from crest to trough. For standing waves, wave height (measured from crest to trough): $H = 2H_i$

H_i wave height of incident wave

H_s significant wave height

h water depth

i imaginary unit

K bulk modulus of elasticity of water

K' apparent bulk modulus of elasticity of water

KC Keulegan–Carpenter number

k coefficient of permeability

k_0 coefficient of lateral earth pressure

L wave length

N number of cycles or waves

N number of stone layers (Chapter 3, Section 3.6)

N SPT blowcount per ft (Appendix C)

N_ℓ number of cycles or waves to cause liquefaction

$N_{1(60)}$ corrected SPT blowcount

$O(\ )$ order of magnitude

p pore-water pressure

p_0 absolute (not excess) pore-water pressure, which can be taken as equal to the initial value of pressure

p_1 pressure induced by waves on the seabed, the bed pressure

p_b maximum value of the bed pressure

p_s surface loading (or surcharge) corresponding to cover stones

q_c cone resistance in CPT testing

Re   Reynolds number

SPT   Standard Penetration Test

S_r Degree of saturation

s specific gravity of sediment (or soil) grains

s_{liq} specific gravity of liquefied soil

s_p specific gravity of pipe

T wave period

T_p peak period of irregular waves

t	time
U	current velocity
U_c	velocity of compaction front
U_m	maximum value of undisturbed orbital velocity at the seabed
u, v, w	soil displacement in x-, y- and z-directions, respectively
V_x, V_y, V_z	components of ground-water velocity in x-, y- and z-directions, respectively
w	fall velocity of sediment (or soil) grains
x, y, z	Cartesian coordinates
z	depth measured from the mudline/seabed downwards
γ	specific weight of water
γ'	submerged specific weight of soil, $\gamma' = \gamma_t - \gamma$
γ_p	specific weight of pipe
γ_{liq}	specific weight of liquefied soil
γ_s	specific weight of soil grains
γ_t	specific weight of soil (not to be confused with the specific weight of soil grains, γ_s)
$\gamma_x, \gamma_y, \gamma_z$	shear (angular) deformations soil deformation in x-, y- and z-directions, respectively
ϵ	volume expansion
η	free-surface elevation
θ	Shields parameter
θ_{cr}	critical value of the Shields parameter corresponding to the initiation of motion at the bed
λ	wave number, $\lambda = 2\pi/L$
ν	Poisson's ratio
ν	kinematic viscosity (mainly Chapter 6)
ρ	water density
ρ_s	density of sediment grains
$\sigma_x, \sigma_y, \sigma_z$	x-, y- and z-components of normal stress, respectively
$\sigma'_x, \sigma'_y, \sigma'_z$	x-, y- and z-components of effective stress, respectively
σ'_0	initial mean normal effective stress
σ'_{v0}	initial vertical effective stress
τ	amplitude of the cyclic shear stress, τ_y, in the soil
τ_x, τ_y, τ_z	x-, y- and z-components of shear stress in the soil, respectively
φ	friction angle
ω	angular frequency of waves, $\omega = 2\pi/T$

Author Index

Subject Index